NEUROBIOLOGY of BODY FLUID HOMEOSTASIS

TRANSDUCTION AND INTEGRATION

NEUROBIOLOGY of BODY FLUID HOMEOSTASIS

TRANSDUCTION AND INTEGRATION

Edited by

Laurival Antonio De Luca Jr.
São Paulo State University
UNESP
Araraquara, Brazil

José Vanderlei Menani
São Paulo State University
UNESP
Araraquara, Brazil

Alan Kim Johnson
University of Iowa
Iowa City, USA

CRC Press
Taylor & Francis Group
Boca Raton London New York

CRC Press is an imprint of the
Taylor & Francis Group, an **informa** business

CRC Press
Taylor & Francis Group
6000 Broken Sound Parkway NW, Suite 300
Boca Raton, FL 33487-2742

First issued in paperback 2019

ISBN-13: 978-1-4665-0693-0 (hbk)
ISBN-13: 978-0-367-37941-4 (pbk)

Library of Congress Cataloging-in-Publication Data

Neurobiology of body fluid homeostasis : transduction and integration / editors, Laurival Antonio De Luca Jr., Jose Vanderlei Menani, Alan Kim Johnson.
p. ; cm. -- (Frontiers in neuroscience)
Includes bibliographical references and index.
ISBN 978-1-4665-0693-0 (hardback : alk. paper)
I. De Luca, Laurival Antonio, Jr., editor of compilation. II. Menani, Jose Vanderlei, editor of compilation. III. Johnson, Alan Kim, editor of compilation. IV. Series: Frontiers in neuroscience (Boca Raton, Fla.)
[DNLM: 1. Sodium, Dietary--pharmacology. 2. Appetite--drug effects. 3. Appetite--physiology. 4. Sodium, Dietary--administration & dosage. 5. Thirst--drug effects. 6. Thirst--physiology. 7. Water-Electrolyte Balance. QV 273]

QP535.N2
572'.5238224--dc23 2013028516

Visit the Taylor & Francis Web site at
http://www.taylorandfrancis.com

and the CRC Press Web site at
http://www.crcpress.com

Dedication

This book is dedicated to Alan Neil Epstein and Miguel Rolando Covian, whose high scientific endeavors and mentoring inspired generations of students in the field of the neurobiology of body fluid homeostasis throughout the world.

Contents

Preface

Life of terrestrial vertebrates depends on the maintenance of water and salt in the extracellular compartment ("internal milieu") within strictly regulated boundaries. Recognition of this fact was the basis upon which Walter Cannon* formulated his definition of the term *homeostasis* almost a century ago. This book, *Neurobiology of Body Fluid Homeostasis: Transduction and Integration*, was inspired by Cannon's definition, and it is the result of an invitation made by the *Frontiers in Neuroscience Series* editors, Sidney Simon and Miguel Nicolelis, to produce a volume based on an international symposium with a similar title.

The symposium was held in Araraquara, São Paulo, Brazil, in September 2011 and was convened as an official satellite of a joint meeting held in Buzios, Rio de Janeiro, of the International Society for Autonomic Neuroscience (ISAN) and the American Autonomic Society (AAS).†,‡ The initial proposal for a satellite symposium on body fluid homeostasis was made by the chair of the Local Organizing Committee, Professor Sérgio Cravo, and it was selected by the International Programming Committee of ISAN–AAS. The symposium attracted speakers from countries across the globe (Argentina, Australia, Brazil, Canada, China, Israel, and United States) and from states representing several different regions of Brazil (Bahia, Goiás, Rio de Janeiro, and São Paulo) and it was sponsored by CAPES, Brazilian Society of Physiology (SBFis), Joint Graduate Program in Physiological Sciences (PIPGCF UFSCar-UNESP), FAPESP, and UNESP,§ which allowed for their support. Such broad international participation generated stimulating discussion among the invited speakers, faculty and young investigators (post-doctoral fellows, graduate and undergraduate students) presenting their work at the symposium and ultimately played a very positive role in the production of this book.

The study of the neural control of body fluid homeostasis provides rich examples of integrative neurobiology and regulatory physiology. The reader will find many examples in this book of research conducted by leading scientists on signal transduction and sensory afferent mechanisms, molecular genetics, perinatal and adult long-term influences on regulation, central neural integrative circuitry, and autonomic/neuroendocrine effector systems. In the spirit of integration, many examples are provided demonstrating the interface between body fluid homeostasis with systems controlling cardiovascular function, temperature, energy balance, and behavior. Hormones and neurotransmitters are at the core of the mechanisms that control body fluid balance, and in this volume they are found to be involved with the control of the hypothalamic–neurohypophysial axis, neuroplasticity, thirst, and sodium appetite.

We thank the contributors for their time and effort to produce excellent chapters. It is important to emphasize their patience when asked to revise their chapters. Each

---

* Cannon, W. B. 1929. Organization for physiological homeostasis. *Physiol Rev* 9:399–31.

† http://200.145.81.21/body_fluid2011/interna_program.php.

‡ http://www.eventus.com.br/isan2011/.

§ Several offices from UNESP sponsored the symposium: FOAr, FUNDUNESP, PROPe, and PROPG.

author also contributed to the enterprise as blind reviewer for other manuscripts to help bring each of the chapters to its final form.

In just one symposium and in a single volume, it is impossible to cover all of the critical research in a field as large as body fluid homeostasis. However, we trust that omissions in coverage in the text will be compensated for, at least in part, by the extensive number of citations of important research in the reference section of each chapter.

We are confident that readers will find many outstanding examples throughout this volume of how an integrative approach to research applies the latest methods and conceptual frameworks to enhance the study and knowledge on the control of body fluid homeostasis.

Laurival A. De Luca Jr.
José V. Menani
Alan Kim Johnson

Editors

Name: Laurival A. De Luca Jr., Ph.D.
Research emphasis: Neurobiology and integrative physiology.
Position: Professor, Department of Physiology and Pathology, School of Dentistry - FOAr, São Paulo State University - UNESP, Araraquara, Brazil.
Administration: Coordinator-President Joint Graduate Program in Physiological Sciences, PIPGCF UFSCar-UNESP (2010-12).
Education: Bachelor in Biology, Institute of Biosciences-IB, University of São Paulo - USP; Undergraduate Research at FCMBB - Botucatu, UNESP; Master and Doctorate in Physiology, School of Medicine at Ribeirão Preto - USP; Doctorate Sandwich Fellowship CAPES-Brazil, University of Pennsylvania, USA.

Name: José Vanderlei Menani, Ph.D.
Research emphasis: Neuroscience and cardiovascular physiology.
Position: Professor, Department of Physiology and Pathology, School of Dentistry - FOAr, São Paulo State University - UNESP, Araraquara, Brazil.
Education: Pharmacy, UNESP-Araraquara; Master and Doctorate in Physiology, School of Medicine at Ribeirão Preto, University of São Paulo - USP; Post-Doctoral Fellow, The University of Iowa, USA.

Name: Alan Kim Johnson, Ph.D.
Research Emphasis: Neural control of body fluid and cardiovascular homeostasis.
Position: Professor, Departments of Psychology, Health and Human Physiology, and Pharmacology, The University of Iowa, Iowa City, USA.
Education: B.S. The Pennsylvania State University; M.A. Temple University; Ph.D. The University of Pittsburgh; Post-Doctoral Fellow, University of Pennsylvania.

Contributors

José Antunes-Rodrigues
School of Medicine of Ribeirão Preto
University of São Paulo
Ribeirão Preto, São Paulo, Brazil

Ximena E. Caeiro
Instituto de Investigación Médica Mercedes y Martín Ferreyra (INIMEC-CONICET-Universidad Nacional de Córdoba)
Córdoba, Argentina

Maria J. Cambiasso
Instituto de Investigación Médica Mercedes y Martín Ferreyra (INIMEC-CONICET-Universidad Nacional de Córdoba)
Córdoba, Argentina

Débora S. A. Colombari
Department of Physiology and Pathology, School of Dentistry
São Paulo State University-UNESP
Araraquara, São Paulo, Brazil

Eduardo Colombari
Department of Physiology and Pathology, School of Dentistry
São Paulo State University-UNESP
Araraquara, São Paulo, Brazil

Sergio L. Cravo
Department of Physiology, Escola Paulista de Medicina
Universidade Federal de São Paulo
São Paulo, São Paulo, Brazil

Andréia L. da Silva
School of Medicine of Ribeirão Preto
University of São Paulo-UNESP
Ribeirão Preto, São Paulo, Brazil

Carolina Dalmasso
Instituto de Investigación Médica Mercedes y Martín Ferreyra (INIMEC-CONICET-Universidad Nacional de Córdoba)
Córdoba, Argentina

Derek Daniels
Department of Psychology, Program in Behavioral Neuroscience
University at Buffalo, The State University of New York
Buffalo, New York

Richard B. David
Department of Physiology and Pathology, School of Dentistry
São Paulo State University-UNESP
and
Joint Graduate Program in Physiological Sciences
PIPGCF/UFSCar-UNESP
Araraquara, São Paulo, Brazil

Iracema G. de Araujo
Laboratory of Behavioral Neuroendocrinology
Department of Physiological Sciences, Institute of Biology
Federal Rural University of Rio de Janeiro
Seropédica, Rio de Janeiro, Brazil

Laurival A. De Luca Jr.
Department of Physiology and Pathology, School of Dentistry
São Paulo State University-UNESP
Araraquara, São Paulo, Brazil

Patrícia M. De Paula
Department of Physiology and Pathology, School of Dentistry
São Paulo State University-UNESP
Araraquara, São Paulo, Brazil

Lucila L. K. Elias
School of Medicine of Ribeirão Preto
University of São Paulo
Ribeirão Preto, São Paulo, Brazil

Alastair V. Ferguson
Department of Biomedical and Molecular Sciences
Queen's University
Kingston, Ontario, Canada

Hilda S. Ferreira
Life Sciences Department
Bahia State University
Salvador, Bahia, Brazil

Fabricia V. Fonseca
Laboratory of Behavioral Neuroendocrinology
Department of Physiological Sciences, Institute of Biology
Federal Rural University of Rio de Janeiro
Seropédica, Rio de Janeiro, Brazil

Josmara B. Fregoneze
Department of Physiology, Health Sciences Institute
Federal University of Bahia
Salvador, Bahia, Brazil

Andrea Godino
Instituto de Investigación Médica Mercedes y Martín Ferreyra (INIMEC-CONICET-Universidad Nacional de Córdoba)
Córdoba, Argentina

Seth W. Hurley
Department of Psychology
University of Iowa
Iowa City, Iowa

Alan Kim Johnson
Departments of Psychology, Health and Human Physiology, Pharmacology and Cardiovascular Center
University of Iowa
Iowa City, Iowa

Micah Leshem
Department of Psychology
University of Haifa
Haifa, Israel

Na Li
Institute for Fetology and Reproductive Medicine Center
First Hospital of Soochow University
Suzhou, China

Wolfgang Liedtke
Departments of Neurology and Neurobiology
Duke University
Durham, North Carolina

Oswaldo U. Lopes
Department of Physiology, Escola Paulista de Medicina
Universidade Federal de São Paulo
São Paulo, São Paulo, Brazil

Carla Patricia N. Luz
Department of Biological Sciences
State University of Southwest Bahia
Jequié, Bahia, Brazil

Ana F. Macchione
Instituto de Investigación Médica Mercedes y Martín Ferreyra (INIMEC-CONICET-Universidad Nacional de Córdoba)
Córdoba, Argentina

Caiping Mao
Institute for Fetology and Reproductive Medicine Center
First Hospital of Soochow University
Suzhou, China

Lisandra O. Margatho
School of Medicine of Ribeirão Preto
University of São Paulo
Ribeirão Preto, São Paulo, Brazil

Michael L. Mathai
Howard Florey Institute, Florey Neuroscience Institutes
University of Melbourne
Parkville, Victoria, Australia

and

School of Biomedical and Health Sciences
Victoria University
St Albans, Victoria, Australia

Michael J. McKinley
Howard Florey Institute, Florey Neuroscience Institutes

and

Department of Physiology
University of Melbourne
Parkville, Victoria, Australia

André S. Mecawi
Laboratory of Neuroendocrinology, Department of Physiology
School of Medicine of Ribeirão Prêto
University of São Paulo
Ribeirão Preto, São Paulo, Brazil

José Vanderlei Menani
Department of Physiology and Pathology, School of Dentistry
São Paulo State University-UNESP
Araraquara, São Paulo, Brazil

Gustavo R. Pedrino
Department of Physiological Science
Universidade Federal de Goiás
Goiânia, Goiás, Brazil

Luis C. Reis
Laboratory of Behavioral Neuroendocrinology
Department of Physiological Sciences, Institute of Biology
Federal Rural University of Rio de Janeiro
Seropédica, Rio de Janeiro, Brazil

Wagner L. Reis
Medical College of Georgia
Augusta, Georgia

Daniel A. Rosa
Department of Physiological Science
Universidade Federal de Goiás
Goiânia, Goiás, Brazil

Silvia G. Ruginsk
School of Medicine of Ribeirão Preto
University of São Paulo
Ribeirão Preto, São Paulo, Brazil

Willis K. Samson
Department of Pharmacological and Physiological Science
Saint Louis University School of Medicine
Saint Louis, Missouri

Lijun Shi
Department of Exercise Physiology
Beijing Sport University
Beijing, People's Republic of China

Robert L. Thunhorst
Department of Psychology and Cardiovascular Center
University of Iowa
Iowa City, Iowa

Renato R. Ventura
Federal University of Alfenas
Alfenas, Minas Gerais, Brazil

Alexandre A. Vieira
Department of Physiology and Pathology, School of Dentistry
São Paulo State University-UNESP
Araraquara, São Paulo, Brazil

Tatiane Vilhena-Franco
School of Medicine of Ribeirão Preto
University of São Paulo
Ribeirão Preto, São Paulo, Brazil

Laura Vivas
Instituto de Investigación Médica Mercedes y Martín Ferreyra (INIMEC-CONICET-Universidad Nacional de Córdoba)
Córdoba, Argentina

Feichao Xu
Institute for Fetology and Reproductive Medicine Center
First Hospital of Soochow University
Suzhou, China

Zhice Xu
Institute for Fetology and Reproductive Medicine Center
First Hospital of Soochow University
Suzhou, China

and

Center for Perinatal Biology
Loma Linda University
Loma Linda, California

Gina L. C. Yosten
Department of Pharmacological and Physiological Science
Saint Louis University School of Medicine
Saint Louis, Missouri

1 The Human Penchant for Deranged Salt Balance

Micah Leshem

CONTENTS

1.1 OVERVIEW

Salt intake is deranged because humans ingest vast quantities of salt with no clear utility—and despite substantial evidence of its deleterious effects. We review the paucity of ideas as to why this occurs and make some novel suggestions.

1.2 CONSEQUENCES AND STUDY OF DERANGED SALT INTAKE

"Habitual high salt intake is one of the quantitatively important, preventable mass exposures causing the unfavorable population-wide pattern that is a major risk factor for epidemic cardiovascular disease," and "...CVD is the world's leading cause of death and excess dietary sodium has convincingly been shown to be a serious public health hazard..." (The Lancet 2011; Stamler 1997). In contradiction, in a human experiment, blood pressure (BP) was not found to increase with sodium load, unlike animal experiments, and recent evidence highlights a protective role of salt intake in cardiovascular disease (Alderman and Cohen 2012; Dong et al. 2010; Taylor et al. 2011; Todd et al. 2012).

The consensus that dietary sodium is toxic, and the contention that it is not, fuel an important debate that, astonishingly, ignores its root cause—why humans eat salt in the first place. What high salt intake causes is well researched, what causes high salt intake is not known at all (Beauchamp 1987; Leshem 2009a; Mattes 1997).

There have been very many and extensive studies on tens of thousands of people over at least 4 decades elucidating and refining the connection between sodium intake and pathology, and there has been considerable public investment in health promotion and legislation to reduce sodium dietary content and intake (Boscoe et al. 2006; Cohen et al. 2006; Institute of Medicine (US) Committee on Strategies to Reduce Sodium Intake 2010; Dietary Guidelines for Americans 2010; He and MacGregor 2010; Millett et al. 2012; Stamler 1997). Despite this tremendous investment, in the same period only three review articles have addressed human sodium appetite. They concluded that "we know of no reason for intake of salt in humans" (Beauchamp 1987), that "the basis for the high, apparently need-free, sodium chloride ingestion ... has not been established" (Mattes 1997), and that "humans do not have a sodium appetite as we know it in animals, leaving us with a rather limited understanding of why humans ingest so much salt" (Leshem 2009a).

Hence, essentially, the efforts to manage sodium consumption, by the individual, the physician, and government, are not evidence-based.

In contrast, we have a solid base of knowledge from animal experiments about the controls of salt intake as a biological homeostatic system—we can halt salt intake when it is crucially required, and we can boost it when it is not, both acutely and for the duration of the animal's life. However, the animal system has primarily evolved to prevent and repair sodium deficit—not to cope with persistent and chronic excess. Analogous arguments have been applied to the obesity epidemic as a failure to regulate ingestion with homeostatic controls evolved to contend with famine rather than feast; however, there is an important difference between salt and food with crucial implications for control—the results of excess caloric intake are visible and persistently troubling even if its long-term implications are not, whereas those of excess sodium are not. In susceptible individuals, its primary pathology is hypertension, for which we have no senses, no defenses, and no warnings. Thus, among hypertensives worldwide, 6%–97% are unaware or uncontrolled (Dreisbach 2011). These are no trivial figures; hypertensives may include 67% of U.S. adults aged 60 years and over (Ostchega et al. 2007).

The study of the causes and determinants of salt appetite will assist in addressing its behavioral regulation, largely sodium overconsumption, but also the increasingly recognized problems of hyponatremia and fluid regulation, neonatally, in exertion, in disease, in mental anguish, and in senescence (Almond et al. 2005; Boscoe et al. 2006; He and MacGregor 2010; Johnson and Grippo 2006; Koleganova et al. 2011; Malaga et al. 2005; Moritz 2008; O'Donnell et al. 2011; Shirazki et al. 2007; Stein et al. 2006; Stolarz-Skrzypek et al. 2011).

1.3 DERANGED INTAKE OF SALT

Some experts point to the benefits of a daily sodium intake of 0.9 mmol (21 mg sodium, 0.53 g salt; Mancilha-Carvalho et al. 2003), some go 5-fold higher to 5 mmol (115 mg, 0.29 g) for healthy adults living in a temperate climate, but in consideration of the wide variation in (U.S.) physical activity and climatic exposure, increase that 4-fold to a safe minimum of 21.7 mmol (500 mg, 1.27 g), tripled to 65 mmol (1500 mg, 3.8 g), not because this is the actual sodium requirement, but because it ensures that the overall diet provides an adequate intake of other important nutrients, as well as to cover possible excessive sodium loss (Dietary Reference Intakes 2005). This is increased by a further 50% to 100 mmol (2300 mg, 5.85 g) by health authorities (e.g., the U.S.) as an achievable interim reduction goal (Dietary Guidelines for Americans 2010; The Seventh Report of the Joint National Committee on Prevention, Detection, Evaluation, and Treatment of High Blood Pressure 2004).

However, 12- to 50-year-old American men consume 196 mmol (4500 mg, 11.5 g) and American women a third less, which averages to the similar intake of people all over the world at 162.4 ± 22.4 mmol (3735 mg, 9.5 g; Bernstein and Willett 2010; Brown et al. 2009; McCarron et al. 2009).

This is 2–200 times the requirement estimates, a difference showing the human penchant for deranged salt balance. This derangement is repeated daily, and for perspective, ponder our ponderosity were our caloric intake similarly apportioned.

We have no inkling as to the reasons for this derangement. It has been sparsely researched, and where so, the conclusion has been unanimous—there is no known reason for the deranged intake of salt (Beauchamp 1987; Leshem 2009a; Mattes 1997).

An alternative approach suggests that this may be no derangement, but an imperative—that although the requirement estimates, based on human physiology, may be adequate for physiological function, there are additional benefits to the persistently higher intake of salt in humans. Such benefits can make salt tasty by conditioning, that is, if salt does us good, it will come to taste good. Taste will therefore regulate its intake, not physiological necessity. On the contrary, physiology is harnessed to cope with this overload, and clearly at a price in bodily resources and health. According to this rationale, and it is presently the best available, to understand the human penchant for deranged salt balance all that remains is to discover how salt benefits us.

In an interesting departure from the reduction proposals, it has recently been argued that, in fact, the current level of intake is the healthiest in terms of cardiovascular mortality—sodium intakes above and below the range of 2.5–6.0 g/day (6.4–15.3 g/day salt) are associated with increased risk (Alderman and Cohen 2012).

However, the long-term benefits of sodium cannot condition sodium preference in the healthy individual—although over generations such increased fitness might.

1.4 EXTANT NOTIONS ABOUT WHY WE HAVE DERANGED INTAKE

We review extant notions about the causes of our salt intake, as a derived derangement of animal salt appetite, the suggestion that much of our salt intake is insensible by virtue of sodium being veiled in industrialized foods, that we are instilled with a love of salt at an early age, or that we are addicted to salt. We end with a review of the single established determinant of salt intake—perinatal sodium loss.

Let us contemplate the nature of the human predilection for salt. We have four, possibly five, basic taste sensors in our mouths. In concert, they inform us of the infinite variety of tastes we enjoy. Yet, of these few sensors, one is dedicated entirely and specifically to respond to sodium, the charged atom that gives salt its unique taste.

1.4.1 Is Our Sodium Appetite Like That of Animals?

What are the implications of such a specialized sodium sensor? Rats—like humans, one of the very rare handful of omnivores—are similarly equipped, and research clearly indicates a role for that sensor in meeting the critical challenge of hyponatremia (life-threatening bodily sodium deficit). It directs the animal to the life-saving sodium it requires (it does the same in vegetarian and herbivorous animals; Denton 1982; Epstein 1990; Schulkin 1991).

It does so by a number of extraordinary means. First, some rats seem to innately and immediately recognize the taste of salt as the appropriate medication for their hyponatremic ailment (Coldwell and Tordoff 1996; Schulkin 1991), although other strains require some experience (Leshem et al. 2004). Second, having once cured hyponatremia, salt becomes even more attractive to the animal, and it will evermore increase its intake (Falk and Lipton 1967; Leshem et al. 2004; Sakai et al. 1989). This is understood as an adaptation to the environmental circumstances that, having precipitated the hyponatremic crisis, may also predict its recurrence. Hence, the resulting enhanced avidity for salt ensures that sodium resources in the surroundings are licked, noted, and remembered more saliently, to be on hand when the crisis recurs (Epstein 1990; Fessler 2003; Leshem 2009a). Long-term changes in the brain may underlie this long-term enhancement of the appetite. After repeated sodium depletions, dopamine D2 receptors are sensitized, dendritic spines and branches in the shell of the nucleus accumbens are more prolific, and FOS immunoreactivity is enhanced in the subfornical organ and basolateral amygdala, loci implicated in mediation of sodium appetite (Clark and Bernstein 2006; Na et al. 2007; Roitman et al. 2002).

Finally, the enhanced avidity ensures a spontaneous increase in the intake of sodium, a strategy that may ameliorate a recurring sodium deficit (Epstein 1990; Fessler 2003; Leshem 2009a). It may even be that higher sodium intake enables storage of sodium reserves to dampen any loss (Heer et al. 2000).

Nevertheless, the differences between human and animal love of salt are profound—human salt intake does not fit the biological model of a regulated sodium appetite. It is similar in that the ion is essential for both men and animals, both have a special sense organ to detect it, in both bodily sodium is regulated by similar physiological systems, given the chance both ingest more salt than is physiologically necessary, and in both perinatal sodium loss enduringly enhances salt intake. On the other hand, humans reject pure salt (e.g., rock salt), which animals seek; humans only eat salt with food and reject it in water, whereas rats prefer sodium in solution (so as to control its intake); humans will only ingest NaCl, whereas animals have a sodium appetite—they will ingest almost any sodium salt. Humans do not respond spontaneously to acute sodium deficit by seeking salt—it is impossible to acutely arouse salt hunger in humans—whereas animals respond with a robust intake of salt. Having been depleted of sodium once, rats will increase their sodium appetite permanently, which does not occur in humans. Humans voluntarily ingest salt every time they eat, and finally—the nub of this chapter—humans eat more salt by far than any other animal.

Most prominently, our human dedicated sodium receptor does not direct us to sodium when we are in deficit. Humans are recorded dying of hyponatremia with salt about them (Almond et al. 2005; Boscoe et al. 2006; Moritz and Ayus 2008). Studies and observations have attempted to demonstrate innate sodium self-medication for hyponatremia in humans, but the findings are consistently unpersuasive (Beauchamp et al. 1990; Leshem 2009a).

Of one heroic woman and two men deprived of salt and heavily sweated for 5 and 11 days, respectively, in an experiment the purpose of which they knew, one man "R.B.N. longed for salt and often went to sleep thinking about it. [The other, the author] R.A.M. felt no specific craving for salt and had difficulty in convincing himself that taking salt would at once make him feel all right again." These 41 words are the only reference to salt in the 812-word description of the symptoms and feelings of these three people during those many days (McCance 1936).

The tragic and classic case is of a little boy who craved and ate much salt. Hospitalized and denied his copious salt, the 3 1/2-year-old child was "reduced to very defective behavior, vomiting, anorexia, and died within a week." The *postmortem* report revealed that he had hyponatremia due to adrenal pathology. Nevertheless, his appetite for salt seems to have been learned (Wilkins and Richter 1940). As an infant, eating salt would have alleviated his condition, and led to a rapid conditioning of salt preference owing to its benefits, in processes well understood (Kochli et al. 2005; Wald and Leshem 2003; Rozin 1965; Yeomans et al. 2004).

In a remarkable experiment designed to replicate as closely as possible the experimental designs that have proved that rats and other animals develop a robust craving for salt when in deficit, over 10 days 10 men were deprived of salt and treated with diuretics to force the loss of sodium through their kidneys and urine. They showed scant evidence of a focused and robust salt craving (Beauchamp et al. 1990).

As to the enhanced palatability of sodium evinced by rats that have been previously sodium depleted, we have no good evidence for that occurring in adult humans after sodium loss. In a series of experiments, we have failed to reliably demonstrate enhanced salt appetite after repeated sodium loss due to hemorrhage (blood donors with up to 12 donations), army veterans who had been dehydrated up to eight times in training, mothers who had up to six births, and hyperhidrosis sufferers,* losing 44 ± 8 mg/min sweat from their palms compared to 6 ± 8 for controls (Leshem 2009b). Thus, unlike animals, humans do not seem to develop some protection to the hazard of hyponatremic crisis. However, we shall see below that, although enhancement may not occur in adult humans, it does occur after perinatal sodium loss, as it does in the rat experimental model (Crystal and Bernstein 1995, 1998; Kochli et al. 2005; Leshem et al. 1998; Shirazki et al. 2007; Stein et al. 1996).

Hence, the primary role of sodium-seeking in animals is irrelevant to humans; we humans will seek salt to please our palate, but not to save our life.

We are thus left with no explanation for the prominence of this sense organ, the sodium detector, or for the powerful influence it exerts on our predilection for salt as the prime condiment and food additive that gives taste and tang to our food, and is of no nutritional necessity, and that we persist in partaking of in defiance of extant public health advisories.

1.4.2 We Do Not Know We Are Ingesting Salt

There is a forcefully argued view that much of our sodium intake occurs unbeknownst because the salt is incorporated into our industrially prepared food such that it largely evades our sodium detector, making us unwitting consumers (Institute of Medicine 2010; He and MacGregor 2010). Indeed, it has long been suggested that there is an insensible component to sodium intake as well as the voluntary one. It is estimated that in the West 80%–90% of our sodium intake thus infiltrates our bodies, with 3%–7% added voluntarily by salting at the table, the balance accounted for by salting in the kitchen (Mattes 1997; Shepherd et al. 1989). It is this view that underpins the public effort to reduce the sodium content of commercially prepared food.

However, although commercially prepared food accounts for much of the nutrient intake in Western countries, insensible intake is less likely to be true of other countries, where salt intake is nevertheless similar or even greater (Brown et al. 2009; He and MacGregor 2010; McCarron et al. 2009). Moreover, just how insensible this sodium intake is, is debatable. We have shown that 10- to 15-year-old Arab and Jewish children who had neonatal hyponatremia had a greater intake of sodium in their diet. The children selected different sodium-rich items in their respective cuisines to attain the same increased intake, suggesting that "untasted" sodium content of food can determine selection, even when no overt sodium preference could be shown (Shirazki et al. 2007).

Stated differently, the postingestive consequences of "insensible" sodium intake from a particular food can condition a preference for that food (Wald and Leshem 2003). Hence, one might argue that whether sodium in industrially prepared foods is tasted or not, it is likely to condition a preference for such foods if it is in some way beneficial for the consumer. How else to explain the persistent selection of such foods? Surveys and sales figures testify that convenience of itself does not engender

* Hyperhidrosis is a medical condition in which a person sweats excessively and unpredictably. People with hyperhidrosis may sweat even when the temperature is cool or when they are at rest.

a preference for bad-tasting foods. In the United States, "taste still prevails as the number one motivator for food and beverage selections in general" (International Food Information Council Foundation 2011; Institute of Medicine 2010).

One can therefore wonder whether food intake determines sodium intake or whether, on the contrary, in part, sodium content determines food intake.

For many people around the world, food and beverage choice is restricted or non-existent, but salt might be the one ubiquitous taste enhancer available.

1.4.3 From a Young Age, We Are Taught to Like Salt

High dietary sodium, early in life and in adulthood, has been proposed to engender the increased salt preference by an indistinct mechanism termed "mere exposure," that is, exposure to a high level of sodium in the diet will make it preferred by force of habit (Institute of Medicine 2010). Many experiments have addressed this possibility, but this investigatory investment has produced scant evidence for what should be a robust phenomenon (cf. Leshem 2009a, Table 4).

On the other hand, there is consistent evidence that the avidity for salt emerges spontaneously in the infant, guides its food preferences, and increases the amount the child eats of those foods (Bernstein 1990; Bouhlal et al. 2011; Harris et al. 1990; Schwartz et al. 2011). This spontaneous avidity may decline in early infancy (Harris et al. 1990), but the subsequent avidity in childhood is greater than in adulthood (Beauchamp and Mennella 2011; Leshem 2009a). There is also consistent evidence that children learn the specific food contexts of salt rigidly, so that it does not generalize to a salt preference *per se* (Beauchamp and Mennella 2011; Birch 1999). Thus, few Americans salt their candies or fruit, yet salted chocolate, sweets, and lemons are delectables in other cuisines.

In reference to mere exposure, and thinking in evolutionary terms, Bernstein (1990) has argued winningly that "infants whose early weaning diets were very low in sodium would have been those most in need of a vigorous salt preference, and therefore, reliance on early diets to set the level for salt preference would seem rather maladaptive."

Also thinking in evolutionary terms, in a pilot experiment designed to show that our physiology evolved to regulate a much lower level of dietary sodium, five medical men changed their diet radically to the "paleolithic salt-free diet," reducing their sodium intake from 150 mEq/L/day (3450 mg Na^+, 8.8 g NaCl) to 10 mEq/L/day (230 mg Na^+, 0.6 g NaCl), consequently losing 1.4 kg body weight in 7 days as 5% of total body saline. Tellingly, despite the fact that they "all felt great" on low sodium, and despite their professional awareness and proximity to the dangers and consequences of high dietary sodium, they reverted to their salt-rich diet after the experiment and re-engorged with saline, observations again suggesting the prodigious attraction of salt-rich food and its scorn for physiological diktat (Klemmer 2010; Klemmer, personal communication). Similarly, in a more formal study, salt policy makers eating at their place of work exposed themselves to a 23%–36% increase in premature cardiovascular mortality compared to their recommended sodium intake, and the authors concluded "that awareness, adoption, power, a sense of urgency, legitimacy to act, or the presence of an institutional policy were not sufficient to adhere to the recommendation to reduce salt intake" (Brewster et al. 2011).

Hence, mere exposure seems a circular proposition inasmuch as it begs the question of why high dietary salt is acceptable or attractive in the first place, and why, if it has no utility, it has endured for all time, for everyone, everywhere, with nary an exception. One is hard put to come up with any such other acquired and universal behavior.

Yet environmental influences obviously influence food preferences—parental guidance and modeling, and in parts of the world child-targeted commercial advertising, have been shown to influence food choice in children (Cornwell and McAlister 2011; Scaglioni et al. 2008). What remains to be determined is whether salt falls into this domain, too, the persistence of these influences into adult preferences, and importantly, the influence on intake, which is but tenuously related to preference: chocolate is preferred to potato in a preference test or questionnaire by most people who nevertheless obtain more of their calories from the potato.

It is important to distinguish between "mere exposure" as a mechanism for a general liking of salt, and exposure in the context of utility. Specific tastes associated with postingestive or contextual stimuli meticulously condition specific food preference, *in utero* and on (Beauchamp and Mennella 2011; Birch 1999; Stein et al. 2012). But the notion of learning as applied to salt intake detached from a specific food item is inadequate. In its absence, it remains unclear what makes salt desirable in infancy, nor why it then defies extinction to persist throughout life.

1.4.4 Salt Craving as an Addiction

There are other molecules of no apparent biological utility that humans seek ardently, such as in the broadly termed addictions. There have been occasional suggestions that our craving for salt may be such (Cocores and Gold 2009; Tekol 2006). I have once or twice come across someone who claimed an inordinate love of salt and to taking it in spoonfuls. Whether an addiction or pica, it is irrelevant to why we *all* seek salt so.

More substantive, if circuitous, evidence for the addictive hypothesis comes from rodents. Enhanced and deficit-induced sodium seeking has been reported to produce cross tolerance with amphetamine, opiates, cocaine, and activate cocaine-seeking genes (Acerbo and Johnson 2011; Clark and Bernstein 2006; Liedtke et al. 2011; McBride et al. 2008; Na et al. 2009). In general, addictions are understood to hijack existing neural systems of great importance to initially engender psychological comfort but then compel pursuit of the substance despite patent revulsion (Robinson and Berridge 2003).

Yet, for a variety of reasons, the human avidity for salt does not fit this model. The receptors for the addictions are buried deep in central neural systems, and the substances craved must penetrate the brain by devious artifactual routes to mimic and surge the endogenous molecular cascades that turn us to them. Not so sodium—its receptor is an external sense organ in the first line of oral vetting of nutrients, and there is no mimicry here; sodium is the unique stimulus for which the receptor is designed, and its access route has evolved biologically, presumably adaptively.

Note, also, that unlike all the other addictive substances, salt *per se* is not craved. There are no reports of any of the salt avid population disrupting their schedule by rushing out to seek salt or facing adversity and degradation to obtain it when it is lacking in a meal.

Indeed, salt alone is never required to satisfy an urgent need—it is invariably a complement in well-structured, mundane, and scheduled situations—meals at mealtime. The craving for salt is exclusively context-embedded; if we miss a meal, we do not crave its salt.

Moreover, arguably, salt as an addiction would violate the very concept. All humans take salt. An addiction can hardly be a norm; for comparison, even for the greatest addiction, tobacco-smoking, addicts rarely exceed a minority of the (male) population, not to mention that the other addictions, destructive as they can be, are proportionally peripheral (WHO 2008).

Indeed, the notion has recently been turned on its head with the suggestion that addiction (at least for cocaine and opiates) has usurped the neural substrates of salt hunger—the drive in animals to replace depleted sodium (Liedtke et al. 2011). However, in the absence of salt hunger in humans, how these substrates may be related to human need-free intake, and to human addiction, awaits inquiry.

1.4.5 The Single Known Determinant of Salt Intake—Perinatal Sodium Loss

In seeking the determinants of high salt intake, studies in rats have shown that sodium depletion *in utero*, neonatally, or in maturity, permanently enhances salt appetite (Alwasel et al. 2012; Nicolaidis et al. 1990; Falk and Lipton 1967; Sakai et al. 1989). In humans, too, salt appetite is permanently enhanced after very early sodium loss, occurring *in utero*, neonatally, and in early infancy (Crystal and Bernstein 1995, 1998; Kochli et al. 2005; Leshem 1998; Shirazki et al. 2007; Stein et al. 1996).

This enduring enhancement of sodium palatability following on putative perinatal sodium privation is the most replicated determinant of the long-term human love of salt (Leshem 1998). It is consequent upon sodium loss *in utero* by maternal vomiting during pregnancy (Crystal and Bernstein 1995, 1998; Kochli et al. 2005; Leshem 1998), infantile vomiting and diarrhea (Kochli et al. 2005; Leshem 1998), electrolyte-deficient infant formula (Stein et al. 1996), neonatal hyponatremia (Shirazki et al. 2007), and possibly neonatal furosemide (Leshem et al. 1998).

In analyzing data from 200 participants in our various studies, we find that prenatal and childhood putative sodium loss (recorded neonatal hyponatremia and recall of pre- and postnatal vomiting and diarrhea) correlates with daily dietary sodium intake ($r = 0.40$, $p < 0.05$) and accounts for some 16% of the variance in children's dietary sodium intake (Leshem 2009a). In one study of adolescents, perinatal sodium loss increased salt preference by 50% (Leshem 1998). If this preference is confirmed to similarly increase salt intake, it would certainly be a dominant determinant of the derangement.

Maternal vomiting has been shown to predict preference in adolescence, as well as sensitivity to the taste of salt, which in turn is inversely related to BP (Crystal and Bernstein 1995; Leshem 1998; Malaga et al. 2005). Neonatal salt preference is also related to BP, and at 2 months to birth weight (Stein et al. 2006; Zinner et al. 2002). This relationship of gestational and neonatal hydromineral challenges, birth weight, and neonatal BP with neonatal and adolescent salt preference, is suggestive of the Barker hypothesis predicting adult cardiovascular disease (Alwasel et al. 2012; Barker 2007; Malaga et al. 2005). However, although salt intake may increase hypertension, it also may reduce cardiovascular and other mortality (Alderman and Cohen 2012; Dong 2010; Stolarz-Skrzypek et al. 2011).

An additional possibility is compromised kidney development *in utero* by maternal insult or sodium intake, low or high (Kett and Denton 2011). The insult can be very brief, and although the effects of maternal vomiting, sodium loss, or dehydration have not been studied directly, maternal angiotensin II can affect fetal kidney development. Such gestational insults can cause alterations in the offspring peripheral renin–angiotensin system, including elevated plasma renin activity and angiotensin-converting enzyme, which, in turn, can increase salt appetite in animal models (Alwasel et al. 2012; Argüelles et al. 2000; Kett and Denton 2011; Koleganova et al. 2011).

However, these are speculations, and the relationship of perinatal challenges to sodium homeostasis, and salt appetite to later disease, remain uncertain.

1.5 NOVEL NOTIONS ABOUT WHY WE HAVE DERANGED INTAKE

In the absence of valid and ascertained causes of the human penchant for salt intake, working hypotheses and research must be imaginative, eclectic, and diverse. We suggest several novel possible causes of the derangement—starting with the possibility that our salt intake is not, after all, deranged, but fulfills some as yet undiscovered needs. We then speculate on what these needs might be, suggesting stress reduction, and individual differences in the need for salt due to genetics, personality, sex, exertion, season, and climate.

1.5.1 Sodium Balance Is Not Deranged

Despite—or because of—its behavioral arm, which persists in apparent excess, sodium balance in humans is exquisitely maintained at various levels of intake. Because of the habitual nature of eating, and the sodium that perforce comes with it, long-term intake is probably constant. In our analysis of the National Health and Nutrition Examination Survey data (NHANES; e.g., Cohen et al. 2006), we find that energy accounts for most of the variance in sodium intake (as it does similarly for macronutrients and minerals; Goldstein and Leshem, unpublished data), and our laboratory data show similar relationships, energy intake correlating with macronutrients and electrolytes (Ca^{2+}, Na^+, K^+) ($n = 80$, $r = 0.44–0.93$, $p < 0.001$; Leshem 2009a; Leshem et al. 2008). Unexpectedly, energy also correlated with salting, sweetening, and adding pepper, ketchup, and mayonnaise.

However, in response to acute salt loads (meals, snacks), sodium in blood is maintained constant by the triplet of banking some in the huge extravascular fluid reservoir, withdrawing water therefrom, and urinating the excess remnant. Thus, constancy is maintained for blood sodium concentration, for body water, to which sodium concentration is of necessity linked, and for body mass too, insofar as it is largely composed of body water (Heer et al. 2000).

In this sense, therefore, "the human penchant for deranged salt balance" is a misnomer; excess sodium intake does not impose imbalance. On the other hand, there is substantial evidence that this persistent enlistment of regulatory resources ultimately brings about their derangement in vulnerable people, and a slew of consequent ailments, primarily hypertension (Zhao et al. 2011). Again, such long-term sequelae of sodium intake cannot condition sodium preference because it antecedes them.

1.5.2 Is Salt Intake Adaptive in Facing Stress?

The possibility that enhanced salt intake in humans protects against stress is an attractive hypothesis resting on the commonality of adrenal corticoids mediating both salt appetite and stress response and deriving from the intertwining endocrine systems regulating sodium concentration, fluid volume, and sympathetic tone. It is a bidirectional hypothesis, positing that stress may increase salt intake, and that sodium intake or loss may respectively reduce stress or aggravate it (Henry 1988).

1.5.2.1 Stress-Increased Sodium Appetite Has Been Demonstrated in Some Animal Experiments but Less so in Humans

In laboratory rodents, and sheep (but not pigs), baboons, or humans, administration of the neurohormone that kick-starts the body's response to stress, adrenocorticotrophic hormone (ACTH), stimulates salt intake (Jankevicius and Widowski 2003; Shade et al. 2002; Wong et al. 1993).

Despite the negative results with humans of exogenous ACTH, endogenous stress might increase salt appetite, because students rated high on hostility report using more salt, but not sugar, than their low-hostility peers, and their hostility score predicts how much salt they add to soup (Miller et al. 1998). Broadly consistent with this, trait anxiety in a sample of British women was positively correlated with cooking salt use, with eating food with high sodium content, and with an estimate of total salt intake (Shepherd and Farleigh 1986), but another study found trait anxiety related to a measure of sweet preference rather than salt (Stone and Pangborn 1990). I have assessed the effect of actual stress, confirmed by the Institute for Personality and Ability Testing Anxiety Scale (IPAT), at examination time for university students and found no relationship to salt or sweet preference (Leshem 2009a). Although these three results are methodologically insubstantial, they are fairly typical. Very few studies, if any, have properly examined stress and specific salt intake, mostly seeking immediate increases in intake to momentary stress in sodium-replete participants, in complete contrast to the methodology of the animal studies (Leshem 2009a; Torres et al. 2010). Furthermore, very many studies have examined the effects of stress on food intake and selection, and where increases rather than reductions are observed, they are generally reported to be for sweet or fat (Gibson 2006).

In a similar vein, depending on the individual, it might well be that stress can either increase or reduce sodium intake, as it does sodium excretion (Rollnik et al. 1995). Tangential evidence that salt appetite may be affected is suggested by the increased autonomic sensitivity to stress in salt-sensitive males (Weber et al. 2008; Zimmermann-Viehoff et al. 2008).

Enduring psychosocial stimulation in mice associated with sympathetic arousal and alterations in the renin–angiotensin system and observations on human societies have suggested that "the salt consumption of society [is] a measure of the social stress to which it is exposed" (Henry 1988). If so, the methodological approach should be to examine the response to stress, exclusively persistent, in people on minimal physiological levels of sodium intake (e.g., 100 mg/day NaCl; Mancilha-Carvalho et al. 2003), and with another nutrient control, because of the tight linkage of energy and sodium intakes (e.g., Leshem et al. 2008).

These are no mean methodological constraints, and this may account for the paucity of solid data. Clearly, the effect of stress on salt preference in humans necessitates more research for resolution.

1.5.2.2 Effects of Sodium Intake or Loss on Stress Have Been Studied Only in Experimental Animals

Sodium-depleted rats, or those with a deoxycorticosterone-stimulated sodium hunger, require respectively greater stimulation current in brain to effect the same rate of self-stimulation and reduce their intake of sweet solution, suggesting reduced hedonics, a model of depression. (Grippo et al. 2006; Morris et al. 2006, 2008). The effects are reversed by blocking aldosterone, which is elevated in depressed humans (Morris et al. 2010; Murck et al. 2003). Our own study on dietary restriction of sodium showed that it increased measures of anxiety in unstressed rats but did not exacerbate either acute or chronic stress (Leshem 2011).

Complementing sodium deficiency–induced depression and anxiety, hypernatremia increases social behavior in rats and decreases ACTH and cortisol levels during stress (Kraus et al. 2011). This is also an interesting complement to the suggestion that salt consumption of society is a measure of its social stress, again consistent with the bidirectional effect (Henry 1988).

Taken together, the animal work suggests a consistent effect of reduced sodium increasing anxiety, stress, or depression, in individual and social milieus, and its prevention by increased salt intake. Oblique evidence in humans for mood impairment comes from sporadic reports of sodium restriction–induced chronic fatigue (Bou-Holaigah et al. 1995; De Lorenzo et al. 1997; Johnson and Grippo 2006; Werbach 2000).

Findings in humans are scrappy, and currently, the case for a substantive role for sodium intake in alleviating stress in humans or for stress promoting our intake of salt, although tantalizing as a potential explanation of great importance, remains unproven.

1.5.3 Individual Differences in Salt Intake

Consideration of individual traits such as gender, genetics, constitution, culture, conditioning, and habit established in individual biographies leading to individual differences in salt intake are of interest, too, and may provide insights into its determinants (Leshem 2009a).

1.5.3.1 Genetics of Salt Avidity

Seventy-four pairs of monozygotic (identical) twins and 35 pairs of dizygotic (fraternal) twins were tested for recognition thresholds of sour and salt tastes. For sour, genetic factors accounted for 53% of the variance, with no environmental component.

This level of heritability is on par with sensitivity to the bitter compounds 6-*n*-propylthiouracil (PROP) and phenylthiocarbamide, and strongly suggests that genetic factors play a larger role than shared environment. In contrast to the genetic contribution to sour and bitter taste, salt taste recognition included an environmental component accounting for 22% of the variance, suggesting that environment plays a larger role than genetics in determining individual differences in recognition thresholds for saltiness (Wise et al. 2007).

The relevance of thresholds for recognition or sensitivity to preference and intake is moot, but a study comparing 663 mono- and dizygotic twin pairs in three countries found heritability estimates for the liking and use frequency of salty fat foods of 45%; however, the authors argue that this may be attributable to a genetic tendency to prefer fat rather than salt taste (Keskitalo et al. 2008).

Another route of inquiry to the possible genetic basis of variation in salt intake has been via the relationship of salt taste and PROP sensitivity. There is a genetic basis to the bitter taste of PROP, with people who cannot taste it being classed as "nontasters," with "supertasters" at the other extreme (each comprising about a quarter of the population). However, there is conflicting evidence on the relationship of salt taste to genetic bitter responding indexed by PROP response. Taste papilla density in the tip of the tongue has also been related to PROP responsivity (Hayes et al. 2008, 2010; Yackinous and Guinard 2002). The important issue raised here is twofold—whether PROP and papilla density predict the preference for salt, and if so, whether together they predict the intake of salt. There is evidence for the first, but less persuasive for the second (Hayes et al. 2008, 2010), much as the preference bias due to PROP sensitivity does not influence body mass index (BMI) (Yackinous and Guinard 2002). In an unpublished study on 82 students, we have found no relationship of any measure of salt (or sweet) preference or intake with either PROP response or papilla density.

A study conducted on sons and daughters of hypertensives and controls showed greater urine sodium in the offspring of hypertensives, interpreted as suggesting a genetic cause for the increased sodium intake (Ukoh et al. 2004).

However, there is no direct relationship between maternal dietary salt intake and offspring salt intake (Beauchamp and Moran 1984; Leshem 1998). The relationship between parental and offspring food preferences seems generally tenuous, although it tends to firm with age. Where found, it is attributed to environment rather than heredity (Birch 1999; Guidetti and Cavazza 2008; Rozin 1991).

In conclusion, there is evidence against a significant genetic basis for variation in salt intake in humans. The equally, if not more, important question of the *extent* to which human salt intake is genetically driven is more difficult to resolve because virtually all humans will ingest salt—just as it is difficult to resolve with language, which all humans use (Stromswold 2006). It is the more important question insofar as it must impinge crucially on strategies to reduce salt intake.

1.5.3.2 Predilection for Salt as a Personality Trait

Salt intake has been examined in relation to personality only twice, and found essentially irrelevant, as it is for other taste preferences such as spicy (Ludy and Mattes 2011; Shepherd and Farleigh 1986; Stone and Panborn 1990).

1.5.3.3 Sex Differences in Salt Intake

In an experiment with some 400 students, we find that women add less salt to pea soup compared to tomato or yam soup, whereas men show no finesse—they plunk the same (higher) amount in all three soups, as others have noted (Hayes et al. 2010; Leshem 2009a). Corrected for body weight, women take less salt than men, respectively 39.0 and 45.4 mg/day/kg (in the United States) and 33.91 and 41.25 mg/kg (in Israel). For a 75-kg Israeli, this is about 1.4 g/day (24 mmol Na^+, 550 mg Na^+) more salt for a man than for a woman. This disparity is equal to a quarter of the daily intake of 2300 mg sodium recommended incongruously equally for both men and women (Leshem 2009a).

When sex differences in fat mass are excluded, caloric intake (and basal metabolic rate) equalizes for the sexes in adults (Johnstone et al. 2005; Lazzer et al. 2010). Hence, because caloric and sodium intake is linked, the sex differences in sodium intake might dissipate, too, when fat mass is excluded. At face value, this suggests that sodium intake is somehow regulated by the lean person within, male or female. Yet, because sodium intake is accounted for by caloric intake (as an index of total food intake), we are returned via an Escherian stairway to the quandary that it is not possible to ignore BMI as a direct correlate of food intake, and hence sodium intake, which is indeed greater in men, even when corrected for body weight (Venezia et al. 2010). Clearly, the relationship of sex and body composition to sodium intake requires empirical study rather than Kabbalistic argumentation.

We have reported that women in Israel add less salt at table than men (Leshem 2009a), but a preliminary analysis of NHANES data indicates the opposite for the United States (Goldstein and Leshem, unpublished). If upheld, this would favor that the quantity of salt added at table is learned or conditioned, consistent with the habitual motor pattern of shaking the salt shaker (Greenfield et al. 1983; Mittelmark and Sternberg 1985), and the inadequate compensation by added salt for dietary sodium reduction (Beauchamp et al. 1987; Huggins et al. 1992; Shepherd and Farleigh 1989). Again, the source of the habit requires elucidation.

For diners in societies where most sodium is ingested via commercial food preparations, the salting habit is believed to be a minor contributor to intake (3%–5%; Institute of Medicine 2010; Mattes 1997; Shepherd and Farleigh 1986). Information about other societies is generally lacking, although we have studied a culture where salting at table, and commercial foods, are absent, yet salt intake is high owing to traditional kitchen food preparation (Leshem et al. 2008).

Ultimately, we cannot divine why men require more sodium than women, and the utility thereof. In rats it is generally, but not invariably, the female that has the higher intake, which is probably linked to reproduction, the dynamics of which do not apply to women, who hardly nurse offspring exceeding their own body weight (Leshem 2009b). Investigation of the determinants of gender differences in patterns of salt intake is necessary for its own interest, and because it may reveal additional determinants of sodium intake.

1.5.3.4 Exertion, Season, and Climate

Unlike perinatal sodium loss (above), sodium loss in adults due to hyperhidrosis, repeated hemorrhage, repeated dehydration, or breastfeeding, does not increase salt preference (Leshem 2009b). The findings contrast with studies in adult rats showing enduring enhancement of salt appetite by sodium deficit depending on sex and strain (Falk and Lipton 1967; Sakai et al. 1989; Leshem et al. 2004).

However, sodium loss in humans may increase salt intake enduringly via another route: because humans do not respond spontaneously to dehydration or sodium loss with a salt craving, it is possible that were other forms of sodium loss, such as copious perspiration, diarrhea, excessive vomiting, dehydration, or hemorrhage, paired with sodium intake, a conditioned increase in salt preference would emerge (Leshem 2009b). In rats this is not necessary, because along with their ability to compensate for sodium deficiency, enduringly enhanced sodium appetite can be acquired without learning (Sakai et al. 1989).

Because rehydration is more rapid with sodium, it could condition a preference for salt, as we have demonstrated with athletes taking untasted salt after exertion (Wald and Leshem 2003). The utility of sodium in maintaining or restoring euhydration in response to challenge may serve to condition increased preference for salt in those given to such challenge, such as workers in heavy physical work and heat, soldiers, and athletes (excluding swimmers) (Leshem 2009b; Leshem et al. 1999; Passe et al. 2009; Takamata et al. 1994; Wald and Leshem 2003). Indeed, recent findings reveal that in the United Kingdom, manual workers have a consistently greater spot urine sodium than nonmanual workers and report greater salting in cooking and at the table (Millett et al. 2012).

It is possible that in hot climes and seasons increased thirst induces sodium intake. It might therefore also be that sodium preference is conditioned by hydration in the heat, as it can be after exertion, and that this generalizes to a salt preference in cooler seasons. Thus, we have found that desert-dwelling Bedouin have a high salt intake even when assessed in the cooler desert spring, and others report a reduced proportion of excreted sodium in summer (Holbrook et al. 1984; Leshem 2009a; Leshem et al. 2008).

These are testable ideas, but the critical question of whether increases in acute sodium intake generalize to an increase in salt preference, and thus contribute to salt intake in the population, is another question that remains to be answered.

1.6 CONCLUSIONS

We are still lacking a basic understanding of why we ingest as much salt as we do and in the many different ways that we do. Salt ameliorates bitterness, which is probably a minor motive for its consumption. Salt does noticeably augment taste, but primarily it is the taste of salt with our food that we seek. Why that should be so remains enigmatic.

We need studies to provide an indication of the sources and determinants of salt intake in humans, and the reasons for excess salt intake. If the causes are understood, evidence-based intervention for controlling salt intake might be more focused, effective, and economical than the current *ad hoc* admonitory and regulatory strategies. By the same token, this can have implications for the health hazards associated with sodium overconsumption or deficiency, tailoring regulation to individual requirement.

As a corollary, the possible adaptive role of sodium intake must be addressed. Although there is a general ethos to reduce salt intake because of its contribution to hypertension and other pathology, there are good reasons for supposing that a blanket reduction could be less beneficial and even harmful to particular groups, and recent important discoveries suggesting increased cardiovascular and end-stage kidney mortality is increased by sodium restriction reinforce the cause of caution (Alderman and Cohen 2012; Dong et al. 2010; Strom et al. 2013; Taylor et al. 2011, 2012).

Furthermore, despite individual success, we have no idea if our population-wide avidity for salt can be at all reduced.

Finally, the general ethos favoring reduction of salt intake must not curtail research into the possibility that the apparently physiologically deranged levels of sodium intake of our society may also be adaptive or beneficial to different individuals in a range of ways as explored herein, but awaiting definitive evidence.

ACKNOWLEDGMENTS

My research was supported by the Israel–US Binational Foundation (to myself and Alan Epstein, and a second grant with Jay Schulkin), the Edelstein Foundation for Population Studies, the Israel Science Foundation, the Salt Institute, and the University of Haifa.

REFERENCES

Acerbo, M. J., and A. K. Johnson. 2011. Behavioral cross-sensitization between DOCA-induced sodium appetite and cocaine-induced locomotor behavior. *Pharmacol Biochem Behav* 98:440–8.

Alderman, M. H., and H. W. Cohen. 2012. Dietary sodium intake and cardiovascular mortality: Controversy resolved? *Am J Hypertens* 25:727–34.

Almond, C. S., A. Y. Shin, E. B. Fortescue et al. 2005. Hyponatremia among runners in the Boston Marathon. *N Engl J Med* 352:1550–6.

Alwasel, S. H., D. J. Barker, and N. Ashton. 2012. Prenatal programming of renal salt wasting resets postnatal salt appetite, which drives food intake in the rat. *Clin Sci (Lond)* 122:281–8.

Argüelles, J., J. I. Brimel, P. López-Selal, C. Perillán, and M. Vijande. 2000. Adult offspring long-term effects of high salt and water intake during pregnancy. *Horm Behav* 37:156–62.

Barker, D. J. 2007. The origins of the developmental origins theory. *J Intern Med* 261:412–7.

Beauchamp, G. K. 1987. The human preference for excess salt. *Am Sci* 75:27–33.

Beauchamp, G. K., and M. Moran. 1984. Acceptance of sweet and salty taste in 2-year-old children. *Appetite* 5:291–305.

Beauchamp, G. K., M. Bertino, and K. Engelman. 1987. Failure to compensate decreased dietary sodium with increased table salt usage. *JAMA* 258:3275–8.

Beauchamp, G. K., M. Bertino, D. Burke, and K. Engelman. 1990. Experimental sodium depletion and salt taste in normal human volunteers. *Am J Clin Nutr* 51:881–9.

Beauchamp, G. K., and J. A. Mennella. 2011. Flavor perception in human infants: Development and functional significance. *Digestion* 83(Suppl 1):1–6.

Bernstein, I. L. 1990. Salt preference and development. *Dev Psychol* 26:552–4.

Bernstein, A. M., and W. C. Willett. 2010. Trends in 24-h urinary sodium excretion in the United States, 1957–2003: A systematic review. *Am J Clin Nutr* 92:1172–80.

Birch, L. L. 1999. Development of food preferences. *Annu Rev Nutr* 19:41–62.

Bou-Holaigah, I., P. C. Rowe, J. Kan, and H. Calkins. 1995. The relationship between neurally mediated hypotension and the chronic fatigue syndrome. *JAMA* 274:961–7.

Boscoe, A., C. Paramore, and J. G. Verbalis. 2006. Cost of illness of hyponatremia in the United States. *Cost Eff Resour Alloc* 4:10.
Bouhlal, S., S. Issanchou, and S. Nicklaus. 2011. The impact of salt, fat and sugar levels on toddler food intake. *Br J Nutr* 105:645–53.
Brewster, L. M., C. A. Berentzen, and G. A. van Montfrans. 2011. High salt meals in staff canteens of salt policy makers: Observational study. *BMJ* 20(343):d7352.
Brown, I. J., I. Tzoulaki, V. Candeias, and P. Elliott. 2009. Salt intakes around the world: Implications for public health. *Int J Epidemiol* 38:791–813.
Clark, J. J., and I. L. Bernstein. 2006. A role for D_2 but not D_1 dopamine receptors in the cross sensitization between amphetamine and salt. *Pharmacol Biochem Behav* 83:277–84.
Cocores, J. A., and M. S. Gold. 2009. The salted food addiction hypothesis may explain overeating and the obesity epidemic. *Med Hypotheses* 73:892–9.
Cohen, H. W., S. M. Hailpern, J. Fang, and M. H. Alderman. 2006. Sodium intake and mortality in the NHANES II follow-up study. *Am J Med* 119:e7–14.
Coldwell, S. E., and M. G. Tordoff. 1996. Immediate acceptance of minerals and HCl by calcium-deprived rats: Brief exposure tests. *Am J Physiol Regul Integr Comp Physiol* 271(1 Pt 2):R11–7.
Cornwell, T. B., and A. R. McAlister. 2011. Alternative thinking about starting points of obesity. Development of child taste preferences. *Appetite* 56:428–39.
Crystal, S., and I. L. Bernstein. 1995. Morning sickness, impact on offspring salt preference. *Appetite* 25:231–40.
Crystal, S. R., and I. L. Bernstein. 1998. Infant salt preference and mother's morning sickness. *Appetite* 30:297–307.
De Lorenzo, F., J. Hargreaves, and V. V. Kakkar. 1997. Pathogenesis and management of delayed orthostatic hypotension in patients with chronic fatigue syndrome. *Clin Auton Res* 7:185–90.
Denton, D. A. 1982. *The Hunger for Salt: An Anthropological, Physiological, and Medical Analysis*. Berlin: Springer-Verlag.
Dietary Guidelines for Americans, U.S. 2010. Department of Agriculture U.S. Department of Health and Human Services www.dietaryguidelines.gov http://health.gov/dietaryguidelines/dga2010/DietaryGuidelines2010.pdf (accessed 7.11.12).
Dietary Reference Intakes for Water, Potassium, Sodium, Chloride, and Sulfate. 2005. Washington, DC: National Academies Press.
Dong, J., Y. Li, Z. Yang, and J. Luo. 2010. Low dietary sodium intake increases the death risk in peritoneal dialysis. *Clin J Am Soc Nephrol* 5:240–7.
Dreisbach, A. W. 2011. Epidemiology of Hypertension. http://emedicine.medscape.com/article/1928048-overview#aw2aab6b6 (accessed 12.1.12).
Epstein, A. N. 1990. Thirst and salt intake, a personal review and some suggestions. In *Thirst—Physiological and psychological aspects*, ed. D. J. Ramsay, and D. A. Booth, 481–501. Berlin: Springer-Verlag.
Falk, J. L., and M. J. Lipton. 1967. Temporal factors in the genesis of NaCl appetite by intraperitoneal dialysis. *J Comp Physiol Psychol* 63:247–51.
Fessler, D. M. 2003. An evolutionary explanation of the plasticity of salt preferences, prophylaxis against sudden dehydration. *Med Hypotheses* 613:412–5.
Gibson, E. L. 2006. Emotional influences on food choice: Sensory, physiological and psychological pathways. *Physiol Behav* 89:53–61.
Greenfield, H., J. Maples, and R. B. H. Wills. 1983. Salting of food – a function of hole size and location of shakers. *Nature* 301:331–2.
Grippo, A. J., J. A. Moffitt, T. G. Beltz, and A. K. Johnson. 2006. Reduced hedonic behavior and altered cardiovascular function induced by mild sodium depletion in rats. *Behav Neurosci* 120:1133–43.
Guidetti, M., and N. Cavazza. 2008. Structure of the relationship between parents' and children's food preferences and avoidances: An explorative study. *Appetite* 50:83–90.

Harris, G., A. Thomas, and D. A. Booth. 1990. Development of salt taste in infancy. *Dev Psychol* 26:534–538.

Hayes, J. E., L. M. Bartoshuk, J. R. Kidd, and V. B. Duffy. 2008. Supertasting and PROP bitterness depends on more than the *TAS2R38* gene. *Chem Senses* 33:255–65.

Hayes, J. E., B. S. Sullivan, and V. B. Duffy. 2010. Explaining variability in sodium intake through oral sensory phenotype, salt sensation and liking. *Physiol Behav* 100:369–80.

He, F. J., and G. A. MacGregor. 2010. Reducing population salt intake worldwide: From evidence to implementation. *Prog Cardiovasc Dis* 52:363–82.

Heer, M., F. Baisch, J. Kropp, R. Gerzer, and C. Drummer. 2000. High dietary sodium chloride consumption may not induce body fluid retention in humans. *Am J Physiol Renal Physiol* 278:F585–95.

Henry, J. P. 1988. Stress, salt and hypertension. *Soc Sci Med* 26:293–302.

Holbrook, J. T., K. Y. Patterson, J. E. Bodner et al. 1984. Sodium and potassium intake and balance in adults consuming self-selected diets. *Am J Clin Nutr* 40:786–93.

Huggins, R. L., R. Di Nicolantonio, and T. O. Morgan. 1992. Preferred salt levels and salt taste acuity in human subjects after ingestion of untasted salt. *Appetite* 18:111–9.

Institute of Medicine (US) Committee on Strategies to Reduce Sodium Intake. 2010. Henney, J. E., C. L. Taylor, and C. S. Boon, editors. Strategies to Reduce Sodium Intake in the United States. Washington (DC): National Academies Press (US). Available from: http://www.ncbi.nlm.nih.gov/books/NBK50956/.

International Food Information Council Foundation. 2011. Food and Health Survey. http://www.foodinsight.org/Resources/Detail.aspx?topic=2011_Food_Health_Survey_Consumer_Attitudes_Toward_Food_Safety_Nutrition_Health (accessed 12.1.12).

Jankevicius, M. L., and T. M. Widowski. 2003. Exogenous adrenocorticotrophic hormone does not elicit a salt appetite in growing pigs. *Physiol Behav* 78:277–4.

Johnson, A. K., and A. J. Grippo. 2006. Sadness and broken hearts: Neurohumoral mechanisms and co-morbidity of ischemic heart disease and psychological depression. *J Physiol Pharmacol* 57(Suppl 11):5–29.

Johnstone, A. M., S. D. Murison, J. S. Duncan, K. A. Rance, and J. R. Speakman. 2005. Factors influencing variation in basal metabolic rate include fat-free mass, fat mass, age, and circulating thyroxine but not sex, circulating leptin, or triiodothyronine. *Am J Clin Nutr* 82:941–948.

Keskitalo, K., H. Tuorila, T. D. Spector et al. 2008. The Three-Factor Eating Questionnaire, body mass index, and responses to sweet and salty fatty foods: A twin study of genetic and environmental associations. *Am J Clin Nutr* 88:263–71.

Kett, M. M., and K. M. Denton. 2011. Renal programming: Cause for concern? *Am J Physiol Regul Integr Comp Physiol* 300:R791–803.

Klemmer, P. J. 2010. Salt appetite. *Am J Kidney Dis* 55:A31–2.

Kochli, A., Rakover, Y., and M. Leshem. 2005. Salt appetite in patients with CAH-21-OH deficiency (congenital adrenal hyperplasia). *Am J Physiol (Regulatory, integrative and comparative physiology)* 288(6):R1673–81.

Koleganova, N., G. Piecha, E. Ritz et al. 2011. Both high and low maternal salt intake in pregnancy alter kidney development in the offspring. *Am J Physiol Renal Physiol* 301:F344–54.

Krause, E. G., A. D. de Kloet, J. N. Flak et al. 2011. Hydration state controls stress responsiveness and social behavior. *J Neurosci* 31:5470–6.

Lazzer, S., G. Bedogni, C. L. Lafortuna et al. 2010. Relationship between basal metabolic rate, gender, age, and body composition in 8,780 white obese subjects. *Obesity (Silver Spring)* 18:71–8.

Leshem, M., M. Maroun, and Z. Weintraub. 1998. Neonatal diuretic therapy may not alter children's preference for salt taste. *Appetite* 30:53–64.

Leshem, M. 1998. Salt preference in adolescence is predicted by common prenatal and infantile mineralofluid loss. *Physiol Behav* 63:699–704.

Leshem, M., A. Abutbul, and R. Eilon. 1999. Exercise increases the preference for salt in humans. *Appetite* 32:251–60.

Leshem, M., A. Kavushansky, J.-M. Devys, and S. Thornton. 2004. Enhancement revisited, the effects of multiple depletions on sodium intake in rats vary with strain, substrain, and gender. *Physiol Behav* 82:571–80.

Leshem, M., A. Saadi, N. Alem, and K. Hendi. 2008. Enhanced salt appetite, diet and drinking in traditional Bedouin women in the Negev. *Appetite* 50:71–82.

Leshem, M. 2009a. Biobehavior of the human love of salt. *Neurosci Biobehav Rev* 33:1–17.

Leshem, M. 2009b. The excess salt appetite of humans is not due to sodium loss in adulthood. *Physiol Behav* 98:331–7.

Liedtke, W. B., M. J. McKinley, L. L. Walker et al. 2011. Relation of addiction genes to hypothalamic gene changes subserving genesis and gratification of a classic instinct, sodium appetite. *Proc Natl Acad Sci USA* 108:12509–14.

Ludy, M. J., and R. D. Mattes. Comparison of sensory, physiological, personality, and cultural attributes in regular spicy food users and non-users. *Appetite* 58:19–27.

Malaga, I., J. Arguelles, J. J. Diaz, C. Perillan, M. Vijande, and S. Malaga. 2005. Maternal pregnancy vomiting and offspring salt taste sensitivity and blood pressure. *Pediatr Nephrol* 20:956–960.

Mancilha-Carvalho J. de J., and N. A. Souza e Silva. 2003. The Yanomami Indians in the INTERSALT Study. *Arq Bras Cardiol* 80:295–300.

Mattes, R. D. 1997. The taste for salt in humans. *Am J Clin Nutr* 65(Suppl.):692S–97S.

McBride, S. M., B. Culver, and F. W. Flynn. 2008. Dietary sodium manipulation during critical periods in development sensitize adult offspring to amphetamines. *Am J Physiol Regul Integr Comp Physiol* 295:R899–905.

McCance, R. A. 1936. Experimental sodium chloride deficiency in man. *Proc Royal Soc* B119:245–68.

McCarron, J. C., J. C. Geerling, A. G. Kazaks, and J. Stern. 2009. Can dietary sodium intake be modified by public policy? *Clin J Am Soc Nephrol* 4:1878–82.

Miller, S. B., M. Friese, L. Dolgoy, A. Sita, K. Lavoie, and T. Campbell. 1998. Hostility, sodium consumption, and cardiovascular response to interpersonal stress. *Psychosom Med* 60:71–77.

Millett, C., A. A. Laverty, N. Stylianou, K. Bibbins-Domingo, and U. J. Pape. 2012. Impacts of a national strategy to reduce population salt intake in England: Serial cross-sectional study. *PLoS One* 7:e29836.

Mittelmark, M. B., and B. Sternberg. 1985. Assessment of salt use at the table: Comparison of observed and reported behavior. *Am J Public Health* 75:1215–6.

Morris, M. J., E. S. Na, A. J. Grippo, and A. K. Johnson. 2006. The effects of deoxycorticosterone induced sodium appetite on hedonic behaviors in the rat. *Behav Neurosci* 120:571–579.

Morris, M. J., E. S. Na, and A. K. Johnson. 2008. Salt craving: The psychobiology of pathogenic sodium intake. *Physiol Behav* 94:709–21.

Morris, M. J., E. S. Na, and A. K. Johnson. 2010. Mineralocorticoid receptor antagonism prevents hedonic deficits induced by a chronic sodium. *Behav Neurosci* 124:211–24.

Moritz, M. L., and J. C. Ayus. 2008. Exercise-associated hyponatremia: Why are athletes still dying? *Clin J Sport Med* 18:379–81.

Murck, H., K. Held, M. Ziegenbein, H. Künzel, K. Koch, and A. Steiger. 2003. The renin–angiotensin–aldosterone system in patients with depression compared to controls—a sleep endocrine study. *BMC Psychiatry* 29:15.

Na, E. S., M. J. Morris, R. F. Johnson, T. G. Beltz, and A. K. Johnson. 2007. The neural substrates of enhanced salt appetite after repeated sodium depletions. *Brain Res* 171:104–10.

Na, E. S., M. J. Morris, and A. K. Johnson. 2009. Behavioral cross-sensitization between morphine-induced locomotion and sodium depletion-induced salt appetite. *Pharmacol Biochem Behav* 93:368–74.

Nicolaidis, S., O. Galaverna, and C. H. Metzler. 1990. Extracellular dehydration during pregnancy increases salt appetite of offspring. *Am J Physiol Regul Integr Comp Physiol* 258(1 Pt 2):R281–3.

O'Donnell, M. J., S. Yusuf, A. Mente et al. 2011. Urinary sodium and potassium excretion and risk of cardiovascular events. JAMA 306:2229–38.

Ostchega, Y., C. F. Dillon, J. P. Hughes, M. Carroll, and S. Yoon. 2007. Trends in hypertension prevalence, awareness, treatment, and control in older U.S. adults: Data from the National Health and Nutrition Examination Survey 1988 to 2004. *J Am Geriatr Soc* 55:1056–65.

Passe, D. H., J. R. Stofan, C. L. Rowe, C. A. Horswill, and R. Murray. 2009. Exercise condition affects hedonic responses to sodium in a sport drink. Appetite 52:561–7.

Robinson, T. E., and K. C. Berridge. 2003. Addiction. *Annu Rev Psychol* 54:25–53.

Roitman, M. F., E. Na, G. Anderson, T. A. Jones, and I. L. Bernstein. 2002. Induction of a salt appetite alters dendritic morphology in nucleus accumbens and sensitizes rats to amphetamine. *J Neurosci* 22:RC225.

Rollnik, J. D., P. J. Mills, and J. E. Dimsdale. 1995. Characteristics of individuals who excrete versus retain sodium under stress. *J Psychosom Res* 39:499–505.

Rozin, P. 1965. Specific hunger for thiamine: Recovery from deficiency and thiamine preference. *J Comp Physiol Psychol* 59:98–101.

Rozin, P. 1991. Family resemblance in food and other domains, the family paradox and the role of parental congruence. *Appetite* 16:93–102.

Sakai, R. R., S. P. Frankmann, W. B. Fine, and A. N. Epstein. 1989. Prior episodes of sodium depletion increase the need-free sodium intake of the rat. *Behav Neurosci* 103:186–92.

Scaglioni, S., M. Salvioni, and C. Galimberti. 2008. Influence of parental attitudes in the development of children eating behaviour. *Br J Nutr* 99(Suppl 1):S22–5.

Schulkin, J. 1991. *Sodium Hunger: The Search for a Salty Taste*. New York: Cambridge University Press.

Schwartz, C., C. Chabanet, C. Lange, S. Issanchou, and S. Nicklaus. 2011. The role of taste in food acceptance at the beginning of complementary feeding. *Physiol Behav* 104:646–52.

Shade, R. E., J. R. Blair-West, K. D. Carey et al. 2002. Ingestive responses to administration of stress hormones in baboons. *Am J Physiol Regul Integr Comp Physiol* 282:R10–8.

Shepherd, R., and C. A. Farleigh. 1986. Attitudes and personality related to salt intake. *Appetite* 7:343–54.

Shepherd, R., C. A. Farleigh, and S. G. Wharf. 1989. Limited compensation by table salt for reduced salt within a meal. *Appetite* 13:193–200.

Shirazki, A., Z. Weintraub, D. Reich, E. Gershon, and M. Leshem. 2007. Lowest neonatal serum sodium predicts sodium intake in low-birthweight children. *Am J Physiol Regul Integr Comp Physiol* 292:R1683–9.

Stamler, J. 1997. The INTERSALT Study: Background, methods, findings, and implications. *Am J Clin Nutr* 65:626S–42S.

Stein, L. J., B. J. Cowart, A. N. Epstein, L. J. Pilot, C. R. Laskin, and G. K. Beauchamp. 1996. Increased liking for salty foods in adolescents exposed during infancy to a chloride deficient feeding formula. *Appetite* 27:65–77.

Stein, L. J., B. J. Cowart, and G. K. Beauchamp. 2006. Salty taste acceptance by infants and young children is related to birth weight: Longitudinal analysis of infants within the normal birth weight range. *Eur J Clin Nutr* 60:272–9.

Stein, L. J., B. J. Cowart, and G. K. Beauchamp. 2012. The development of salty taste acceptance is related to dietary experience in human infants: A prospective study. *Am J Clin Nutr* 94:123–9.

Stolarz-Skrzypek, K., T. Kuznetsova, L. Thijs et al. 2011. European Project on Genes in Hypertension (EPOGH) Investigators. Fatal and nonfatal outcomes, incidence of hypertension, and blood pressure changes in relation to urinary sodium excretion. *JAMA* 305:1777–85.

Stone, L. J., and P. M. Pangborn. 1990. Preferences and intake measures of salt and sugar, and their relation to personality traits. *Appetite* 15:63–79.

Strom, B. L., A. L. Yaktine, and M. Oria (Eds). Sodium intake in populations: Assessment of evidence. Institute of Medicine of The National Academes. http://books.nap.edu/openbook.php?record_id=18311&;page=R1 accessed 21st May 2013.

Stromswold, K. 2006. Why aren't identical twins linguistically identical? Genetic, prenatal and postnatal factors. *Cognition* 101:333–84.

Takamata, A., G. W. Mack, C. M. Gillen, and E. R. Nadel. 1994. Sodium appetite, thirst, and body fluid regulation in humans during rehydration without sodium replacement. *Am J Physiol Regul Integr Comp Physiol* 266:R1493–502.

Taylor, R. S., K. E. Ashton, T. Moxham, L. Hooper, and S. Ebrahim. 2011. Reduced dietary salt for the prevention of cardiovascular disease (Review). *Cochrane Database Syst Rev* 7:CD009217.

Tekol, Y. 2006. Salt addiction: A different kind of drug addiction. *Med Hypotheses* 67:1233–4.

The Lancet. 2011. Editorial: Salt and cardiovascular disease mortality. *Lancet* 377: May 14 2011.

The Seventh Report of the Joint National Committee on Prevention, Detection, Evaluation, and Treatment of High Blood Pressure. 2004. U.S. Department of Health and Human Services National Institutes of Health, National Heart, Lung, and Blood Institute, National High Blood Pressure Education Program, NIH Publication No. 04-5230 August.

Todd, A. S., R. J. Macginley, J. B. Schollum, S. M. Williams, W. H. Sutherland, J. I. Mann, and R. J. Walker. 2012. Dietary sodium loading in normotensive healthy volunteers does not increase arterial vascular reactivity or blood pressure. *Nephrology (Carlton)* 17:249–56.

Torres, S. J., A. I. Turner, and C. A. Nowson. 2010. Does stress induce salt intake? *Br J Nutr* 103:1562–8.

Ukoh, V. A., G. C. Ukoh, R. E. Okosun, and E. Azubike. 2004. Salt intake in first degree relations of hypertensive and normotensive Nigerians. *East Afr Med J* 81:524–8.

Venezia, A., G. Barba, O. Russo et al. 2010. Dietary sodium intake in a sample of adult male population in southern Italy: Results of the Olivetti Heart Study. *Eur J Clin Nutr* 64:518–24.

Wald, N., and M. Leshem. 2003. Salt conditions a flavor preference or aversion after exercise depending on NaCl dose and sweat loss. *Appetite* 40:277–84.

Weber, C. S., J. F. Thayer, M. Rudat et al. 2008. Salt-sensitive men show reduced heart rate variability, lower norepinephrine and enhanced cortisol during mental stress. *J Hum Hypertens* 22:423–31.

Werbach, M. R. 2000. Nutritional strategies for treating chronic fatigue syndrome. *Altern Med Rev* 5:93–108.

WHO Report on the Global Tobacco Epidemic 2008: The MPOWER Package. http://www.who.int/tobacco/mpower/mpower_report_full_2008.pdf (accessed11.12.2011).

Wilkins, L., and C. P. Richter. 1940. A great craving for salt by a child with corticoadrenal insufficiency. *JAMA* 114:866–8.

Wise, P. M., J. L. Hansen, D. R. Reed, and P. A. Breslin. 2007. Twin study of the heritability of recognition thresholds for sour and salty taste. *Chem Senses* 32:749–54.

Wong, K. S., P. M. Williamson, M. A. Brown, V. C. Zammit, D. A. Denton, and J. A. Whitworth. 1993. Effects of cortisol on blood pressure and salt preference in normal humans. *Clin Exp Pharmacol Physiol* 20:121–6.

Yackinous, C. A., and J. X. Guinard. 2002. Relation between PROP (6-*n*-propylthiouracil) taster status, taste anatomy and dietary intake measures for young men and women. *Appetite* 38:201–9.

Yeomans, M. R., J. E. Blundell, and M. Leshem. 2004. Palatability, response to nutritional need or need-free stimulation of appetite? *Br J Nutr* 92(Suppl. 1):S3–S14.

Zhao, D., Y. Qi, Z. Zheng et al. 2011. Dietary factors associated with hypertension. *Nat Rev Cardiol* 8:456–65.

Zimmermann-Viehoff, F., C. S. Weber, M. Merswolken, M. Rudat, and H. C. Deter. 2008. Low anxiety males display higher degree of salt sensitivity, increased autonomic reactivity, and higher defensiveness. *Am J Hyperten* 21:1292–7.

Zinner, S. H., S. T. McGarvey, L. P. Lipsitt, and B. Rosner. 2002. Neonatal blood pressure and salt taste responsiveness. *Hypertension* 40:280–5.

2 Circumventricular Organs

Integrators of Circulating Signals Controlling Hydration, Energy Balance, and Immune Function

Alastair V. Ferguson

CONTENTS

2.1 INTRODUCTION: FEEDBACK CONTROL PATHWAYS IN CENTRAL AUTONOMIC CONTROL

The dominant role of the brain in the hierarchical control of the autonomic nervous system demands that it receive extensive afferent information regarding the "milieu interieur." This information is derived from two primary sources: (1) peripheral and visceral sensory systems that transmit information through classical sensory pathways into the central nervous system (CNS) and (2) sensory systems in the brain that monitor the constituents of the circulation to assess the physiological status of the individual. However, in view of its protected position behind the blood–brain barrier (BBB), the CNS is theoretically unable to monitor many of the most significant controlled variables that constitute this internal environment (osmolarity, glucose, calcium, lipophobic amino acids, and peptide concentrations). Thus, although the BBB acts to protect the brain from large shifts in these variables (their fine control is a prerequisite for normal CNS function), in theory, it also precludes CNS monitoring of such essential information regarding the physiological status of the internal environment. The logic of such a system is clear, providing the brain can gain access to essential sensory information through alternative mechanisms. A specialized group of CNS structures, which lack the normal BBB, the circumventricular organs (CVOs), provide such an alternative, and they thus play a pivotal role in blood–brain communication. Within these structures, circulating substances can directly influence individual CNS neurons, which—through efferent projections to autonomic control centers in the hypothalamus and medulla (Figure 2.1) —transmit this information for integration leading to appropriate modulation of autonomic outputs, and thus maintenance of "healthy autonomic state." In addition, considerable recent information suggests that the CVOs sense multiple signals traditionally thought to be of importance in separate physiological systems, including fluid balance (angiotensin II, natriuretic peptides, osmolarity), metabolic control (amylin, ghrelin, leptin), reproduction (relaxin), and immune regulation [interleukin-1 β (IL-1 β), interleukin-6 (IL-6)], although the boundaries delimiting these specific functional roles are rapidly disappearing.

The purpose of this chapter is to consider the potential mechanisms through which single CVO neurons can sense and integrate the critical information from multiple circulating signals of autonomic status. I will specifically describe the literature suggesting that single subfornical organ (SFO) neurons respond to separate signals relaying information regarding body fluid and metabolic status with a focus on nonselective cationic conductances as a site at which such integration may occur. I will first consider some of the primary circulating molecules that provide key sensory information to the CNS in each of these systems.

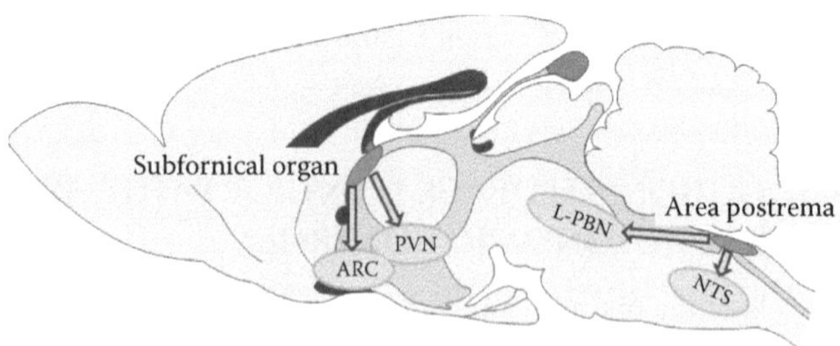

FIGURE 2.1 (See color insert.) This schematic midsagittal section through the rat brain illustrates location of two of circumventricular organs, subfornical organ and area postrema. It also highlights some of their primary efferent connections to arcuate (ARC) and paraventricular (PVN) nuclei for the subfornical organ, or nucleus tractus solitarius (NTS) and lateral parabrachial nucleus (L-PBN) for the area postrema.

2.2 FLUID BALANCE

Historically, our recognition of the CVOs as important sensory structures at the blood–brain interface, probably owes more to the description of the role of SFO in sensing circulating angiotensin II in the late 1970s than any other single observation (Mangiapane and Simpson 1980; Simpson and Routtenberg 1975). The role of the renin–angiotensin system in the regulation of cardiovascular function as a result of angiotensin II actions in controlling vascular tone and sodium reabsorption by the kidney was already well established. However, in addition to these classical endocrine effects, circulating angiotensin II was shown to have behavioral (drinking), neuroendocrine [adrenocorticotrophic hormone, oxytocin (OT), vasopressin (VP) secretion], and autonomic (sympathetic activation) effects as a direct consequence of actions in the CNS. The densest aggregations of angiotensin II receptors are found within the sensory CVOs (Mendelsohn et al. 1984), with the majority of such binding occurring in the core region (Song et al. 1992). SFO neurons show c-*fos* activation after intravenous angiotensin II (McKinley et al. 1992), and direct injection of angiotensin II into the SFO increases arterial pressure (Mangiapane and Simpson 1980), effects that are blocked by section of the ventral stalk of the SFO (Lind et al. 1983). At the cellular level, angiotensin II depolarizes more than 60% of SFO neurons (Li and Ferguson 1993a). These observations suggest the SFO is a critical relay center through which the CNS is able to monitor circulating angiotensin II and influence fluid balance and cardiovascular function.

Stimulation of the SFO by chemical or electrical means results in the secretion of both VP and OT (Ferguson and Kasting 1986, 1987, 1988; Holmes et al. 1987; Simpson 1981) and activation of sympathetic outflow (Ciriello and Gutman 1991; Ferguson et al. 1984). The hypertensive effect of stimulation of the SFO is prevented by ablation of the hypothalamic paraventricular nucleus (PVN) (Ferguson and Renaud 1984), and systemic angiotensin II acts at the SFO to control the activity of PVN neurons (Ferguson and Renaud 1986; Li and Ferguson 1993a). Intriguingly, these SFO neurons projecting to the PVN have also been shown to utilize angiotensin II as a neurotransmitter (Li and Ferguson 1993b). Other regulatory peptides including VP (Smith and Ferguson 1997; Washburn et al. 1999), atrial natriuretic peptide (Saavedra et al. 1986), and relaxin (Mumford et al. 1989) have also been shown to influence cardiovascular function as a result of actions in the SFO.

2.3 METABOLIC CONTROL

Since the discovery of leptin in the early 1990s, a number of other neuroactive peptides and adipokines associated with metabolic syndrome, including adiponectin, angiotensin II, OT, alpha-melanocyte stimulating hormone, and peptideYY/neuropeptide Y (Buijs 1990; Cowley et al. 1999; Ferguson and Washburn 1998; Kadowaki et al. 2006), have been shown to act not only as peripheral hormones, but also as neural signaling molecules in critical autonomic control centers of the brain. Importantly, the majority of these neuroactive signaling molecules exert what are apparently diverse physiological effects on cardiovascular (blood pressure, heart rate, baroreflex sensitivity), metabolic (food intake, metabolic rate), immune, and reproductive functions by

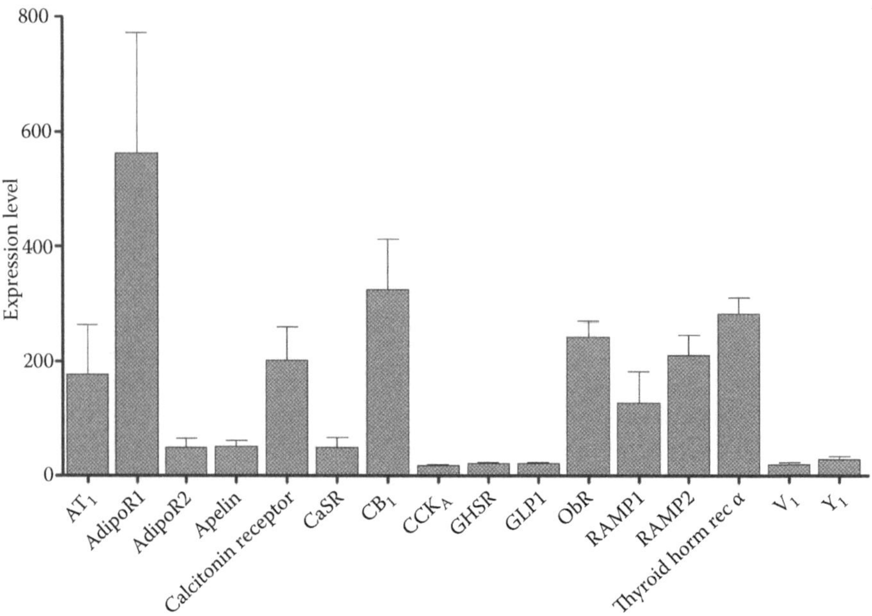

FIGURE 2.2 Summary data from our microarray studies showing relative expression levels of a number of different receptors in the SFO.

acting in these autonomic control centers in the hypothalamus and medulla. Although such commonality suggests these CNS centers are potential sites at which pathological changes may underlie all of the comorbidities associated with the metabolic syndrome, an understanding of the CNS circuitry through which this may occur has yet to emerge. Single cell recordings have shown the effects of metabolic signals such as calcitonin (Schmid et al. 1998) and amylin (Riediger et al. 1999a,b, 2001), whereas during the current funding period we have also shown ghrelin (Pulman et al. 2006) actions on SFO neurons. Our recent genomic analysis of SFO (Hindmarch et al. 2008) has also identified adiponectin (AdipoR1, AdipoR2), apelin, CB_1, ObR (leading to our recent work demonstrating that leptin influences the excitability of SFO neurons; Smith et al. 2009), and Y_1 receptors (Figure 2.2), as well as the potential presence of adiponectin, apelin, CART, CCK, orexin, NPY, OT, pro-melanin-concentrating hormone (PMCH), and VP neurons (or glial cells) in the SFO. Intriguingly, this genomic analysis also showed that 48-h food deprivation results in changes in CART and PMCH expression, adding support to the hypothesis that SFO plays important roles in regulation of energy balance (Hindmarch et al. 2008).

2.4 ANATOMICAL CONNECTIONS OF SFO

The SFO is a midline structure located on the dorsal floor of the third ventricle primarily known for its well-established roles in cardiovascular and neuroendocrine regulation (Ferguson and Bains 1997; McKinley et al. 1998). SFO sends dense projections to the PVN, supraoptic nucleus (SON), median preoptic nucleus, and the organum vasculosum of the lamina terminalis (OVLT; Lind et al. 1982; Miselis 1982; Miselis et al. 1979), and more minor projections to zona incerta, raphé nuclei, infralimbic cortex, rostral and ventral portions of the bed nucleus of the stria terminalis, lateral preoptic area, lateral hypothalamus/dorsal perifornical region, and the ARC (Gruber et al. 1987; Lind 1986; Miselis 1982). SFO neurons have compact dendritic trees and receive limited neural inputs (Dellman and Simpson 1979), which in most cases originate from the same areas that receive SFO efferents (Cottrell and Ferguson 2004). Specific excitatory projections have been found to VP and OT neurons in the SON and PVN, as well as to parvocellular areas of the PVN that in turn project either to the median eminence, the medulla, or the spinal cord (Ferguson and Bains 1996).

2.5 SFO—A PLURIPOTENT SENSOR OF CIRCULATING SIGNALS

The literature described above has, over a number of years, developed a database that makes it clear that the SFO has the ability to sense a large number of circulating signals, with physiological functions traditionally viewed as being distinct, all of which, however, can be argued to play important roles in the regulation of "integrated autonomic function." This is specifically emphasized by our recent gene chip analysis of the transcriptome of the SFO, which identified the presence of 745 transcripts associated with the descriptor "receptor" (Hindmarch et al. 2008), of which a number related to fluid balance and metabolism are highlighted in Figure 2.2. These observations—when considered in association with the literature showing that many of these signaling molecules influence the excitability of at least 30% of SFO neurons—lead to the unavoidable conclusion that single SFO neurons likely sense and integrate information from multiple signals. We have used patch clamp techniques to determine the effects of many of these different signaling molecules on isolated single neurons, an approach that not only allows us to determine the effects on resting membrane potential (depolarization or hyperpolarization) (Ferguson et al. 1997), but also allows direct assessment of the specific ion channels (Ferguson and Bains 1996), the modulation of which results in such effects on membrane potential and spike frequency. We have also used retrogradely transported fluorescent beads microinjected into the PVN before dissociation to label SFO neurons, techniques that have allowed us to determine the electrophysiological fingerprints of PVN (I_A) and non-PVN (I_K) projecting neurons (Anderson et al. 2001). In the remainder of this chapter, I will highlight nonselective cation channels (most likely TRPV1) as a common point for modulation and integration through which a number of circulating signals may act on SFO neurons to regulate central autonomic processing.

2.5.1 Angiotensin II

The development of both dissociated cell (Ferguson et al. 1997), and slice preparations (Ferguson and Li 1996; Xu et al. 2000) of the SFO for the first time presented the opportunity to directly assess the membrane events through which circulating signals influenced the excitability of these neurons. Not surprisingly, in view of the previous focus on the SFO as a sensor of circulating angiotensin II, initial studies focused on the mechanisms through which this peptide caused depolarization of the majority of SFO neurons, and identified the inhibitory effects of angiotensin II on transient potassium conductances (Ferguson and Li 1996; Ono et al. 2005), in our studies primarily on SFO neurons with the I_A fingerprint. Although such actions would not explain angiotensin II's ability to depolarize SFO neurons, they would lead to increases in spike frequency as a direct result of reducing the period of hyperpolarization following action potentials. Intriguingly, later the work of Ono et al. (2001) suggested that angiotensin II activated a nonselective cation channel with a reversal potential of –30 mV, actions that would result in the observed depolarization in response to angiotensin II, and also show that angiotensin II modulates multiple conductances in the same neurons. This picture became even more intriguing with our demonstration that the potentiation of N-type calcium channels by angiotensin II represents yet another ionic mechanism through which circulating ANG is involved in the excitation of SFO neurons (Washburn and Ferguson 2001). Although these high voltage activated Ca^{2+} currents would not be expected to be activated at rest (and therefore probably do not trigger the initial depolarization), it would be predicted that potentiation of these currents that would be activated during depolarization (resulting from activation of nonselective cation channels) could contribute to the overall spike patterning produced during excitation. In bursting SFO neurons, for example, angiotensinergic recruitment of an inward current could contribute to either burst maintenance (inward current would support the depolarization) or termination (Ca^{2+} influx could trigger Ca^{2+}-activated K^+ channels). These potential mechanisms would suggest important roles for these channels in shaping the bursts of action potentials that we have suggested may contribute to burst-mediated angiotensinergic neurotransmission by SFO neurons (Washburn et al. 2000).

2.5.2 Calcium

The localization of the calcium sensing receptor (CaSR) in the SFO (Rogers et al. 1997) identified this as a region at which changes in circulating calcium concentrations might have impact on the regulation of autonomic output. Using allosteric agonists of this receptor, we were able to show that the increases in blood pressure and neuronal excitability that occur after activation of CaSR appeared to be the result of activation of both nonselective cation channels (Washburn et al. 1999) and sodium channels underlying a persistent sodium current (Washburn et al. 2000). These combined effects would be expected to cause depolarization and increased bursting in SFO neurons; the latter effect may be of particular relevance to the ability of these putative angiotensinergic neurons to release angiotensin II as a neurotransmitter at the nerve terminals in the hypothalamus.

2.5.3 Osmolarity

We have obtained recordings from dissociated SFO neurons demonstrating that these cells exhibit an intrinsic osmosensitivity (the majority of which demonstrate the I_A fingerprint) over a wide range of osmolalities (270–330 mOsm) in the physiological range (Anderson et al. 2000). This osmosensitivity is characterized by a direct relationship between the extracellular osmolality and firing frequency of the SFO neurons in the iso-osmotic range (±10 mOsm). Although our studies were not able to identify the specific conductances responsible for this osmosensitivity, previous work in SON and OVLT has shown that such intrinsic osmosensitivity results from the activation of nonselective cation currents, a suggestion confirmed in preliminary recordings from SFO neurons (Bourque et al. 1994).

2.5.4 Ghrelin

Although ghrelin is well recognized as one of the few, if not the only, true orexigenic signaling molecules (Yoshihara et al. 2002), a smaller body of work has also suggested roles in the regulation of fluid intake (Hashimoto and Ueta 2011; Mietlicki and Daniels 2011). The majority of information has suggested that this effect is a result of direct actions at the arcuate nucleus (Cone et al. 2001). However, our recent work showing ghrelin receptor (GHRS) expression in the SFO and direct depolarizing actions of this peptide on SFO neurons has suggested this as an alternative potential site for such actions (Pulman et al. 2006). We also examined the ion channels responsible for such depolarizing actions of ghrelin and were able to show activation of nonselective cation channels in response to concentrations of ghrelin in the high physiological range as illustrated in Figure 2.3a.

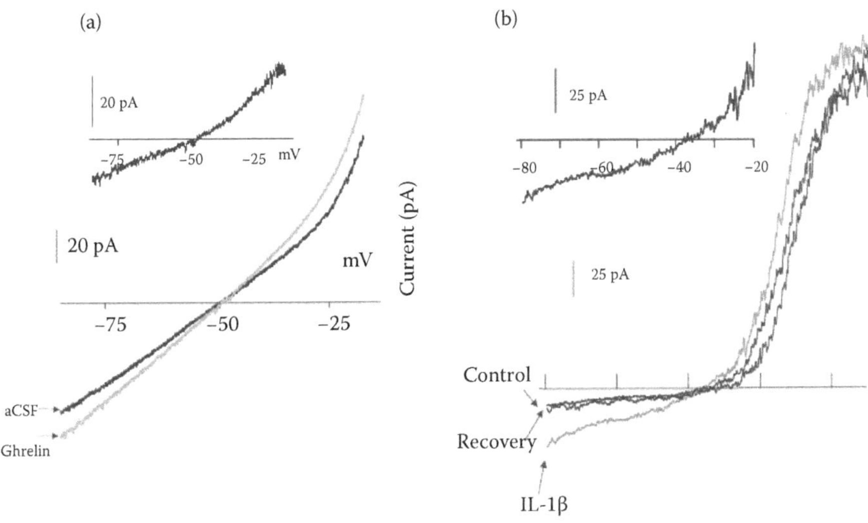

FIGURE 2.3 Voltage clamp data from two different SFO neurons illustrating current recorded in response to slow voltage ramps illustrating activation of voltage-independent current in response to (a) ghrelin and (b) interleukin-1 β. In each case, inset represents the difference (i.e., peptide induced) current obtained by subtracting control currents from those recorded during peptide administration.

2.5.5 Amylin

Amylin, an inhibitor of food intake, was perhaps the first signaling molecule with primary actions in the regulation of energy balance to be shown to have specific physiological actions in the SFO. These initial reports linked the effects of this "metabolic" hormone to actions in the regulation of drinking, which were associated with the excitatory effects on SFO neurons (Riediger et al. 1999a,b). The fact that ghrelin and amylin, which exert opposite effects on food intake, were shown to have similar excitatory effects on SFO neurons led us to studies that demonstrated that these two peptides exerted these excitatory effects on completely separate subpopulations of SFO neurons (Pulman et al. 2006). We were also able to show that amylin exerted these excitatory effects through actions on conductances characteristic of persistent sodium currents, while exerting no effect on nonselective cationic conductances (Pulman et al. 2006).

2.5.6 Glucose

Although not a "traditional hormone," circulating glucose represents a critical circulating signal that provides critical information regarding energy status to the CNS, yet does not diffuse freely across the BBB. We have recently shown that SFO neurons are glucose sensitive [separate SFO neurons were classified as glucose excited (GE) or glucose inhibited (GI); Medeiros et al. 2012], identifying this CVO as a potential site where such information may be collected and integrated by CNS neurons, which in turn project to critical hypothalamic autonomic control centers. We also obtained data suggesting that GI neurons primarily demonstrated the I_A fingerprint, and were backfilled by fluorescent dyes injected into the PVN before the dissociation process, identifying this nucleus as their most likely projection site. These studies also identified activation of nonselective cation and K_{ATP} channels as the most likely conductances mediating GE and GI neuron responsiveness, respectively.

2.5.7 Interleukin 1-β

The SFO has been identified as a potential target for the central actions of circulating cytokines including IL-6, IL-1 β, and tumor necrosis factor-α (Cartmell et al. 1999; Plata-Salaman and Ilyin 1997; Takahashi et al. 1997; Vallieres and Rivest 1997). IL-1 β, in particular, when administered centrally appears to have inhibitory effects on ingestive behaviors (Plata-Salaman 1994; Plata-Salaman and Ffrench-Mullen 1992) in a similar manner to those observed in response to amylin. We have again examined the ion channels underlying the actions of one of these cytokines, IL-1 β, on SFO neurons and identified effects on multiple conductances. Our studies have shown that primarily in SFO I_K neurons IL-1 β causes the activation of a nonselective cationic conductance (Figure 2.3b) and the inhibition of sustained voltage activated potassium conductances characteristic of I_K (Desson and Ferguson 2003).

2.6 INTEGRATION BY NONSELECTIVE CATION CHANNELS

Previous studies have shown that in the supraoptic nucleus of the hypothalamus stretch, inactivated cation channels represent not only a critical part of the intrinsic osmosensory abilities of magnocellular neurons, but also represent a common target for modulatory actions of the peptides cholecystokinin and neurotensin on the same neuron (Chafke and Bourque 2000). These studies elegantly demonstrated that the same single channel in a single neuron was influenced by all three sensory signals; the work summarized above has not addressed the issue of convergence of multiple signals in such a precise and definitive manner. The information above suggests a similar potential point for integration at nonselective cation conductances in SFO neurons. The data summarized in Table 2.1 suggest that in SFO cells with the I_A fingerprint (primarily projecting to PVN), angiotensin II, osmolarity, calcium, ghrelin, and glucose may all modulate this same conductance. Such a point of integration would lead to simple hypotheses regarding how such signals would interact in regulating different behaviors:

- During hyperosmolarity (thirst), other signals would be ineffective as the conductance would be maximally activated.
- During high ghrelin (hunger), the remaining signals would also be ineffective for the same reason.
- Under normal conditions, both would contribute to the normal drive to eat and drink.

Importantly, a potentially central role for this conductance in the regulation of autonomic function is particularly intriguing in view of the fact that activation will tend to always regulate the membrane potential around the reversal potential (–40

TABLE 2.1
Summary Data from a Number of Studies Highlighted in This Chapter Indicating the Effects of Different Signaling Molecules on SFO Neurons with I_A and I_K Fingerprints of a Variety of Different Ion Channels

	Behavior		Ion channels					
Signal	Drinking	Feeding	NSCC	I_A	I_K	I_{NaP}	I_{Ca}	SFO cell type
Angiotensin	↑		↑	↓			↑	I_A
Osmolarity	↑		↑					I_A
Calcium			↑			↑		?
Ghrelin		↑	↑					I_A (?)
Amylin		↓	→			↑		I_K (?)
Glucose (GE)			↑					I_A
Il-1 β	↓	↓	↑	→	↓			I_K

Note: Blank boxes indicate a lack of definitive information.

to –55 mV) for the channels. Thus, all of these signals will tend to regulate/maintain the membrane potential of SFO neurons at levels close to normal resting membrane potential. Another interesting point demonstrated by these observations relates to the fact that nearly all of these signals modulate multiple conductances, thus ultimately offering multiple points for such integration.

In conclusion, individual SFO neurons clearly have the ability to sense multiple circulating signals involved in the regulation of what we often think of as separate physiological systems (cardiovascular, fluid, metabolic, immune, etc.). This suggests critical roles for SFO neurons in integrating information from these multiple (perhaps not so diverse?) signals and providing integrated "autonomic information" to hypothalamic control centers. What is also apparent is that each of these signals influences excitability of SFO neurons as a consequence of the modulation of multiple ion channels. In this chapter, I have emphasized the potential ability of one such conductance to integrate these signals, although it should be stressed that each of these separate conductances has the ability to contribute to the total integration of circulating signals. The immediate challenges thus center around gaining a clearer understanding of not only how different conductances truly perform such roles, but also about how different subsets of SFO neurons play different roles in contributing to appropriate regulation of physiological autonomic state.

REFERENCES

Anderson, J. W., P. M. Smith, and A. V. Ferguson. 2001. Subfornical organ neurons projecting to paraventricular nucleus: Whole-cell properties. *Brain Res* 921:78–85.

Anderson, J. W., D. L. Washburn, and A. V. Ferguson. 2000. Intrinsic osmosensitivity of subfornical organ neurons. *Neuroscience* 100:539–47.

Bourque, C. W., S. H. R. Oliet, and D. Richard. 1994. Osmoreceptors, osmoreception, and osmoregulation. *Front Neuroendocrinol* 15:231–74.

Buijs, R. M. 1990. Vasopressin and oxytocin localization and putative functions in the brain. *Acta Neurochir Suppl (Wien)* 47:86–9.

Cartmell, T., G. N. Luheshi, and N. J. Rothwell. 1999. Brain sites of action of endogenous interleukin-1 in the febrile response to localized inflammation in the rat. *J Physiol* 518:585–94.

Chafke, Y., and C. W. Bourque. 2000. Excitatory peptides and osmotic pressure modulate mechanosensitive cation channels in concert. *Nat Neurosci* 3:572–9.

Ciriello, J., and M. B. Gutman. 1991. Functional identification of central pressor pathways originating in the subfornical organ. *Can J Physiol Pharmacol* 69:1035–45.

Cone, R. D., M. A. Cowley, A. A. Butler, W. Fan, D. L. Marks, and M. J. Low. 2001. The arcuate nucleus as a conduit for diverse signals relevant to energy homeostasis. *Int J Obes Relat Metab Disord* 25(Suppl 5):S63–7.

Cottrell, G. T., and A. V. Ferguson. 2004. Sensory circumventricular organs: Central roles in integrated autonomic regulation. *Regul Pept* 117:11–23.

Cowley, M. A., N. Pronchuk, W. Fan, D. M. Dinulescu, W. F. Colmers, and R. D. Cone. 1999. Integration of NPY, AGRP, and melanocortin signals in the hypothalamic paraventricular nucleus: Evidence of a cellular basis for the adipostat. *Neuron* 24:155–63.

Dellman, H. D., and J. B. Simpson. 1979. The subfornical organ. *Int Rev Cytol* 58:333–421.

Desson, S. E., and A. V. Ferguson. 2003. Interleukin 1beta modulates rat subfornical organ neurons as a result of activation of a non-selective cationic conductance. *J Physiol* 550:113–22.

Ferguson, A. V., and J. S. Bains. 1996. Electrophysiology of the circumventricular organs. *Front Neuroendocrinol* 17:440–75.

Ferguson, A. V., and J. S. Bains. 1997. Actions of angiotensin in the subfornical organ and area postrema: Implications for long term control of autonomic output. *Clin Exp Pharmacol Physiol* 24:96–101.

Ferguson, A. V., R. J. Bicknell, M. A. Carew, and W. T. Mason. 1997. Dissociated adult rat subfornical organ neurons maintain membrane properties and angiotensin responsiveness for up to 6 days. *Neuroendocrinology* 66:409–15.

Ferguson, A. V., T. A. Day, and L. P. Renaud. 1984. Subfornical organ stimulation excites paraventricular neurons projecting to the dorsal medulla. *Am J Physiol Regul Integr Comp Physiol* 247:R1088–92.

Ferguson, A. V., and N. W. Kasting. 1986. Electrical stimulation in the subfornical organ increases plasma vasopressin concentrations in the conscious rat. *Am J Physiol Regul Integr Comp Physiol* 251:R425–8.

Ferguson, A. V., and N. W. Kasting. 1987. Activation of subfornical organ efferents stimulates oxytocin secretion in the rat. *Regul Pept* 18:93–100.

Ferguson, A. V., and N. W. Kasting. 1988. Angiotensin acts at the subfornical organ to increase plasma oxytocin concentrations in the rat. *Regul Pept* 23:343–52.

Ferguson, A. V., and Z. Li. 1996. Whole cell patch recordings from forebrain slices demonstrate angiotensin II inhibits potassium currents in subfornical organ neurons. *Regul Pept* 66:55–58.

Ferguson, A. V., and L. P. Renaud. 1984. Hypothalamic paraventricular nucleus lesions decrease pressor responses to subfornical organ stimulation. *Brain Res* 305:361–4.

Ferguson, A. V., and L. P. Renaud. 1986. Systemic angiotensin acts at subfornical organ to facilitate activity of neurohypophysial neurons. *Am J Physiol Regul Integr Comp Physiol* 251:R712–17.

Ferguson, A. V., and D. L. S. Washburn. 1998. Angiotensin II: A peptidergic neurotransmitter in central autonomic pathways. *Progr Neurobiol* 54:169–92.

Gruber, K., A. McRae-Degueurce, L. D. Wilkin, L. D. Mitchell, and A. K. Johnson. 1987. Forebrain and brainstem afferents to the arcuate nucleus in the rat: Potential pathways for the modulation of hypophyseal secretions. *Neurosci Lett* 75:1–5.

Hashimoto, H., and Y. Ueta. 2011. Central effects of ghrelin, a unique peptide, on appetite and fluid/water drinking behavior. *Curr Protein Pept Sci* 12:280–7.

Hindmarch, C., M. Fry, S. T. Yao, P. M. Smith, D. Murphy, and A. V. Ferguson. 2008. Microarray analysis of the transcriptome of the subfornical organ in the rat: Regulation by fluid and food deprivation. *Am J Physiol Regul Integr Comp Physiol* 295:R1914–20.

Holmes, M. C., K. J. Catt, and G. Aguilera. 1987. Involvement of vasopressin in the down-regulation of pituitary corticotropin-releasing factor receptors after adrenalectomy. *Endocrinology* 121:2093–8.

Kadowaki, T., T. Yamauchi, N. Kubota, K. Hara, K. Ueki, and K. Tobe. 2006. Adiponectin and adiponectin receptors in insulin resistance, diabetes, and the metabolic syndrome. *J Clin Invest* 116:1784–92.

Li, Z., and A. V. Ferguson. 1993a. Angiotensin II responsiveness of rat paraventricular and subfornical organ neurons *in vitro*. *Neuroscience* 55:197–207.

Li, Z., and A. V. Ferguson. 1993b. Subfornical organ efferents to the paraventricular nucleus utilize angiotensin as a neurotransmitter. *Am J Physiol Regul Integr Comp Physiol* 265:R302–9.

Lind, R. W. 1986. Bi-directional, chemically specified neural connections between the subfornical organ and the midbrain raphe system. *Brain Res* 384:250–61.

Lind, R. W., L. E. Ohman, M. B. Lansing, and A. K. Johnson. 1983. Transection of subfornical organ neural connections diminishes the pressor response to intravenously infused angiotensin II. *Brain Res* 275:361–4.

Lind, R. W., G. W. Van Hoesen, and A. K. Johnson. 1982. An HRP study of the connections of the subfornical organ of the rat. *J Comp Neurol* 210:265–77.

Mangiapane, M. L., and J. B. Simpson. 1980. Subfornical organ lesions reduce the pressor effect of systemic angiotensin II. *Neuroendocrinology* 31:380–4.

McKinley, M. J., A. M. Allen, P. Burns, L. M. Colvill, and B. J. Oldfield. 1998. Interaction of circulating hormones with the brain: The roles of the subfornical organ and the organum vasculosum of the lamina terminalis. *Clin Exp Pharmacol Physiol Suppl* 25:S61–7.

McKinley, M. J., E. Badoer, and B. J. Oldfield. 1992. Intravenous angiotensin II induces Fos-immunoreactivity in circumventricular organs of the lamina terminalis. *Brain Res* 594:295–300.

Medeiros, N., L. Dai, and A. V. Ferguson. 2012. Glucose-responsive neurons in the subfornical organ of the rat—a novel site for direct CNS monitoring of circulating glucose. *Neuroscience* 201:157–65.

Mendelsohn, F. A., R. Quirion, J. M. Saavedra, G. Aguilera, and K. J. Catt. 1984. Autoradiographic localization of angiotensin II receptors in rat brain. *Proc Natl Acad Sci USA* 81:1575–9.

Mietlicki, E. G., and D. Daniels. 2011. Ghrelin reduces hypertonic saline intake in a variety of natriorexigenic conditions. *Exp Physiol* 96:1072–83.

Miselis, R. R. 1982. The subfornical organ's neural connections and their role in water balance. *Peptides* 3:501–2.

Miselis, R. R., R. E. Shapiro, and P. J. Hand. 1979. Subfornical organ efferents to neural systems for control of body water. *Science* 205:1022–5.

Mumford, A. D., L. J. Parry, and A. J. Summerlee. 1989. Lesion of the subfornical organ affects the haemotensive response to centrally administered relaxin in anaesthetized rats. *J Endocrinol* 122:747–55.

Ono, K., E. Honda, and K. Inenaga. 2001. Angiotensin II induces inward currents in subfornical organ neurones of rats. *J Neuroendocrinol* 13:517–23.

Ono, K., T. Toyono, E. Honda, and K. Inenaga. 2005. Transient outward K^+ currents in rat dissociated subfornical organ neurones and angiotensin II effects. *J Physiol* 568:979–91.

Plata-Salaman, C. R. 1994. Meal patterns in response to the intracerebroventricular administration of interleukin-1 beta in rats. *Physiol Behav* 55:727–33.

Plata-Salaman, C. R., and J. M. Ffrench-Mullen. 1992. Intracerebroventricular administration of a specific IL-1 receptor antagonist blocks food and water intake suppression induced by interleukin-1 beta. *Physiol Behav* 51:1277–9.

Plata-Salaman, C. R., and S. E. Ilyin. 1997. Interleukin-1beta (IL-1beta)-induced modulation of the hypothalamic IL-1beta system, tumor necrosis factor-alpha, and transforming growth factor-beta1 mRNAs in obese (fa/fa) and lean (Fa/Fa) Zucker rats: Implications to IL-1beta feedback systems and cytokine-cytokine interactions. *J Neurosci Res* 49:541–50.

Pulman, K. J., W. M. Fry, G. T. Cottrell, and A. V. Ferguson. 2006. The subfornical organ: A central target for circulating feeding signals. *J Neurosci* 26:2022–30.

Riediger, T., M. Rauch, and H. A. Schmid. 1999a. Actions of amylin on subfornical organ neurons and on drinking behavior in rats. *Am J Physiol Regul Integr Comp Physiol* 276:R514–21.

Riediger, T., H. A. Schmid, T. Lutz, and E. Simon. 2001. Amylin potently activates AP neurons possibly via formation of the excitatory second messenger cGMP. *Am J Physiol Regul Integr Comp Physiol* 281:R1833–43.

Riediger, T., H. A. Schmid, A. A. Young, and E. Simon. 1999b. Pharmacological characterisation of amylin-related peptides activating subfornical organ neurones. *Brain Res* 837:161–8.

Rogers, K. V., C. K. Dunn, S. C. Hebert, and E. M. Brown. 1997. Localization of calcium receptor mRNA in the adult rat central nervous system by in situ hybridization. *Brain Res* 744:47–56.

Saavedra, J. M., A. Israel, M. Kurihara, and E. Fuchs. 1986. Decreased number and affinity of rat atrial natriuretic peptide (6–33) binding sites in the subfornical organ of spontaneously hypertensive rats. *Circ Res* 58:389–92.

Schmid, H. A., M. Rauch, and J. Koch. 1998. Effect of calcitonin on the activity of ANG II-responsive neurons in the rat subfornical organ. *Am J Physiol Regul Integr Comp Physiol* 274:R1646–52.

Simpson, J. B. 1981. The circumventricular organs and the central actions of angiotensin. *Neuroendocrinology* 32:248–56.

Simpson, J. B., and J. B. Routtenberg. 1975. Subfornical organ lesions reduce intravenous angiotensin-induced drinking. *Brain Res* 88:154–61.

Smith, P. M., A. P. Chambers, C. J. Price et al. 2009. The subfornical organ: A CNS site for actions of circulating leptin. *Am J Physiol Regul Integr Comp Physiol* 296:512–20.

Smith, P. M., and A. V. Ferguson. 1997. Vasopressin acts in the subfornical organ to decrease blood pressure. *Neuroendocrinology* 66:130–5.

Song, K., A. M. Allen, G. Paxinos, and F. A. O. Mendelsohn. 1992. Mapping of angiotensin II receptor subtype heterogeneity in rat brain. *J Comp Neurol* 316:467–84.

Takahashi, Y., P. M. Smith, A. V. Ferguson, and Q. J. Pittman. 1997. Circumventricular organs and fever. *Am J Physiol Regul Integr Comp Physiol* 273:R1690–5.

Vallieres, L., and S. Rivest. 1997. Regulation of the genes encoding interleukin-6, its receptor, and gp130 in the rat brain in response to the immune activator lipopolysaccharide and the proinflammatory cytokine interleukin-1 beta. *J Neurochem* 69:1668–83.

Washburn, D. L., and A. V. Ferguson. 2001. Selective potentiation of N-type calcium channels by angiotensin II in rat subfornical organ neurones. *J Physiol* 536:667–75.

Washburn, D. L., P. M. Smith, and A. V. Ferguson. 1999. Control of neuronal excitability by an ion-sensing receptor (correction of anion-sensing). *Eur J Neurosci* 11:1947–54.

Washburn, D. L. S., J. W. Anderson, and A. V. Ferguson. 2000. A subthreshold persistent sodium current mediates bursting in rat subfornical organ neurones. *J Physiol* 529:359–71.

Washburn, D. L. S., A. M. Beedle, and A. V. Ferguson. 1999. Inhibition of subfornical organ neuronal potassium channels by vasopressin. *Neuroscience* 93:349–59.

Xu, S. H., K. Inenaga, E. Honda, and H. Yamashita. 2000. Glutamatergic synaptic inputs activate neurons in the subfornical organ through non-NMDA receptors. *J Auton Nerv Syst* 78:177–80.

Yoshihara, F., M. Kojima, H. Hosoda, M. Nakazato, and K. Kangawa. 2002. Ghrelin: A novel peptide for growth hormone release and feeding regulation. *Curr Opin Clin Nutr Metab Care* 5:391–5.

3 Neuroendocrinology of Hydromineral Homeostasis

José Antunes-Rodrigues, Silvia G. Ruginsk, André S. Mecawi, Lisandra O. Margatho, Wagner L. Reis, Renato R. Ventura, Andréia L. da Silva, Tatiane Vilhena-Franco, and Lucila L. K. Elias

CONTENTS

3.1 SENSORY AND NEUROENDOCRINE INTEGRATION IN CONTROL OF FLUID BALANCE

The maintenance of extracellular fluid (ECF) osmolality within narrow limits of variation is essential for the functioning of cells. The constant composition of the internal environment is primarily regulated by the balance between the intake and renal excretion of both sodium and water (Denton et al. 1996).

Sodium is the most abundant ion in the ECF, and it is also the solute that most effectively creates an osmotic gradient, which allows the movement of water between intra- and extracellular compartments. The total content of sodium and its accompanying anions (Cl^- and HCO_3^-) determines plasma osmolality.

Increased plasma osmolality is one of the most common disturbances of fluid osmolality, for which there are two major causes. A decrease in body water content could be induced by water deprivation or by insensible losses, such as after excessive perspiration and evaporation through the skin or the respiratory system, or after intense diarrhea and vomiting. On the other hand, an increase in plasma concentrations of solutes can occur after excessive sodium ingestion. Accordingly, the excessive intake of sodium in the diet is one of the emerging health problems in modern society, and this ingestive behavior may be associated with the development of several chronic debilitating diseases, such as cardiovascular disorders, including hypertension.

The mechanism for the correction of body fluid tonicity is so efficient that an increase in plasma osmolality as small as 1%–2% can trigger neuroendocrine responses such as vasopressin (AVP) release and, subsequently, the behavioral search for water (Johnson and Edwards 1991). Vasopressin acts to reduce urinary water excretion. At the same time, an inhibition of sodium consumption is observed in parallel with increased renal sodium excretion, primarily mediated by oxytocin (OT) and atrial natriuretic peptide (ANP) (Antunes-Rodrigues et al. 1997; McCann et al. 1989). These neuroendocrine responses result from an increase in effective osmolality that activates sensors located in the central nervous system (CNS) as well as in the periphery.

The concept of effective osmolality (i.e., osmolality resulting from the accumulation of nonpermeable solutes) was first introduced by Verney (1947), who proposed the existence of osmoreceptors as a mechanism related to the release of AVP in response to increased extracellular osmolality. "Osmosensors" are highly specialized cells responsible for translating changes in the extracellular osmotic pressure into electrical signals that activate CNS areas involved in the control of water and salt intake and excretion through the release of acetylcholine or angiotensin II (ANG II) as neurotransmitters (Chowdrey and Bisset 1988). Electrophysiological studies using hypothalamic magnocellular cells from the supraoptic nucleus (SON) showed that these neurons are depolarized and hyperpolarized by increases and decreases in the extracellular osmolality, respectively, and that these responses result from variations in the activity of cation channels (Bourque et al. 2002). This physiological response to hyperosmolality results in increased release of AVP and OT from the neurohypophysis and from parvocellular neurons of the paraventricular nucleus (PVN) projecting to other brain areas (Antunes-Rodrigues et al. 2004; Bourque et al. 2002; Verney 1947).

In addition, Andersson and McCann postulated that sodium sensors located at brain regions devoid of the blood–brain barrier could control sodium appetite and excretion in response to changes in the concentration of this solute in the cerebrospinal fluid (Andersson 1977; Andersson and McCann 1955, 1956). The infusion of hypertonic NaCl solution can induce a more effective increase in water intake if compared to equimolar solutions of nonionic substances, demonstrating the functional role of this ion. Blackburn et al. 1995 definitively demonstrated the dissociation between Na^+- and osmolality-induced effects on fluid intake. However, more related to the osmoreceptor hypothesis, recent work revealed that hyperosmolality activates intrinsic stretch-inactivated cation channels, as a result of taurine release from surrounding glia cells and changes in the expression of gene encoding sodium channels (Bourque 2008).

Current studies corroborate the hypothesis that cells sensitive to changes in osmolality or in the extracellular concentration of sodium are located in the lamina terminalis, which lies in the anterior wall of the third ventricle, which is composed of the subfornical organ (SFO), median preoptic nucleus (MnPO), and organum vasculosum of the lamina terminalis (OVLT). These brain regions, and the area postrema (AP) in the brainstem, are strategically located very close to the cerebral ventricles. They establish direct or multisynaptic connections with other areas controlling the cardiovascular function, such as the nucleus of the solitary tract (NTS) and several hypothalamus nuclei. This neuroanatomical organization is essential for the initiation of integrated responses to changes in plasma osmolality and sodium content.

In addition to their location in the CNS, osmoreceptors are also present in the afferent nerve endings adjacent to the hepatic vessels, kidney, and bowel. It has been suggested that osmoreceptors in the hepatoportal or mesenteric area are involved in the regulation of AVP release from the neurohypophysis (Arsenijevic and Baertschi 1985; Choi-Kwon and Baertschi 1991). The osmosensitive hepatic and splanchnic receptors are located in the mesentery of the upper small intestine and possibly in the portal vein. Inputs coming from the portal vein area, via the vagal afferents and both the right and left major splanchnic nerves, activate neurons located in the NTS, which in turn, stimulate the neural structures of the brainstem, leading to natriuresis and inhibition of intestinal absorption of sodium (Hosomi and Morita 1996). In addition to controlling sodium ingestion, the gastrointestinal tract participates in the restoration of normal plasma osmolality by controlling water absorption. When water is provided to an animal that was kept for 24–48 h under water restriction, the more intense behavioral response in water intake occurs in the following period of 3–10 min. In the meantime, thirst will be satiated without, however, completely restoring plasma osmolality. Stimuli generated in the mouth, pharynx, or stomach are converted into afferent impulses to CNS structures involved in the integrative response inducing the inhibition of thirst.

Aside from the regulation of osmolality, the existence of an adequate circulating volume is crucial to homeostatic regulation because, under these conditions, blood pressure must be maintained at an appropriate level to allow tissue perfusion. The volume of the extracellular compartment is largely determined by the balance between the intake and renal excretion of sodium. Variations in the circulating volume (detected by volume receptors or baroreceptors) activate the structures of the CNS responsible for integrating information and triggering a set of neural, endocrine, cardiovascular, and renal responses to correct these deviations.

Changes in blood volume or pressure lead to functional changes in the renal fluid and electrolyte excretion through neural and endocrine adaptive responses. Hypovolemia induces AVP release from magnocellular neurons, which acts by increasing the reabsorption of water in the distal nephron by stimulating the insertion of aquaporin-2 into the luminal membranes of the nephron. The threshold for stimulation of AVP release induced by hypovolemia is generally reported to be between 10% and 20% of the total blood volume in several species (Share 1988). On the other hand, isotonic expansion of the extracellular volume results in reduction of plasma AVP concentrations (Haanwinckel et al. 1995; Johnson et al. 1970; Ledsome et al. 1985; Leng et al. 1999; Shade and Share 1975; Share 1988).

The release of AVP from the neurohypophysial terminals of the hypothalamic magnocellular neurosecretory system is also controlled by peripheral baroreceptors, cardiopulmonary volume receptors, and circulating ANG II concentrations (Thrasher 1994). An increase in arterial pressure sufficient to activate baroreceptors is associated with a transient and selective γ-aminobutyric acid-mediated (GABAergic) inhibition of these neurosecretory neurons, which is achieved through a multisynaptic pathway that involves ascending catecholaminergic projections from neurons in the diagonal band of Broca (DBB). Baroreceptor activation induces a consistent increase in the firing rate of the DBB neurons, which project to the hypothalamic SON, potentially mediating baroreceptor-induced inhibition of hypothalamic vasopressinergic neurons (Chiodera et al. 1996; Jhamandas and Renaud 1986a,b; Kimura et al. 1994; Otake et al. 1991).

Afferent nerve impulses from stretch receptors in the left atrium, aortic arch, and carotid sinus tonically inhibit AVP secretion, and a reduction in their discharge leads to AVP release (Chowdrey and Bisset 1988). This information is relayed through the vagal and glossopharyngeal nerves to the NTS in the brainstem, respectively, from which postsynaptic pathways connect with the magnocellular neurons of the SON and PVN (Share 1988; Duan et al. 1997). Low-pressure receptors in the atrium tonically inhibit AVP release via a pathway involving the NTS, and AVP release induced by hypovolemia occurs through a reduction in the activity of this inhibitory input (Chowdrey and Bisset 1988; Share 1988).

A decrease in the arterial pressure, in turn, activates peripheral low-volume receptors in the great veins, atria, and lungs. Moreover, neural inputs coming from these peripheral regions induce an increase in the excitability of AVP-secreting neurons, achieved via pathways that include direct projections from the A1 neurons in the caudal ventrolateral medulla. Severe stimuli appear to involve the activation of both A1 projections and an additional AVP stimulatory pathway that bypasses the A1 region (Smith and Johnson 1995).

The main behavioral response observed after hypovolemia is the increase in specific appetite for sodium (Denton 1982; Johnson and Thunhorst 1997; McKinley et al. 1986; Muller et al. 1983; Weisinger et al. 1982, 1985), which is mediated by the activation of the renin–angiotensin–aldosterone system (RAAS) (Fitzsimons 1998). This effector system, together with an increase in sympathetic activity, also stimulates sodium reabsorption by the kidneys. A decrease in blood volume or pressure as large as 10% is also an important stimulus for water intake, but at a lesser extent, if compared to the hyperosmolality-mediated water ingestion.

3.2 NEUROCHEMICAL PATHWAYS REGULATING EXTRACELLULAR VOLUME AND OSMOLALITY

The pioneer studies to determine the actions of neurotransmitters in synaptic hypothalamic structures that regulate body fluid homeostasis were published by Grossman in the 1960s (Grossman 1960). He demonstrated that stimulation of the hypothalamus with noradrenergic and cholinergic agonists increased the intake of both food and water. In parallel, studies completed at the end of decade showed that intracranial injection of ANG II induced a rapid increase in water intake (Fitzsimons 1998) and that injection of carbachol (a cholinergic agonist) into the anterior and ventral portion of the third cerebral ventricle—or intracerebroventricular (icv) injection—resulted not only in thirst, but also in intense natriuresis, kaliuresis, and antidiuresis (Antunes-Rodrigues and McCann 1970), effects very similar to those observed after central injections of hypertonic saline (Dorn and Porter 1970). Subsequent studies showed that the effects of carbachol injected (icv) were mediated by muscarinic receptors and that they could interact with noradrenergic systems (Camargo et al. 1976, 1979; da Rocha et al. 1985; Dorn et al. 1970; Franci et al. 1980, 1983; Menani et al. 1984; Morris et al. 1976, 1977; Perez et al. 1984; Saad et al. 1975, 1976).

Research on brain control of renal sodium excretion paved the way for the demonstration of a central control of the secretion of ANP. Originally discovered as a peptide that induces natriuresis, ANP is released by blood volume expansion, possibly due to stretch receptors located in the atrium (Antunes-Rodrigues et al. 1991; Haanwinckel et al. 1995). However, carbachol injected icv also induced the release of ANP into the circulation (Baldissera et al. 1989). Moreover, ANP release induced by volume expansion could be blocked by previous administration of a muscarinic receptor blocker, atropine (Antunes-Rodrigues et al. 1993). Thus, muscarinic signaling in the CNS controls the release of ANP into the systemic circulation.

In addition to cholinergic and adrenergic pathways present in the forebrain, an important serotonergic pathway (also see Chapters 9, 10 and 11) emerges from the raphe system and AP to the lateral parabrachial nucleus (LPBN) (Armstrong and Stern 1998; Berger et al. 1972). The LPBN receives signals from peripheral volume receptors and, through the activation of serotonergic circuits, activates several neuroendocrine responses after extracellular volume expansion (EVE) (Godino et al. 2005). Margatho et al. (2007) published the first study demonstrating that the LPBN serotonergic mechanisms participate in the modulation of urine volume, and sodium and potassium excretion, and also regulates the plasma levels of ANP and OT in response to isotonic EVE. Subsequently, Margatho and colleagues (2008) demonstrated that isotonic EVE also induced an increase in c-Fos immunoreactivity in the central amygdaloid nucleus (CeA), bed nucleus of the stria terminalis, and in the PVN. These structures are monosynaptically connected with the LPBN and, considering that the CeA is rich in GABA receptors, these studies supported a role for GABAergic mechanisms within the CeA in the control of urine electrolyte excretion and hormone release induced by isotonic and hypertonic increases in extracellular volume. Accordingly, it has been demonstrated that the increment in renal sodium excretion as well as in OT and ANP plasma levels observed in response to isotonic and hypertonic EVE were attenuated in rats pretreated with the agonist of GABA-A receptors (muscimol) injected into the CeA, suggesting that the GABAergic inhibitory drive coming from this area is able to regulate the activity of oxytocinergic magnocellular neurons of the PVN and SON. Curiously, bicuculline (an antagonist of GABA-A receptors) injected into the CeA specifically regulated urinary volume and did not affect sodium excretion in response to the same stimuli (Margatho et al. 2009). Taken together, these results represented the first evidence to suggest that the GABAergic system of the CeA modulates the physiological mechanisms involved in the control of urine electrolyte excretion and hormone release in rats submitted to changes in extracellular volume and osmolality.

Besides being considered key sites for the regulation of drinking behavior, the LPBN and the CeA have been also implicated in the regulation of hormone secretion from neurohypophysis. However, it is still unclear which central pathways are responsible for connecting the serotonergic system of the LPBN with the oxytocinergic neurons in the PVN and SON under isotonic hypervolemic conditions. Additionally, it is reasonable to postulate that the serotonergic system, together with the CeA GABAergic cells/receptors, are part of this functional pathway that modulates sodium appetite, plasma OT and ANP release, and consequently, the renal responses elicited by volume expansion.

Physiologic responses to changes in the extracellular osmolality or volume must operate synergistically to modulate renal water and sodium excretion. These changes are detected by receptors, known to operate through stretch-activated channels, and are communicated to specific areas of the CNS, which are responsible for integrating the effector responses. In addition, the lamina terminalis (where the SFO and OVLT are located), PVN, SON, and several brainstem nuclei, among others, seem to be a part of this complex circuitry. Such structures, when stimulated, may trigger responses that include the modulation of (1) the renal environment, through the control of vasculature tonus and RAAS activity; (2) AVP secretion; (3) the release of natriuretic hormones ANP and OT; (4) thirst, salt appetite, or both; and (5) sympathetic activity.

Over the past decades, attempts were made to identify the specific brain areas responsible for the regulation of plasma osmolality, in particular, the control of ingestion and excretion of water and electrolytes. Neurons produce and release a substantial number of chemical mediators (neurotransmitters and neuromodulators) into the interstitial space, which act locally as paracrine and/or autocrine signals, or via the systemic circulation as neurohormones, exerting their effects at distant targets. Several locally produced mediators have been implicated in the regulation of endocrine function. This list of substances is still growing and includes not only classic neurotransmitters such as acetylcholine, amines (noradrenaline, dopamine, serotonin, and histamine), and amino acids (glutamate, GABA, taurine, and glycine), but also purinergic (adenosine, monophosphate of adenosine, and triphosphate of adenosine), gaseous [nitric oxide (NO), carbon monoxide and hydrogen sulfide], ionic (calcium), and lipid-derived (endocannabinoids) mediators. Peripheral hormones such as the adrenal (glucocorticoids and mineralocorticoids) and gonadal steroids (estrogen, progesterone, and testosterone) are also involved in the control of neuronal activity.

3.3 GASEOUS NEUROMODULATORS AND HYDROMINERAL BALANCE

Soluble gases are autocrine or paracrine signaling molecules with short half-lives. The best-studied gaseous neuromodulator, NO, is synthesized from L-arginine. Until the 1980s, NO was known as an environmental gas pollutant that was able to exert a variety of effects in biological systems. However, Ignarro et al.'s (1987) discovery that NO is an endothelium-derived relaxing factor improved our understanding of the physiological role of NO signaling in different peripheral and central functions. Subsequently, a series of studies that showed the vasodilatory effects of NO on the cardiovascular system led Dr. Ignarro and colleagues to be awarded with the Nobel Prize in Physiology and Medicine in 1998 (Huang et al. 1995; Moncada and Higgs 1993; Palmer et al. 1987). Since then, however, NO, as well as endogenous carbon monoxide (CO) and hydrogen sulfide (H_2S) have been shown to participate in several neuroendocrine regulatory mechanisms (Brann et al. 1997; Costa et al. 1996; Grossman et al. 1997; Mancuso et al. 2010; Rivier 1998).

Once synthesized, these highly membrane-permeable molecules are rapidly converted into ineffective products by specific chemical reactions, thus restricting their actions in an autocrine and/or paracrine manner (Suematsu 2003). The short half-life of these molecules obligates a colocalization or a close spatial proximity between these molecules and the enzymatic machinery responsible for their synthesis.

3.3.1 Nitric Oxide

NO is synthesized from the amino acid L-arginine by the enzyme nitric oxide synthase (NOS), and this reaction provides equimolar quantities of L-citruline. Three distinct NOS enzymes have been identified with different cellular distributions and regulatory mechanisms. Neuronal (nNOS) and endothelial NOS (eNOS) are both constitutively expressed in several tissues, and they are activated by increased intracellular Ca^{2+} concentrations, whereas inducible NOS (iNOS) is Ca^{2+}-independent (Salter et al. 1991). The cellular mechanisms of NO signaling involve not only the activation of soluble guanylate cyclase (sGC), which in turn increases cGMP intracellular levels (Moncada et al. 1991), but also a direct effect via protein *S*-nitrosylation (Gow et al. 2002). NO generated locally exerts numerous actions, including its modulatory effect on the hypothalamo-neurohypophysial tract, which is the principal neuroendocrine axis involved in body fluid homeostasis (Forsling and Grossman 1998).

NOS is expressed in several neuronal clusters, including the lamina terminalis (SFO, MnPO, OVLT), magnocellular neurosecretory cells of the SON and PVN nuclei, and the neurohypophysis (Bredt et al. 1990, 1991; Ueta et al. 2002). These structures form a neural network that participates in the control of drinking behavior as well as OT and AVP secretion in response to osmotic stimulation (Bourque et al. 2002; Verney 1947). NO has been shown to have either a stimulatory (Calapai and Caputi 1996; Reis et al. 2010) or an inhibitory (Liu et al. 1996) effect on water intake. In addition, rats centrally or peripherally injected with L-NAME (a nonselective inhibitor of NOS) have higher levels of plasma AVP and OT than the vehicle-treated animals (Kadekaro et al. 1997; Reis et al. 2007, 2010). Interestingly, up-regulation of NOS messenger RNA (mRNA) and of NOS activity has been demonstrated in rats subjected to chronic salt loading, a condition known to elevate plasma AVP and OT levels (Kadowaki et al. 1994; Serino et al. 2001; Ventura et al. 2002, 2005). Thus, NO has a tonic inhibitory effect on neurohypophysial secretion, which raises the intriguing question of why does NO production increase during sustained AVP and OT increased secretion? Some studies have attributed this response to a mechanism responsible for preventing the hormonal stores depletion (Srisawat et al. 2004; Ventura et al. 2005).

These findings indicate that the nitrergic system responds to changes in plasma volume or extracellular tonicity and that NO fluctuations could be involved in the fine-tuned control that regulates neurohypophysial secretion, not only at basal levels but also during high hormonal demand. It is well established that AVP and OT are synthesized in the soma of the magnocellular neurons and, via axonal transport, they reach the neurosecretory terminals in the neurohypophysis (Gainer et al. 1977). Thus, changes in the neuronal firing rate and/or in the rate of AVP/OT synthesis have been considered key points to the nitrergic modulatory effects on magnocellular cells.

Several studies using extracellular and intracellular electrodes have demonstrated that the precursor for NO synthesis L-arginine decreases, whereas L-NAME increases the firing frequency of the magnocellular secretory cells, suggesting a tonic inhibitory effect on neuronal activity (Liu et al. 1997; Stern and Ludwig 2001; Ventura et al. 2008). Interestingly, Stern and Ludwig (2001) reported an increase in GABA inhibitory postsynaptic currents in brain slices containing magnocellular cells bathed with NO donors. Moreover, the demonstration that cGMP is not synthesized in magnocellular neurons but in surrounding GABAergic neurons strongly suggested that NO could regulate the release of AVP and OT indirectly by modulating the activity of the afferent inputs to magnocellular cells (Vacher et al. 2003). However, in NOS-positive cells, the proposal that a gaseous molecule is required to pass through the cell membrane to exert its effect is not intuitive because some membrane proteins are directly affected by NO, for example, by inducing the nitrosylation of ion channels (Gonzalez et al. 2009). Thus, it appears also plausible that NO can exert its effects intracellularly, and not only by activating extracellular signaling pathways. This hypothesis has been confirmed by the demonstration that L-arginine decreases the firing rate of magnocellular cells even in the presence of synaptic blockers (Ventura et al. 2008).

In addition to its central actions, NO exerts actions in the periphery influencing body fluid homeostasis. In the heart, the secretion of ANP, a potent hormone that induces renal sodium and water excretion, is modulated by local NO fluctuations. Studies have demonstrated that the basal release of ANP is increased in rat-isolated atria bathed with methylene blue, a NO scavenger (Sanchez-Ferrer et al. 1990). Similarly, the presence of L-NAME increases (Leskinen et al. 1995; Skvorak and Dietz 1997), whereas L-arginine decreases the basal levels of plasma ANP (Melo and Sonnenberg 1996). In addition, the increase in ANG II-induced ANP release and natriuresis is potentiated by pretreatment with a central injection of L-NAME (Reis et al. 2007). These findings indicate that NO produced by endothelial cells and/or originated from brain tissue has a tonic inhibitory effect on ANP release. Moreover, an abnormal regulation of this NO/ANP modulatory pathway may be involved in the genesis of heart failure (Dietz 2005).

Finally, all the neuroendocrine mechanisms described above converge at the kidneys, the final common pathway that regulates sodium and water excretion. The demonstration that both NOS are present in vascular and tubular elements of the kidneys has supported the NO effects on renal function, particularly its role in the control of renin secretion (Kakoki et al. 2001; Mattson and Wu 2000; Reid 1994; Zou and Cowley 1999). For example, it has been demonstrated that bradykinin and acetylcholine induces renal vasodilatory effects by increasing NO synthesis, which contributes to an increase in diuresis and natriuresis (Bachmann and Mundel 1994). Thus, NO produced in the macula densa is involved in the regulation of glomerular capillary pressure (Wilcox et al. 1992), in the renin release by the justaglomerular cells (Schnackenberg et al. 1997) and in the ANG II–mediated tubular reabsorption (De Nicola et al. 1992). Additionally, sustained changes in NO synthesis have been implicated in some pathological conditions such as glomerulonephritis and chronic kidney disease with its cardiovascular complications (Bachmann and Mundel 1994; Baylis 2011).

3.3.2 Carbon Monoxide

The enzyme heme oxygenase (HO) was first identified in the spleen and liver, where it metabolizes heme derived from the breakdown of aging blood cells (Tenhunen et al. 1968, 1969). For several years, the HO signaling system was considered only as a "molecular wrecking ball" for the degradation of heme and formation of toxic waste products, such as carbon monoxide (CO) and bile pigments (Maines and Kappas 1974). Since the early 1990s, however, after the discovery of high levels of HO in the brain (Sun et al. 1990), a role for HO signaling in central physiologic functions has been proposed.

HO catalyzes the oxidative cleavage of the heme ring with the resulting formation of biologically active molecules, such as the cellular antioxidant precursor biliverdin, free iron, and CO (Stocker et al. 1987; McDonagh 1990). Apart from controlling heme levels, the HO system has additional cellular functions, including cellular defense mechanisms, regulation of gene expression, and synaptic transmission (Shinomura et al. 1994).

There are at least three different isoforms of HO that have been described to date: (1) HO-1, the first identified metal-inducible enzyme (Maines and Kappas 1974, 1977), is the stress-induced isoform and has also been referred to as a heat shock protein (Keyse and Tyrrell 1987; Shibahara et al. 1987). The inducible HO-1 isoform is up-regulated by a vast number of chemicals and stimuli ranging from infection, oxidative stress (Dwyer et al. 1995), hypoxia/hyperthermia (Ewing and Maines 1991; Ewing et al. 1992), ischemic injury, hemorrhage and trauma (Geddes et al. 1996; Turner et al. 1998), as well as hyperosmotic stimulation (Richmon et al. 1998; Reis et al. 2012). (2) HO-2, wherein the constitutive isoform is expressed in almost all cell types and corresponds to most of the brain HO activity (Sun et al. 1990; Ewing and Maines 1992; Verma et al. 1993). This isoform is responsive to glucocorticoids (Weber et al. 1994). (3) The constitutive isoform HO-3 has also been identified (McCoubrey et al. 1997); however, its role in physiological and pathological conditions has yet to be clarified.

The two most important isoenzymes, HO-1 and HO-2, display vast differences in tissue distribution. Under basal conditions, the HO-1 mRNA and protein are generally undetectable in the whole brain, being only highly expressed in a few select neuronal populations, such as those in the dentate gyrus, cerebellum, hypothalamus, and the brainstem (Ewing and Maines 1991, 1992; Ewing et al. 1992; Maines et al. 1998; Vincent et al. 1994). On the other hand, HO-2 is expressed in high densities in the brain, and it is much more widely expressed in neuronal populations under basal conditions, including in the endothelial lining of the blood vessels (Ewing et al. 1994; Ewing and Maines 1992; Verma et al. 1993). Because of the predominance of neurons that express HO-2, the activity of this isoenzyme is commonly responsible for the expressive CO-generating capacity of the brain (Dawson and Snyder 1994).

CO is one of the most toxic molecules in nature. This biologically active diatomic gas may act as an intracellular messenger binding to the iron in the heme group of the soluble guanylyl cyclase (sGC), whose activation determines cGMP production (Ignarro et al. 1986; Maines 1993; Marks et al. 1991; Verma et al. 1993; Zhuo et al. 1993). HO and sGC are colocalized in several brain regions, and HO inhibitors were shown to deplete cGMP levels (Verma et al. 1993). There is growing evidence that supports a role for CO as a biologically active signaling molecule. Verma and colleagues (1993) demonstrated the physiological importance of CO in synaptic plasticity, followed by several reports demonstrating the importance of CO as an endogenous neuromodulator in the CNS (Snyder et al. 1998; Zakhary et al. 1996).

Within the hypothalamus, magnocellular neurosecretory neurons of the PVN and SON show a moderate distribution of HO-1 immunoreactivity, in contrast with the high density of HO-2 staining (Maines 1997; Reis et al. 2012; Vincent et al. 1994). Moreover, the hypothalamus has been shown to have one of the highest CO production rates in the brain (Laitinen and Juvonen 1995; Maines 1997; Mancuso et al. 1997). Activation of HO resulting in the formation of CO plays a primary role in the central control of AVP and OT (Antunes-Rodrigues et al. 2004).

Very few studies have demonstrated that the endogenously produced CO plays a role in fluid/electrolyte homeostasis via the modulation of magnocellular neurosecretory neurons. A recent study showed that HO-1 is expressed in the OT and AVP neurons, suggesting that both neuronal cell types have the capacity to produce CO within the SON and the PVN. Moreover, in animals subjected to water deprivation, HO-1 immunoreactivity in both vasopressinergic and oxytocinergic neurons is increased (Reis et al. 2012). Using different experimental approaches, Richmon and colleagues (1998) showed that the hyperosmotic challenge induced in the rat brain after intracarotid administration of hyperosmolar mannitol promoted HO-1 expression in astrocytes and microglia/macrophages in the ipsilateral hemisphere. Collectively, these data indicate that HO-1 is activated by osmotic challenges, consequently increasing endogenous CO levels. Furthermore, the inhibition of HO was shown to decrease the hyperosmolality-induced release of OT and ANP into the incubation medium by the medial basal hypothalamus *in vitro* (Gomes et al. 2004, 2010). In the same way, inhibition of this enzyme also induced a strong hyperpolarization and a decrease in the firing rate of magnocellular neurosecretory cells in water-deprived animals (Reis et al. 2012). Taken together, these results suggest that CO has a stimulatory effect on AVP and OT secretion. However, CO may have opposing effects on OT and AVP release induced by endotoxemia (Giusti-Paiva et al. 2005; Mancuso et al. 1999; Moreto et al. 2006) and K^+ stimulation (Kostoglou-Athanassiou et al. 1996, 1998; Mancuso et al. 1997). Thus, differences in the experimental conditions, including the concentration and type of drugs used, among others, may account for these reported discrepancies.

Therefore, CO actions on AVP and OT secretion may be dependent on the type of challenge. However, the observation that endogenous CO interferes with neurohypophysial hormone release may be directly associated with hydromineral imbalances. Furthermore, in addition to acting individually, CO has been shown to interact with the nitrergic system in the control of hydromineral homeostasis. It was previously demonstrated that the use of an inhibitor of HO increases the production of L-citruline, a co-product of NO synthesis, by the medial basal hypothalamus *in vitro* in response to hyperosmolality (Gomes et al. 2010). Furthermore, it has also been shown that CO can bind to the heme group of NOS and inhibit its enzymatic activity, consequently controlling NO production (White and Marletta 1992). On the other hand, NO donors were shown to increase both the expression and the activity of HO (Foresti et al. 1997).

Therefore, CO seems to play a predominately facilitatory role on AVP and OT release, although most of the studies agree that NO has an opposite effect, inhibiting AVP and OT release. Based on the evidence that the final products of both NOergic and COergic systems can reciprocally affect enzymatic activity, it can be hypothesized that these two systems integrate a feedback loop implicated in the local control of hypothalamic neurosecretory system output.

3.4 PURINERGIC SYSTEM AND CONTROL OF FLUID BALANCE

Adenosine triphosphate (ATP) has been implicated in the regulation of several biological functions, including neurotransmission (Ribeiro 1978). ATP can act as an autocrine and paracrine regulator preferentially stimulating AVP secretion from the neurohypophysial terminals (Troadec et al. 1998). According to this study and those of others (Knott et al. 2008), the exocytosis of neurosecretory granules containing neuropeptides and ATP would produce a positive feedback for the release of AVP and more ATP from neighboring vasopressinergic terminals. The presence of ATP receptors in AVP- and OT-expressing neurons of the hypothalamus has been well characterized (Guo et al. 2009). These cells types expressed different subunits of the P_{2X} receptor, which could provide the neuroanatomical basis for why different responses are elicited by ATP in the secretion of these neuropeptides.

Finally, ATP actions at terminals are regulated by the presence of ectonucleases at presynaptic sites in neurohypophysis (Thirion et al. 1996), which rapidly hydrolyzes ATP to adenosine 5′-diphosphate, adenosine 5′-monophosphate, and adenosine (Gordon et al. 1989). Adenosine, the final metabolite of ATP, is abundant in the mammalian brain and acts at specific receptors at both the pre- and postsynaptic terminals to inhibit hormone release, terminating the effects initiated by ATP at the synaptic cleft (Rathbone et al. 1999; White 1988). Although both inhibitory and excitatory actions have been attributed to adenosine (Cunha 2001), it is believed that the majority of the reported effects is mediated by inhibitory A_1 receptors (Ponzio and Hatton 2005; Reppert et al. 1991).

3.5 HYPOTHALAMIC–PITUITARY–ADRENAL AXIS AND CONTROL OF FLUID BALANCE

Both isotonic and hypertonic EVE induce an increase in corticosterone secretion (Durlo et al. 2004; Ruginsk et al. 2007), suggesting a role for the hypothalamic–pituitary–adrenal (HPA) axis in the modulation of neurohypophysial hormone secretion. Indeed, studies have demonstrated that the administration of dexamethasone, a synthetic glucocorticoid, inhibits the secretion of OT, but not of AVP, induced by a hypertonic EVE (Ruginsk et al. 2007) or central cholinergic and angiotensinergic stimulation (Lauand et al. 2007). These endocrine responses were correlated with neuronal changes, because the activity of the magnocellular neurons of the PVN and SON was also decreased in dexamethasone-treated rats subjected to the same stimuli (Ruginsk et al. 2007). Thus, glucocorticoids can act centrally to negatively modulate neurohypophysial hormone synthesis and secretion. Activation of the HPA axis under stress conditions is known to exert a predominantly inhibitory effect on most of neuroendocrine functions, including gonadal and growth hormone axes (for review, see McCann et al. 2000).

Under basal conditions, glucocorticoids are bound predominantly to mineralocorticoid receptors (MRs). However, glucocorticoid receptors (GRs) are recruited to mediate the evoked increase in the circulating levels of cortisol or corticosterone by stress stimuli or during the ultradian secretion peak (Reul and de Kloet 1985; Windle et al. 1998). These two types of receptors not only present different affinities depending on ligand concentrations but are also differentially distributed in the CNS. The GRs are diffusely expressed, whereas MRs are restricted to hypothalamic areas, such as the PVN, and the limbic system (hippocampus, amygdala, lateral septal nucleus, and some cortical areas) (de Kloet et al. 2005). GRs are expressed in the parvocellular neurons of the PVN and in the SON, which contains only magnocellular neurons, but not in the magnocellular subdivision of the PVN (Han et al. 2005; Kiss et al. 1988). However, subsequent studies have reported an inhibitory effect of dexamethasone administration on the number of single (c-Fos) and double labeled (c-Fos/OT and c-Fos/AVP) magnocellular neurons of the PVN in response to diverse types of stimulation (Lauand et al. 2007; Ruginsk et al. 2007). We can also speculate that glucocorticoids act not only on neurons, but also on glial cells because both GRs and MRs are expressed in this nonneuronal population (Sierra et al. 2008).

Once activated, the complex formed by the ligand and its receptor can form homo- or heterodimers and bind to glucocorticoid responsive elements in the promoter region of target genes or directly interact with transcription factors as monomers (Beato and Sanchez-Pacheco 1996). This genomic action of glucocorticoids was the only known pathway described until the end of the past century. However, more recently, some responses evoked by glucocorticoids independently of the transcriptional pathway have been described (Limbourg et al. 2002; Mikics et al. 2004; Orchinik et al. 1991; Sandi et al. 1996). These effects are observed within a few minutes after hormone secretion and cannot be blocked by the inhibition of protein synthesis, what characterizes them as nongenomic actions (Newton 2000).

Consistent with this finding, Limbourg and Liao (2003) demonstrated that high doses of glucocorticoids can prevent ischemic tissue injury. This effect was attributed to the rapid glucocorticoid-mediated production and release of NO from vascular endothelial cells. Although some studies have suggested that these actions may be initiated by the glucocorticoid binding to different types of receptors, such the ones coupled to G proteins, the participation of the classic cytoplasmic receptors in these responses could not be ruled out because receptors with structural homology to GRs and MRs were also identified in neuronal and nonneuronal membranes (Gametchu et al. 1993; Liposits and Bohn 1993), suggesting that the same proteins may have very distinct properties depending on the subcellular compartment. Growing evidence suggests that this fast nongenomic pathway is involved in the regulation of hypothalamic function and recruits other signaling molecules, such as endogenously produced cannabinoids (Di et al. 2003), as discussed in the following section.

3.5.1 Endocannabinoids

Recently, endogenously produced cannabinoids have emerged as important signaling molecules mediating several physiological processes. The main endocannabinoids (ECBs) described thus far are anandamide (AEA) and 2-arachidonoilglycerol (2-AG), which are derived from membrane phospholipids. They are produced and metabolized by independent enzymatic pathways and have distinct pharmacological properties (Freund et al. 2003).

A receptor for cannabis-like substances (CB1R) was identified for the first time in 1988. This isoform has a widespread expression in the CNS and its mRNA is found in the PVN and SON, as well as in the external layer of the median eminence in rodents (Herkenham et al. 1991; Wittmann et al. 2007). In humans, although the posterior pituitary presents only a poor CB1R immunoreactivity, this receptor subtype is densely expressed in both the anterior and intermediate lobes (Pagotto et al. 2001). Another receptor subtype (CB2R), characterized few years later, exhibits 44% homology with CB1R and a very distinct expression profile. The physiological actions of the ECB system are produced mostly by the interaction of endogenous ligands with CB1R and/or CB2R. However, the involvement of orphan receptors such as GPR55 in the mediation of some responses cannot be ruled out.

Several studies have suggested that the ECB system may be active under basal conditions as a result not only of constant production and release of ECBs, but also of the presence of constitutively activated CB1R, even in the presence or absence of relatively low concentrations of the endogenous ligands (Pertwee 2005). AEA binds to CB1R with high affinity and regulates the activity of intracellular effector proteins as a partial agonist (Bouaboula et al. 1995). More recently, AEA was shown to act as a promiscuous ligand, with high-affinity binding to the vanilloid type 1 receptor (TRPV1). According to Tóth and colleagues (2005), AEA simultaneously activates ECB and vanilloid systems in several areas of the CNS. Accordingly, some effects of AEA cannot be mimicked by the synthetic cannabinoid agonist, WIN55,212-2 (Al-Hayani et al. 2001), and some responses that were primarily attributed to AEA, such as antinociception, are still conserved in CB1R knockout mice (Di Marzo et al. 2000). In contrast, 2-AG, the most abundant ECB found in the CNS, has a low affinity for CB1R when compared to AEA but can stimulate the same intracellular cascade as a full agonist (Mechoulam et al. 1995).

ECBs may mediate the actions of glucocorticoids, particularly in the CNS. Activation of the HPA axis, with the consequent release of glucocorticoids, is known to exert an inhibitory effect on the activation of most neuroendocrine functions, including the synthesis/secretion of the neurohypophysial peptides, in particular, OT (Durlo et al. 2004; Lauand et al. 2007; Ruginsk et al. 2007). Nongenomic signaling pathways activated by the binding of glucocorticoids to conserved cell membrane sites have already been characterized (Avanzino et al. 1987; Evans et al. 2000; Suyemitsu and Terayama 1975), and this mechanism has been implicated in the rapid changes observed in neurotransmission after an increase in circulating levels of glucocorticoids.

The local production and release of ECBs within the PVN and SON in the presence of glucocorticoids result in decreased glutamatergic and increased GABAergic inputs to both parvocellular and magnocellular neurons *in vitro* (Di et al. 2003, 2005, 2009). In addition to directly stimulating GABA neurotransmission, ECBs have also been shown to stimulate NO release, which potentiates the negative drive to these cell groups. The CB1R appears as the main isoform involved in these responses because rimonabant, a CB1R antagonist, potentiated the secretion of AVP and OT induced by hypertonic EVE in whole animals (Ruginsk et al. 2010) and reversed the inhibitory effects of dexamethasone on hormone release under the same experimental conditions (Ruginsk et al. 2012). Recently, these mechanisms of synaptic regulation have also been implicated in the fast feedback regulation of the HPA axis induced by restraint stress (Evanson et al. 2010) and also in the desensitization of endogenous glucocorticoid responses after repeated exposures to a stress paradigm (Hill et al. 2010).

The ECB system produces a well-known orexigenic effect (Di Marzo et al. 2001), but its participation in the control of water intake remains elusive. It is well known that food consumption induces acute changes in osmolality and that both fluid homeostasis and energy metabolism share common central pathways. The PVN, NTS, and LPBN are involved in reactive and hedonic control of energy balance (Elias et al. 2000) and also integrate information regarding extracellular volume and osmolality (Antunes-Rodrigues et al. 2004; Godino et al. 2010; Margatho et al. 2007, 2008).

3.6 GONADAL STEROIDS AND CONTROL OF FLUID BALANCE

The first studies that proposed that gonadal hormones are involved in the control of hydromineral balance stemmed from data showing that humoral changes that normally occur during the estrous and/or menstrual cycle are followed by changes in both behavioral and humoral responses related to fluid balance. Consistent with previous reports, it is known that water intake and sodium appetite can vary along the estrous cycle (Danielsen and Buggy 1980; Tarttelin and Gorski 1971). In addition, pioneer studies from Antunes-Rodrigues and Covian (1963) reported for the first time the spontaneous decrease in sodium intake during estrus, which is the stage when the female reproductive cycle undergoes high estrogen plasma concentrations, and increase in sodium intake during diestrus, when low plasma levels of estrogen are detected.

Results from different groups have demonstrated that water intake is enhanced in ovariectomized (OVX) rats and that these responses are prevented or partially attenuated by estrogen replacement. Estrogen attenuates water intake of OVX rats during treatment with hyperoncotic colloid (Vijande et al. 1978), isoproterenol, and hypertonic saline (Carlberg et al. 1984) and after water deprivation (Krause et al. 2003).

Several groups have reported that estrogen therapy decreases not only water intake but also abolishes salt ingestion induced by icv injection of ANG II in OVX rats (Do-Vale et al. 1995; Mecawi et al. 2007). In addition, an enhanced appetite for NaCl in adult OVX rats under long-term sodium deprivation was observed by Stricker and colleagues (1991), and this response was blunted by estrogen therapy. Conversely, data reported by Kensicki and colleagues (2002) have demonstrated that increasing plasma levels of estrogen promoted by systemic administration were positively correlated with daily sodium intake in OVX rats, and OVX rats ingested less salt than intact females in response to ANG II microinjection into the MnPO (De-Angelis et al. 1996).

More recently, the participation of the RAAS in the mediation of the behavioral responses induced by estradiol has been demonstrated. Estradiol was shown to inhibit ANG II AT1 receptors and angiotensin-converting enzyme expression in the SFO (Dean et al. 2006; Kisley et al. 1999; Krause et al. 2006). Furthermore, our group demonstrated that the central AT1 blockade inhibited: (1) the dipsogenic response induced by water deprivation, osmotic stimulation, chronic sodium depletion, and furo + cap protocol and (2) the natriorexigenic response induced by sodium depletion in OVX rats (Mecawi et al. 2007, 2008). These studies also demonstrated that estrogen administration significantly attenuated the losartan-induced antidipsogenic and antinatriorexigenic actions driven by sodium depletion and the furo + cap experimental models. Consistent with recent evidence from literature suggesting a possible protective effect of estradiol on the cardiovascular system, these observations further support a role for this hormone on the behavioral control of water and salt intake, because a clear influence of estrogenic status on the AT1-mediated behavioral responses has been demonstrated through different experimental protocols.

In addition, estrogen also modulates the inhibitory serotonergic mechanisms controlling salt intake. Several studies have demonstrated that estrogen modulates serotonin synthesis and release (Robichaud and Debonnel 2005; Rubinow et al. 1998), induces an increase in the basal firing rate of serotonergic neurons in female rats (Peysner and Forsling 1990), and enhances serotonin transporter mRNA expression in the dorsal raphe nucleus (DRN) of OVX animals (Sumner et al. 1999; McQueen et al. 1997). Furthermore, when estrogen is chronically administered to OVX rats, an increase in the mRNA expression for tryptophan hydroxylase-2 was observed in the DRN (Donner and Handa 2009). Estrogen-dependent inhibition of sodium appetite in both normally cycling rats and OVX animals with estradiol replacement may involve an interaction between excitatory neurons from the OVLT and the inhibitory serotonergic drive from the DRN, which are key pathways that underlie the responses to hyponatremia and hypovolemia (Dalmasso et al. 2011).

Further effects of estrogens on fluid balance involve the control of neurohypophysial hormones and ANP. Changes in plasma AVP and OT during the reproductive cycle in both animal models and humans have been previously demonstrated (Forsling and Peysner 1988; Forsling et al. 1981; Mitchell et al. 1981; Windle and Forsling 1993). Estradiol receptor beta (ER-β) is expressed in the SON and PVN, where it colocalizes with AVP and OT (Alves et al. 1998; Hrabovszky et al. 2004; Laflamme et al. 1998), and estradiol receptor alpha (ER-α) is expressed in afferent osmosensitive neurons located in the SFO and OVLT that project to the SON and the PVN (Shughrue et al. 1997; Somponpun et al. 2004; Voisin et al. 1997). Consistent with findings for AVP and OT neurons, studies also demonstrated the colocalization of ERs and ANP in cardiomyocytes (Back et al. 1989; Jankowski et al. 2001).

Corroborating the colocalization studies, several others have demonstrated the stimulatory effects of estradiol on OT and ANP secretion (Amico et al. 1981; Belo et al. 2004; Caligioni and Franci 2002; Karjalainen et al. 2004; Xu et al. 2008). However, regarding the control of AVP secretion, whereas some have reported a stimulatory effect for estradiol (Barron et al. 1986; Crowley and Amico 1993; Hartley et al. 2004; Skowsky et al. 1979), other studies have shown that estradiol does not induce any changes in this parameter (Crofton et al. 1985; Swenson and Sladek 1997).

Recent data demonstrated the contribution of estradiol on AVP, OT, and ANP secretion as well as on vasopressinergic and oxytocinergic neuronal activation in response to EVE and hypovolemic shock (Mecawi et al. 2011; Vilhena-Franco et al. 2011). In addition to what has been demonstrated by previous studies, these data suggest that estradiol modulates AVP, OT, and ANP secretion in response to hyperosmolality and hypovolemia. It was observed that estradiol replacement did not alter hormonal secretion in response to isotonic EVE, but it increased AVP secretion and potentiated plasma OT and ANP concentrations in response to hypertonic EVE (Vilhena-Franco et al. 2011). It has been also demonstrated that estradiol replacement potentiated AVP and OT hormonal secretion in response to hypovolemic shock, but did not alter ANP secretion induced by the same stimulus (Mecawi et al. 2011). The same studies also showed that the activity of both oxytocinergic and vasopressinergic neurons in the SON and PVN induced by hypertonic EVE and hypovolemic shock may be modulated by estradiol (Mecawi et al. 2011; Vilhena-Franco et al. 2011). These findings support a previous report from Hartley and colleagues (2004) that demonstrated an increase in neuronal activation as assessed by c-Fos immunolabeling in the SON induced by estradiol.

Collectively, these data indicate that estradiol participates in the control of fluid balance by (1) increasing the responsiveness of vasopressinergic and oxytocinergic neurons in the PVN and SON and the secretion of the neuropeptides, AVP, and OT; (2) increasing the release of natriuretic factors (ANP and OT) in response to osmotic stimulation; (3) decreasing brain ANG II responsiveness through the modulation of AT1-mediated signaling; and (4) increasing the DRN serotonergic activity.

3.7 CONCLUDING REMARKS

Figure 3.1 summarizes the main systems implicated in the control of hydromineral homeostasis, as well as the main reports of the literature supporting each mechanism. Cholinergic (Ach), adrenergic (Adr), and angiotensinergic pathways provide inputs from the lamina terminalis to the hypothalamic neurohypophysial system (HNS). Several *in vivo* and *in vitro* studies provided strong evidence showing that neuromodulators such as NO, CO and ECBs, and peptides such as ANP, produced

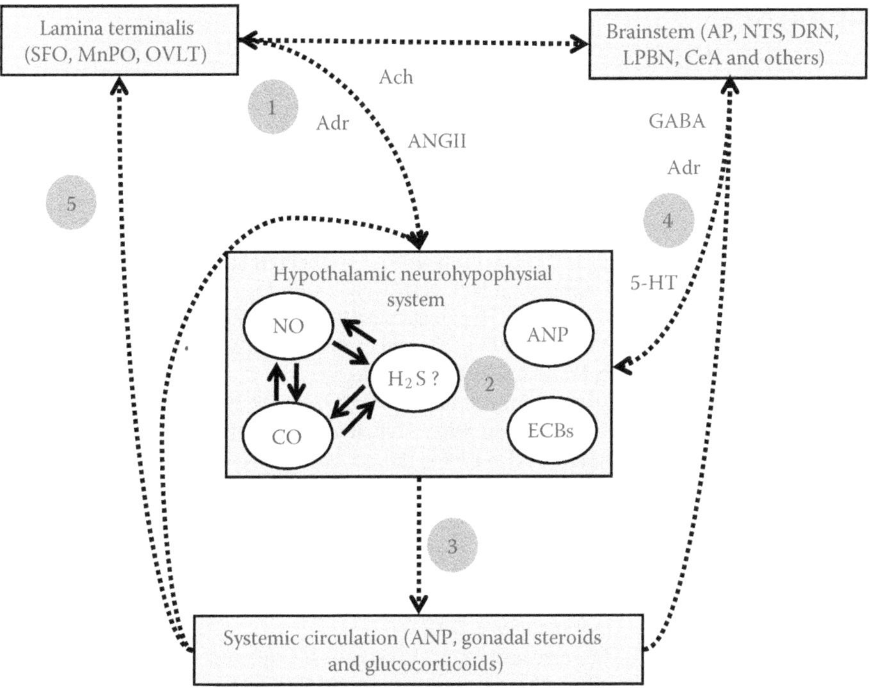

FIGURE 3.1 **(See color insert.)** Schematic illustration of hypothalamic–neurohypophysial system, its main connections and central/peripheral control. Each number indicates a mechanism discussed in the literature: Mechanism 1 (Antunes-Rodrigues et al. 2004), Mechanism 2 (Baldissera et al. 1989; Gomes et al. 2004, 2010; Reis et al. 2007; Ruginsk et al. 2010; Ventura et al. 2002, 2005), Mechanism 3 (Antunes-Rodrigues et al. 1991; Haanwinkel et al. 1995), Mechanism 4 (Godino et al. 2005, 2010; Margatho et al. 2007, 2008, 2009), and Mechanism 5 (Durlo et al. 2004; Lauand et al. 2007; Mecawi et al. 2007, 2008, 2011; Ruginsk et al. 2007; Vilhena-Franco et al. 2011).

locally within the HNS, modulate hormone release. The participation of other locally produced mediators, such as H_2S, however, remains unclear. Within the CNS, Adr and serotonergic (5-HT) pathways from the brainstem establish connections with the HNS, modulating neuroendocrine output. The consequent secretion of hypothalamic neuropeptides to the systemic circulation also modulates the release of other circulating factors, such as ANP, also implicated in the integrated renal responses to altered fluid volume and osmolality. Other circulating factors, such as estrogen and glucocorticoids, have a very prominent effect on the activity of the HNS, controlling both central and peripheral aspects of hydromineral balance. Although much progress has been made during the past few years to the understanding of how the body responds to hydroelectrolytic imbalances, a growing effort is being made to further elucidate the molecular aspects implicated in the regulation of hydromineral homeostasis, which may also contribute in the near future to an integrated view of this homeostatic function.

REFERENCES

Al-Hayani, A., K. N. Wease, R. A. Ross, R. G. Pertwee, and S. N. Davies. 2001. The endogenous cannabinoid anandamide activates vanilloid receptors in the rat hippocampal slice. *Neuropharmacology* 41:1000–5.

Alves, S. E., V. Lopez, B. S. McEwen, and N. G. Weiland. 1998. Differential colocalization of oestrogen receptor beta (ERbeta) with oxytocin and vasopressin in the paraventricular and supraoptic nuclei of the female rat brain: An immunocytochemical study. *Proc Natl Acad Sci USA* 95:3281–6.

Amico, J. A., S. M. Seif, and A. G. Robinson. 1981. Oxytocin in human plasma: Correlation with neurophysin and stimulation with oestrogen. *J Clin Endocrinol Metab* 52:988–93.

Andersson, B., and S. M. McCann. 1955. Hypothalamic control of water intake. *J Physiol* 129:44P.

Andersson, B., and S. M. McCann. 1956. The effect of hypothalamic lesions on the water intake of the dog. *Acta Physiol Scand* 35:312–20.

Andersson, B. 1977. Regulation of body fluids. *Annu Rev Physiol* 39:185–200.

Antunes-Rodrigues, J., and M. R. Covian. 1963. Hypothalamic control of sodium chloride and water intake. *Acta Physiol Lat Am* 13:94–100.

Antunes-Rodrigues, J., and S. M. McCann. 1970. Chemical stimulation of water, sodium chloride and food intake by injections of cholinergic and adrenergic drugs into the third brain ventricle. *Proc Soc Exp Biol Med* 138:1464–70.

Antunes-Rodrigues, J., M. J. Ramalho, L. C. Reis et al. 1991. Lesions of the hypothalamus and pituitary inhibit volume-expansion-induced release of atrial natriuretic peptide. *Proc Natl Acad Sci USA* 88:2956–60.

Antunes-Rodrigues, J., U. Marubayashi, A. L. Favaretto, J. Gutkowska, and S. M. McCann. 1993. Essential role of hypothalamic muscarinic and alpha-adrenergic receptors in atrial natriuretic peptide release induced by blood volume expansion. *Proc Natl Acad Sci USA* 90:10240–4.

Antunes-Rodrigues, J., A. L. Favaretto, J. Gutkowska, and S. M. McCann. 1997. The neuroendocrine control of atrial natriuretic peptide release. *Mol Psychiatry* 2:359–67.

Antunes-Rodrigues, J., M. Castro, L. L. K. Elias, M. Valença, and S. M. McCann. 2004. Neuroendocrine control of body fluid metabolism. *Physiol Rev* 84:169–208.

Armstrong, W. E., and J. E. Stern. 1998. Electrophysiological distinctions between oxytocin and vasopressin neurons in the supraoptic nucleus. *Adv Exp Med Biol* 449:67–77.

Arsenijevic, Y., and A. J. Baertschi. 1985. Activation of the hypothalamo-neurohypophysial system by hypertonic superfusion of the rat mesentery. *Brain Res* 347:169–72.

Avanzino, G. L., R. Ermirio, P. Ruggeri, and C. E. Cogo. 1987. Effects of corticosterone on neurons of reticular formation in rats. *Am J Physiol Regul Integr Comp Physiol* 253: R25–30.

Bachmann, S., and P. Mundel. 1994. Nitric oxide in the kidney: Synthesis, localization, and function. *Am J Kidney Dis* 24:112–29.

Back, H., W. G. Forssmann, and W. E. Stumpf. 1989. Atrial myoendocrine cells (cardiodilatin/atrial natriuretic polypeptide-containing myocardiocytes) are target cells for oestradiol. *Cell Tissue Res* 255:673–4.

Baldissera, S., J. V. Menani, L. F. dos Santos et al. 1989. Role of the hypothalamus in the control of atrial natriuretic peptide release. *Proc Natl Acad Sci USA* 86:9621–5.

Barron, W. M., J. Schreiber, and M. D. Lindheimer. 1986. Effect of ovarian sex steroids on osmoregulation and vasopressin secretion in the rat. *Am J Physiol Endocrinol Metab* 250:E352–61.

Baylis, C. 2011. Nitric oxide synthase derangements and hypertension in kidney disease. *Curr Opin Nephrol Hypertens* 21:1–6.

Beato, M., A. Sánchez-Pacheco. 1996. Interaction of steroid hormone receptors with the transcription initiation complex. *Endocr Rev* 17:587–609.

Belo, N. O., J. Silva-Barra, E. C. Carnio, J. Antunes-Rodrigues, J. Gutkowska, and A. M. Dos Reis. 2004. Involvement of atrial natriuretic peptide in blood pressure reduction induced by oestradiol in spontaneously hypertensive rats. *Regul Pept* 117:53–60.

Berger, J. M., J. C. Sosa-Lucero, F. A. D. L. Iglesia, A. G. Lumb, and S. A. Benecosme. 1972. Relationship of atrial catecholamines to cytochemistry and fine structure of atrial specific granules. *Recent Adv Studies Cardiac Struct Metab* 1:340–50.

Blackbum, R. E., W. K. Samson, R. J. Fulton, E. M. Stricker, and J. G. Verbalis. 1995. Central oxytocin and ANP receptors mediate osmotic inhibition of salt appetite in rats. *Am J Physiol Regul Integ Comp Physiol* 269:R245–51.

Bouaboula, M., C. Poinot-Chazel, B. Bourrié et al. 1995. Activation of mitogen-activated protein kinases by stimulation of the central cannabinoid receptor CB1. *Biochem J* 312:637–41.

Bourque, C. W. 2008. Central mechanisms of osmosensation and systemic osmoregulation. *Nat Rev Neurosci* 9:519–31.

Bourque, C. W., D. L. Voisin, and Y. Chakfe. 2002. Stretch-inactivated cation channels: Cellular targets for modulation of osmosensitivity in supraoptic neurons. *Prog Brain Res* 139:85–94.

Brann, D. W., G. K. Bhat, C. A. Lamar, and V. B. Mahesh. 1997. Gaseous transmitters and neuroendocrine regulation. *Neuroendocrinol* 65:385–95.

Bredt, D. S., P. M. Hwang, and S. H. Snyder. 1990. Localization of nitric oxide synthase indicating a neural role for nitric oxide. *Nature* 347:768–70.

Bredt, D. S., C. E. Glatt, P. M. Hwang, M. Fotuhi, T. M. Dawson, and S. H. Snyder. 1991. Nitric oxide synthase protein and mRNA are discretely localized in neuronal populations of the mammalian CNS together with NADPH diaphorase. *Neuron* 7:615–24.

Calapai, G., and A. P. Caputi. 1996. Nitric oxide and drinking behaviour. *Regul Pept* 66: 117–21.

Caligioni, C. S., and C. R. Franci. 2002. Oxytocin secretion induced by osmotic stimulation in rats during the estrous cycle and after ovariectomy and hormone replacement therapy. *Life Sci* 71:2821–31.

Camargo, L. A. A., W. A. Saad, C. R. Silva-Netto, C. G. Gentil, J. Antunes-Rodrigues, and M. R. Covian. 1976. Effects of catechol-amines injected into the septal area of the rat brain on natriuresis, kaliuresis and diuresis. *Can J Physiol Pharmacol* 54:219–28.

Camargo, L. A. A., W. A. Saad, C. R. Silva-Netto, J. Antunes-Rodrigues, and M. Covian. 1979. Effect of beta-adrenergic stimulation of the septal area on renal excretion of electrolytes and water in the rat. *Pharmacol Biochem Behav* 11:141–4.

Carlberg, K. A., M. J. Fregly, and M. Fahey. 1984. Effects of chronic oestrogen treatment on water exchange in rats. *Am J Physiol Endocrinol Metab* 247:E101–10.

Chiodera, P., R. Volpi, L. Capretti, and V. Coiro. 1996. Gamma-aminobutyric acid mediation of the inhibitory effect of nitric oxide on the arginine vasopressin and oxytocin responses to insulin-induced hypoglycemia. *Regul Pept* 67:21–5.

Choi-Kwon, S., and A. J. Baertschi. 1991. Splanchnic osmosensation and vasopressin: Mechanisms and neural pathways. *Am J Physiol Endocrinol Metab* 261:E18–25.

Chowdrey, H. S., and G. W. Bisset. 1988. Central inhibition by gamma-aminobutyric acid of the release of vasopressin by carbachol in the rat. *Br J Pharmacol* 93:349–56.

Costa, A., A. Poma, P. Navarra, M. L. Forsling, and A. Grossman. 1996. Gaseous transmitters as new agents in neuroendocrine regulation. *J Endocrinol* 149:199–207.

Crofton, J. T., P. G. Baer, L. Share, and D. P. Brooks. 1985. Vasopressin release in male and female rats: Effects of gonadectomy and treatment with gonadal steroid hormones. *Endocrinol* 117:1195–200.

Crowley, R. S., and J. A. Amico. 1993. Gonadal steroid modulation of oxytocin and vasopressin gene expression in the hypothalamus of the osmotically stimulated rat. *Endocrinol* 133:2711–8.

Cunha, R. A. 2001. Adenosine as a neuromodulator and as a homeostatic regulator in the nervous system: Different roles, different sources and different receptors. *Neurochem Int* 38:107–25.

da Rocha, M. J., C. R. Franci, and J. Antunes-Rodrigues. 1985. Participation of cholinergic and adrenergic synapses of the medial septal area (MSA) in the natriuretic and kaliuretic responses to intraventricular hypertonic saline (NaCl). *Physiol Behav* 34:23–8.

Dalmasso, C., J. L. Amigone, and L. Vivas. 2011. Serotonergic system involvement in the inhibitory action of estrogen on induced sodium appetite in female rats. *Physiol Behav* 104:398–407.

Danielsen, J., and J. Buggy. 1980. Depression of ad lib and angiotensin-induced sodium intake at oestrus. *Brain Res Bull* 5:501–4.

Dawson, T. M., and S. H. Snyder. 1994. Gases as biological messengers: Nitric oxide and carbon monoxide in the brain. *J Neurosci* 14:5147–59.

de Kloet, E. R., M. Joëls, and F. Holsboer. 2005. Stress and the brain: From adaptation to disease. *Nat Rev Neurosci* 6:463–75.

De Nicola, L., R. C. Blantz, and F. B. Gabbai. 1992. Nitric oxide and angiotensin II. Glomerular and tubular interaction in the rat. *J Clin Invest* 89:1248–56.

Dean, S. A., J. Tan, R. White, E. R. O'Brien, and F. H. H. Leenen. 2006. Regulation of components of the brain and cardiac renin-angiotensin systems by 17β-oestradiol after myocardial infarction in female rats. *Am J Physiol Regul Integr Comp Physiol* 291:R155–62.

De-Angelis, P. R., V. R. Antunes, G. M. Camargo, W. A. Saad, A. Renzi, A., and L. A. A. Camargo. 1996. Central interaction between atrial natriuretic peptide and angiotensin II in the control of sodium intake and excretion in female rats. *Braz J Med Biol Res* 29: 1671–4.

Denton, D. A. 1982. *The Hunger for Salt.* Heidelberg, Germany: Springer.

Denton, D. A., M. J. McKinley, and R. S. Weisinger. 1996. Hypothalamic integration of body fluid regulation. *Proc Natl Acad Sci USA* 93:7397–404.

Di Marzo, V., C. S. Breivogel, Q. Tao et al. 2000. Levels, metabolism, and pharmacological activity of anandamide in CB(1) cannabinoid receptor knockout mice: Evidence for non-CB(1), non-CB(2)receptor-mediated actions of anandamide in mouse brain. *J Neurochem* 75:2434–44.

Di Marzo, V., S. K. Goparaju, L. Wang et al. 2001. Leptin-regulated endocannabinoids are involved in maintaining food intake. *Nature* 410:822–5.

Di, S., R. Malcher-Lopes, K. C. Halmos, and J. Tasker. 2003. Nongenomic glucocorticoid inhibition via endocannabinoid release in the hypothalamus: A fast feedback mechanism. *J Neurosci* 23:4850–7.

Di, S., R. Malcher-Lopes, V. L. Marcheselli, N. G. Bazan, and J. G. Tasker. 2005. Rapid glucocorticoid-mediated endocannabinoid release and opposing regulation of glutamate and γ-aminobutiric acid inputs to hypothalamic magnocellular neurons. *Endocrinology* 145:4292–301.

Di, S., M. M. Maxson, A. Franco, and J. G. Tasker. 2009. Glucocorticoids regulate glutamate and GABA synapse-specific retrograde transmission via divergent nongenomic signaling pathways. *J Neurosci* 29:393–401.

Dietz, J. R. 2005. Mechanisms of atrial natriuretic peptide secretion from the atrium. *Cardiovasc Res* 68:8–17.

Donner, N., and R. J. Handa. 2009. Estrogen receptor beta regulates the expression of tryptophan-hydroxylase 2 mRNA within serotonergic neurons of the rat dorsal raphe nuclei. *Neurosci* 163:705–18.

Dorn, J., J. Antunes-Rodrigues, and S. M. McCann. 1970. Natriuresis in the rat following intraventricular carbachol. *Am J Physiol* 219:1292–98.

Dorn, J., and J. C. Porter. 1970. Diencephalic involvement in sodium excretion in the rat. *Endocrinology* 86:1112–7.

Do-Vale, C. F., W. A. Saad, A. Renzi et al. 1995. Progesterone administration to ovariectomized rats reduces water and salt intake induced by central administration of angiotensin II. *Braz J Med Biol Res* 28:999–1002.

Duan, Y. F., W. Winters, P. M. McCabe, and E. N. Schneiderman. 1997. Cardiorespiratory components of the defense reaction elicited from the paraventricular nucleus. *Physiol Behav* 61:325–30.

Durlo, F. V., M. Castro, L. L. K. Elias, and J. Antunes-Rodrigues. 2004. Interaction of prolactin, ANPergic, oxytocinergic and adrenal systems in response to extracellular volume expansion in rats. *Exp Physiol* 89:541–48.

Dwyer, B. E., R. N. Nishimura, and S. Y. Lu. 1995. Differential expression of heme oxygenase-1 in cultured cortical neurons and astrocytes determined by the aid of a new heme-oxygenase antibody: Response to oxidative stress. *Brain Res Mol Brain Res* 30:37–47.

Elias, C. F., J. F. Kelly, C. E. Lee et al. 2000. Chemical characterization of leptin-activated neurons in the rat brain. *J Comp Neurol* 423:261–81.

Evans, S. J., T. F. Murray, and F. L. Moore. 2000. Partial purification and biochemical characterization of a membrane glucocorticoid receptor from an amphibian brain. *J Steroid Biochem Mol Biol* 72:209–21.

Evanson, N. K., J. G. Tasker, M. N. Hill, C. J. Hillard, and J. P. Herman. 2010. Fast feedback inhibition of the HPA axis by glucocorticoids is mediated by endocannabinoid signaling. *Endocrinology* 151:4811–9.

Ewing, J. F., S. N. Haber, and M. D. Maines. 1992. Normal and heat-induced patterns of expression of heme oxygenase-1 (HSP32) in rat brain: Hyperthermia causes rapid induction of mRNA and protein. *J Neurochem* 58:1140–9.

Ewing, J. F., and M. D. Maines. 1991. Rapid induction of hemeoxygenase 1 mRNA and protein by hyperthermia in rat brain: Hemeoxygenase 2 is not a heat shock protein. *Proc Natl Acad Sci USA* 88:5364–8.

Ewing, J. F., and M. D. Maines. 1992. In situ hybridization and immunohistochemical localization of heme oxygenase-2 mRNA and protein in normal rat brain: Differential distribution of isozyme 1 and 2. *Mol Cell Neurosci* 3:559–70.

Ewing, J. F., V. S. Raju, and M. D. Maines. 1994. Induction of heart heme oxygenase -1 (HSP32) by hyperthermia: Possible role in stress-mediated elevation of cyclic 3′:5″-guanosine monophosphate. *J Pharmacol Exp Ther* 271:408–14.

Fitzsimons, J. T. 1998. Angiotensin, thirst, and sodium appetite. *Physiol Rev* 78:583–686.

Foresti, R., J. E. Clark, C. J. Green, and R. Motterlini, R. 1997. Thiol compounds interact with nitric oxide in regulating heme oxygenase-1 induction in endothelial cells. Involvement of superoxide and peroxynitrite anions. *J Biol Chem* 272:18411–7.

Forsling, M. L., M. Akerlund, and P. Strömberg. 1981. Variations in plasma concentrations of vasopressin during the menstrual cycle. *J Endocrinol* 89:263–6.

Forsling, M. L., and K. Peysner. 1988. Pituitary and plasma vasopressin concentrations and fluid balance throughout the oestrous cycle of the rat. *J Endocrinol* 117:397–402.

Forsling, M. L., and A. Grossman. 1998. Gaseous neurotransmitter modulation of vasopressin and oxytocin release. *Adv Exp Med Biol* 449:191–2.

Franci, C. R., J. Antunes-Rodrigues, C. R. Silva-Netto, L. A. A. Camargo, and W. A. Saad. 1983. Identification of pathways involved in the natriuretic, kaliuretic and diuretic responses induced by cholinergic stimulation of the medial septal area (MSA). *Physiol Behav* 30:65–71.

Franci, C. R., C. R. Silva-Netto, W. A. Saad, L. A. A. Camargo, and J. Antunes-Rodrigues. 1980. Interaction between the lateral hypothalamic area (LHA) and the medial septal area (MAS) in the control of sodium and potassium excretion in rat. *Physiol Behav* 25:801–6.

Freund, T. F., I. Katona, and D. Piomelli. 2003. Role of endogenous cannabinoids in synaptic signaling. *Physiol Rev* 83:1017–66.

Gainer, H., Y. Sarne, and M. J. Brownstein. 1977. Biosynthesis and axonal transport of rat neurohypophysial proteins and peptides. *J Cell Biol* 73:366–81.

Gametchu, B., C. S. Watson, and S. Wu. 1993. Use of receptor antibodies to demonstrate membrane glucocorticoid receptor in cells from human leukemic patients. *FASEB J* 7:1283–92.

Geddes, J. W., L. C. Pettigrew, M. L. Holtz, S. D. Craddock, and M. D. Maines. 1996. Permanent focal and transient global cerebral ischemia increase glial and neuronal expression of heme oxygenase-1, but not heme oxygenase-2, protein in rat brain. *Neurosci Lett* 210:205–8.

Giusti-Paiva, A., L. L. K. Elias, and J. Antunes-Rodrigues. 2005. Inhibitory effect of gaseous neuromodulators in vasopressin and oxytocin release induced by endotoxin in rats. *Neurosci Lett* 381:320–4.

Godino, A., A. Giusti-Paiva, J. Antunes-Rodrigues, and L. Vivas. 2005. Neurochemical brain groups activated after an isotonic blood volume expansion in rats. *Neurosci* 133:493–505.

Godino, A., L. O. Margatho, X. E. Caeiro, J. Antunes-Rodrigues, and L. Vivas. 2010. Activation of lateral parabrachial afferent pathways and endocrine responses during sodium appetite regulation. *Exp Neurol* 221:275–84.

Gomes, D. A., W. L. Reis, R. R. Ventura et al. 2004. The role of carbon monoxide and nitric oxide in hyperosmolality-induced atrial natriuretic peptide release by hypothalamus in vitro. *Brain Res* 1016:33–9.

Gomes, D. A., A. Giusti-Paiva, R. R. Ventura, L. L. K. Elias, F. Q. Cunha, and J. Antunes-Rodrigues. 2010. Carbon monoxide and nitric oxide modulate hyperosmolality-induced oxytocin secretion by the hypothalamus in vitro. *Biosci Rep* 30:351–7.

Gonzalez, D. R., A. Treuer, Q. A. Sun, J. S. Stamler, and J. M. Hare. 2009. *S*-Nitrosylation of cardiac ion channels. *J Cardiovasc Pharmacol* 54:188–95.

Gordon, E. L., J. D. Pearson, E. S. Dickinson, D. Moreau, and L. L. Slakey. 1989. The hydrolysis of extracellular adenine nucleotides by arterial smooth muscle cells: Regulation of adenosine production at the cell surface. *J Biol Chem* 264:8986–95.

Gow, A. J., Q. Chen, D. T. Hess, B. J. Day, H. Ischiropoulos, and J. S. Stamler. 2002. Basal and stimulated protein *S*-nitrosylation in multiple cell types and tissues. *J Biol Chem* 277:9637–40.

Grossman, A., A. Costa, M. L. Forsling et al. 1997. Gaseous neurotransmitters in the hypothalamus. The roles of nitric oxide and carbon monoxide in neuroendocrinology. *Horm Metab Res* 29:477–82.

Grossman, S. P. 1960. Eating or drinking elicited by direct adrenergic or cholinergic stimulation of hypothalamus. *Science* 132:301–2.

Guo, W., J. Sun, X. Xu, G. Burnstock, C. He, and Z. Xiang. 2009. P2X receptors are differentially expressed on vasopressin- and oxytocin-containing neurons in the supraoptic and paraventricular nuclei of rat hypothalamus. *Histochem Cell Biol* 131:29–41.

Haanwinckel, M. A., L. L. K. Elias, A. L. Favaretto, J. Gutkowska, S. M. McCann, and J. Antunes-Rodrigues. 1995. Oxytocin mediates atrial natriuretic peptide release and natriuresis after volume expansion in the rat. *Proc Natl Acad USA* 92:7902–6.

Han, F., H. Ozawa, K. Matsuda, M. Nishi, and M. Kawata. 2005. Colocalization of mineralocorticoid receptor and glucocorticoid receptor in the hippocampus and hypothalamus. *Neurosci Res* 51:371–81.

Hartley, D. E., S. L. Dickson, and M. L. Forsling. 2004. Plasma vasopressin concentrations and Fos protein expression in the supraoptic nucleus following osmotic stimulation or hypovolaemia in the ovariectomized rat: Effect of oestradiol replacement. *J Neuroendocrinol* 16:191–7.

Herkenham, M., A. B. Lynn, M. R. Johnson, L. S. Melvin, B. R. de Costa, and K. C. Rice. 1991. Characterization and localization of cannabinoid receptors in rat brain: A quantitative in vitro autoradiographic study. *J Neurosci* 11:563–83.

Hill, M. N., R. J. McLaughlin, B. Bingham et al. 2010. Endogenous cannabinoid signaling is essential for stress adaptation. *Proc Natl Acad Sci USA* 107:9406–11.

Hosomi, H., and H. Morita. 1996. Hepatorenal and hepatointestinal reflexes in sodium homeostasis. *News Physiol Sci* 11:103–7.

Hrabovszky, E., I. Kalló, A. Steinhauser et al. 2004. Oestrogen receptor-beta in oxytocin and vasopressin neurons of the rat and human hypothalamus: Immunocytochemical and in situ hybridization studies. *J Comp Neurol* 473:315–33.

Huang, P. L., Z. Huang, H. Mashimo et al. 1995. Hypertension in mice lacking the gene for endothelial nitric oxide synthase. *Nature* 377:239–42.

Ignarro, L. J., J. B. Adams, P. M. Horwitz, and K. S. Wood. 1986. Activation of soluble guanylatecyclase by NO-hemoproteins involves NO-heme exchange. Comparison of heme-containing and heme-deficient enzyme forms. *J Biol Chem* 261:4997–5002.

Ignarro, L. J., G. M. Buga, K. S. Wood, R. E. Byrns, and G. Chaudhuri. 1987. Endothelium-derived relaxing factor produced and released from artery and vein is nitric oxide. *Proc Natl Acad Sci USA* 84:9265–9.

Jankowski, M., G. Rachelska, W. Donghao, S. M. McCann, and J. Gutkowska. 2001. Oestrogen receptors activate atrial natriuretic peptide in the rat heart. *Proc Natl Acad Sci USA* 98:11765–70.

Jhamandas, J. H., and L. P. Renaud. 1986a. A gamma-aminobutyric-acid mediated baroreceptor input to supraoptic vasopressin neurones in the rat. *J Physiol* 381:595–606.

Jhamandas, J. H., and L. P. Renaud. 1986b. Diagonal band neurons may mediate arterial baroreceptor input to hypothalamic vasopressin secreting neurons. *Neurosci Lett* 65:214–48.

Johnson, A. K., and G. L. Edwards. 1991. *Thirst: Physiological Aspects.* London: Springer-Verlag.

Johnson, A. K., and R. L. Thunhorst. 1997. The neuroendocrinology of thirst and salt appetite: Visceral sensory signals and mechanisms of central integration. *Front Neuroendocrinol* 18:292–353.

Johnson, J. A., J. E. Zehr, and W. W. Moore. 1970. Effects of separate and concurrent osmotic and volume stimuli on plasma ADH in sheep. *Am J Physiol* 218:1273–80.

Kadekaro, M., H. Liu, M. L. Terrell, S. Gestl, V. Bui, and J. Y. Summy-Long. 1997. Role of NO on vasopressin and oxytocin release and blood pressure responses during osmotic stimulation in rats. *Am J Physiol Regul Integr Comp Physiol* 273:R1024–30.

Kadowaki, K., J. Kishimoto, G. Leng, and P. C. Emson. 1994. Up-regulation of nitric oxide synthase (NOS) gene expression together with NOS activity in the rat hypothalamo-hypophysial system after chronic salt loading: Evidence of a neuromodulatory role of nitric oxide in arginine vasopressin and oxytocin secretion. *Endocrinology* 134:1011–7.

Kakoki, M., A. P. Zou, and D. L. Mattson. 2001. The influence of nitric oxide synthase 1 on blood flow and interstitial nitric oxide in the kidney. *Am J Physiol Regul Integr Comp Physiol* 281:R91–7.

Karjalainen, A. H., H. Ruskoaho, O. Vuolteenaho et al. 2004. Effects of oestrogen replacement therapy on natriuretic peptides and blood pressure. *Maturitas* 47:201–8.

Kensicki, E., G. Dunphy, and D. Ely. 2002. Oestradiol increases salt intake in female normotensive and hypertensive rats. *J Appl Physiol* 93:479–83.

Keyse, S. M., and R. M. Tyrrell. 1987. Both near ultraviolet radiation and the oxidizing agent hydrogen peroxide induce a 32-kDa stress protein in normal human skin fibroblasts. *J Biol Chem* 262:14821–25.

Kimura, T., T. Funyu, M. Ohta et al. 1994. The role of GABA in the central regulation of AVP and ANP release and blood pressure due to angiotensin and carbachol, and central GABA release due to blood pressure changes. *J Auton Nerv Syst* 50:21–9.

Kisley, L. R., R. R. Sakai, L. Y. Ma, and S. J. Fluharty. 1999. Oestrogen decreases hypothalamic angiotensin II AT1 receptor binding and mRNA in the female rat. *Brain Res* 844:34–42.

Kiss, J. Z., J. A. Van Eekelen, J. M. Reul, H. M. Westphal, and E. R. De Kloet. 1988. Glucocorticoid receptor in magnocellular neurosecretory cells. *Endocrinology* 122:444–9.

Knott, T. K., H. G. Marrero, E. E. Custer, and J. R. Lemos. 2008. Endogenous ATP potentiates only vasopressin secretion from neurohypophysial terminals. *J Cell Physiol* 217:155–61.

Kostoglou-Athanassiou, I., M. L. Forsling, P. Navarra, and A. B. Grossman. 1996. Oxytocin release is inhibited by the generation of carbon monoxide as a neuromodulator. *Mol Brain Res* 42:301–6.

Kostoglou-Athanassiou, I., A. Costa, P. Navarra, G. Nappi, M. L. Forsling, and A. B. Grossman. 1998. Endotoxin stimulates an endogenous pathway regulating corticotropin-releasing hormone and vasopressin release involving the generation of nitric oxide and carbon monoxide. *J Neuroimmunol* 86:104–9.

Krause, E. G., K. S. Curtis, L. M. Davis, J. R. Stowe, and R. J. Contreras. 2003. Oestrogen influences stimulated water intake by ovariectomized female rats. *Physiol Behav* 79:267–74.

Krause, E. G., K. S. Curtis, T. L. Stincic, J. P. Markle, and R. J. Contreras. 2006. Oestrogen and weight loss decrease isoproterenol-induced Fos immunoreactivity and angiotensin type 1 mRNA in the subfornical organ of female rats. *J Physiol* 573:251–62.

Laflamme, N., R. E. Nappi, G. Drolet, C. Labrie, and S. Rivest. 1998. Expression and neuropeptidergic characterization of oestrogen receptors (ERalpha and ERbeta) throughout the rat brain: Anatomical evidence of distinct roles of each subtype. *J Neurobiol* 36:357–78.

Laitinen, J. T., and R. O. Juvonen. 1995. A sensitive microassay reveals marked regional differences in the capacity of rat brain to generate carbon monoxide. *Brain Res* 694:246–52.

Lauand, F., S. G. Ruginsk, H. L. Rodrigues et al. 2007. Glucocorticoid modulation of atrial natriuretic peptide, oxytocin, vasopressin and Fos expression in response to osmotic, angiotensinergic and cholinergic stimulation. *Neurosci* 147:247–57.

Ledsome, J. R., N. Wilson, and C. A. Courneya. 1985. Plasma vasopressin during increases and decreases in blood volume in anaesthetized dogs. *Can J Physiol Pharmacol* 63:224–9.

Leng, G., C. H. Brown, and J. A. Russell. 1999. Physiological pathways regulating the activity of magnocellular neurosecretory cells. *Prog Neurobiol* 57:625–55.

Leskinen, H., O. Vuolteenaho, J. Leppäluoto, and H. Ruskoaho. 1995. Role of nitric oxide on cardiac hormone secretion: Effect of N^{G}-nitro-L-arginine methyl ester on atrial natriuretic peptide and brain natriuretic peptide release. *Endocrinology* 136:1241–9.

Limbourg, F. P., and J. K. Liao. 2003. Nontranscriptional actions of the glucocorticoid receptor. *J Mol Med* 81:168–74.

Limbourg, F. P., Z. Huang, J. C. Plumier et al. 2002. Rapid nontranscriptional activation of endothelial nitric oxide synthase mediates increased cerebral blood flow and stroke protection by corticosteroids. *J Clin Invest* 110:1729–38. Erratum in: *J Clin Invest* 2003 111:759.

Liposits, Z., and M. C. Bohn. 1993. Association of glucocorticoid receptor immunoreactivity with cell membrane and transport vesicles in hippocampal and hypothalamic neurons of the rat. *J Neurosci Res* 35:14–9.

Liu, H., M. L. Terrell, J. Y. Summy-Long, and M. Kadekaro. 1996. Drinking and blood pressure responses to central injection of L-NAME in conscious rats. *Physiol Behav* 59:1137–45.

Liu, Q. S., Y. S. Jia, and G. Ju. 1997. Nitric oxide inhibits neuronal activity in the supraoptic nucleus of the rat hypothalamic slices. *Brain Res Bull* 43:121–5.

Maines, M. D. 1993. Carbon monoxide: An emerging regulator of cGMP in the brain. *Mol Cell Neurosci* 4:389–97.

Maines, M. D. 1997. The heme oxygenase system: A regulator of second messenger gases. *Annu Rev Pharmacol Toxicol* 37:517–54.

Maines, M. D., and A. Kappas. 1974. Cobalt induction of hepatic hemeoxygenase; with evidence that cytochrome P-450 is not essential for this enzyme activity. *Proc Natl Acad Sci USA* 71:4293–7.

Maines, M. D., and A. Kappas. 1977. Regulation of cytochrome P-450-dependent microsomal drug-metabolizing enzymes by nickel, cobalt, and iron. *Clin Pharmacol Ther* 22:780–90.

Maines, M. D., B. Polevoda, T. Coban et al. 1998. Neuronal overexpression of heme oxygenase-1 correlates with an attenuated exploratory behavior and causes an increase in neuronal NADPH diaphorase staining. *J Neurochem* 70:2057–69.

Mancuso, C., I. Kostoglou-Athanassiou, I., M. L. Forsling et al. 1997. Activation of heme oxygenase and consequent carbon monoxide formation inhibits the release of arginine vasopressin from rat hypothalamic explants: Molecular linkage between heme catabolism and neuroendocrine function. *Mol Brain Res* 50:267–6.

Mancuso, C., E. Ragazzoni, G. Tringali et al. 1999. Inhibition of hemeoxygenase in the central nervous system potentiates endotoxin-induced vasopressin release in the rat. *J Neuroimmunol* 99:189–94.

Mancuso, C., P. Navarra, and P. Preziosi. 2010. Roles of nitric oxide, carbon monoxide, and hydrogen sulfide in the regulation of the hypothalamic–pituitary–adrenal axis. *J Neurochem* 113:563–75.

Margatho, L. O., A. Giusti-Paiva, J. V. Menani, L. L. K. Elias, L. M. Vivas, and J. Antunes-Rodrigues. 2007. Serotonergic mechanisms of the lateral parabrachial nucleus in renal and hormonal responses to isotonic blood volume expansion. *Am J Physiol Regul Integr Comp Physiol* 292:R1190–7.

Margatho, L. O., A. Godino, A., F. R. Oliveira, L. Vivas, and J. Antunes-Rodrigues. 2008. Lateral parabrachial afferent areas and serotonin mechanisms activated by volume expansion. *J Neurosci Res* 86:3613–21.

Margatho, L. O., L. L. Elias, and J. Antunes-Rodrigues. 2009. GABA in the central amygdaloid nucleus modulates the electrolyte excretion and hormonal responses to blood volume expansion in rats. *Braz J Med Biol Res.* 42:114–21.

Marks, G. S., J. F. Brien, K. Nakatsu, and B. E. McLaughlin. 1991. Does carbon monoxide have a physiological function? *Trends Pharmacol Sci* 12:185–8.

Mattson, D. L., and F. Wu. 2000. Control of arterial blood pressure and renal sodium excretion by nitric oxide synthase in the renal medulla. *Acta Physiol Scand* 168:149–54.

McCann, S. M., J. Antunes-Rodrigues, C. R. Franci, A. Anselmo-Franci, S. Karanth, and V. Rettori. 2000. Role of the hypothalamic pituitary adrenal axis in the control of the response to stress and infection. *Braz J Med Biol Res* 33:1121–31.

McCann, S. M., C. R. Franci, and J. Antunes-Rodrigues. 1989. Hormonal control of water and electrolyte intake and output. *Acta Physiol Scand* 583:97–104.

McCoubrey, W. K. Jr., T. J. Huang, and M. D. Maines. 1997. Isolation and characterization of a cDNA from the rat brain that encodes hemoprotein heme oxygenase-3. *Eur J Biochem* 247:725–32.

McDonagh, A. F. 1990. Is bilirubin good for you? *Clin Perinatol* 17:359–69.

McKinley, M. J., D. A. Denton, M. Leventer et al. 1986. *The Physiology of Thirst and Sodium Appetite*. New York: Plenum Press.

McQueen, J. K., H. Wilson, and G. Fink. 1997. Estradiol-17 beta increases serotonin transporter (SERT) mRNA levels and the density of SERT-binding sites in female rat brain. *Brain Res Mol Brain Res* 45:13–23.

Mecawi, A. S., A. Lepletier, I. G. Araujo, E. L. Olivares, and L. C. Reis. 2007. Assessment of brain AT1-receptor on the nocturnal basal and angiotensin-induced thirst and sodium appetite in ovariectomised rats. *J Renin Angiotensin Aldosterone Syst* 8:169–75.

Mecawi, A. S., A. Lepletier, I. G. Araujo, F. V. Fonseca, and L. C. Reis. 2008. Oestrogenic influence on brain AT1 receptor signalling on the thirst and sodium appetite in osmotically stimulated and sodium-depleted female rats. *Exp Physiol* 93:1002–10.

Mecawi, A. S., T. Vilhena-Franco, I. G. Araujo, L. C. Reis, L. L. Elias, and J. Antunes-Rodrigues. 2011. Oestradiol potentiates hypothalamic vasopressin and oxytocin neuron activation and hormonal secretion induced by hypovolemic shock. *Am J Physiol Regul Integr Comp Physiol* 301:905–15.

Mechoulam, R., S. Ben-Shabat, L. Hanus et al. 1995. Identification of an endogenous 2-monoglyceride, present in canine gut, that binds to cannabinoid receptors. *Biochem Pharmacol* 50:83–90.

Melo, L. G., and H. Sonnenberg. 1996. Effect of nitric oxide inhibition on secretion of atrial natriuretic factor in isolated rat heart. *Am J Physiol Heart Circ Physiol* 270:H306–11.

Menani, J. V., W. A. Saad, L. A. A. Camargo, J. Antunes-Rodrigues, M. R. Covian, and W. Abrão-Saad. 1984. Effect of cholinergic and adrenergic stimulation of the subfornical organ in water intake. *Pharmacol Biochem Behav* 20:301–6.

Mikics, E., M. R. Kruk, and J. Haller. 2004. Genomic and non-genomic effects of glucocorticoids on aggressive behavior in male rats. *Psychoneuroendocrinology* 29:618–35.

Mitchell, M. D., P. J. Haynes, A. B. Anderson, and A. C. Turnbull. 1981. Plasma oxytocin concentrations during the menstrual cycle. *Eur J Obstet Gynecol Reprod Biol* 12:195–200.

Moncada, S., R. M. Palmer, and E. A. Higgs. 1991. Nitric oxide: Physiology, pathophysiology, and pharmacology. *Pharmacol Rev* 43:109–42.

Moncada, S., and E. A. Higgs. 1993. The L-arginine-nitric oxide pathway. *N Engl J Med* 329: 2002–12.

Moreto, V., A. M. Stabile, J. Antunes-Rodrigues, and E. C. Carnio. 2006. Role of heme-oxygenase pathway on vasopressin deficiency during endotoxemic shock-like conditions. *Shock* 26:472–6.

Morris, M., S. M. McCann, and R. Orias. 1976. Evidence for hormonal participation in the natriuretic and kaliuretic responses to intraventricular hypertonic saline and norepinephrine. *Proc Soc Exp Biol Med* 152:95–8.

Morris, M., S. M. McCann, and R. Orias. 1977. Role of transmitters in mediating hypothalamic control of electrolyte excretion. *Can J Physiol Pharmacol* 55:1143–54.

Muller, A. F., D. A. Denton, M. J. McKinley, E. Tarjan, and R. S. Weisinger. 1983. Lowered cerebrospinal fluid sodium antagonizes effect of raised blood sodium on salt appetite. *Am J Physiol* 244:810–4.

Newton, R. 2000. Molecular mechanisms of glucocorticoid action: What is important? *Thorax* 55:603–13.

Orchinik, M., T. F. Murray, and F. L. Moore. 1991. A corticosteroid receptor in neuronal membranes. *Science* 252:1848–51.

Otake, K., K. Kondo, and Y. Oiso. 1991. Possible involvement of endogenous opioid peptides in the inhibition of arginine vasopressin release by gamma-aminobutyric acid in conscious rats. *Neuroendocrinol* 54:170–4.

Pagotto, U., G. Marsicano, F. Fezza et al. 2001. Normal human pituitary gland and pituitary adenomas express cannabinoid receptor type 1 and synthesize endogenous cannabinoids: First evidence for a direct role of cannabinoids on hormone modulation at the human pituitary level. *J Clin Endocrinol Metab* 86:2687–96.

Palmer, R. M. J., A. G. Ferrige, and S. Moncada. 1987. Nitric oxide release accounts for the biological activity of endothelium-derived relaxing factor. *Nature* 327:524–6.

Perez, S. E., C. R. Silva-Netto, W. A. Saad, L. A. A. Camargo, and J. Antunes-Rodrigues. 1984. Interaction between cholinergic and osmolar stimulation of the lateral hypothalamic area (LHA) on sodium and potassium excretion. *Physiol Behav* 32:191–4.

Pertwee, R. G. 2005. Inverse agonism and neutral antagonism at cannabinoid CB1 receptors. *Life Sci* 76:1307–24.

Peysner, K., and M. L. Forsling. 1990. Effect of ovariectomy and treatment with ovarian steroids on vasopressin release and fluid balance in the rat. *J Endocrinol* 124:277–84.

Ponzio, T. A., and G. I. Hatton. 2005. Adenosine postsynaptically modulates supraoptic neuronal excitability. *J Neurophysiol* 93:535–47.

Rathbone, M. P., P. J. Middlemiss, J. W. Gysbers et al. 1999. Trophic effects of purines in neurons and glia cells. *Prog Neurobiol* 59:663–90.

Reid, I. A. 1994. Role of nitric oxide in the regulation of renin and vasopressin secretion. *Front Neuroendocrinol* 15:351–83.

Reis, W. L., A. Giusti-Paiva, R. R. Ventura et al. 2007. Central nitric oxide blocks vasopressin, oxytocin and atrial natriuretic peptide release and antidiuretic and natriuretic responses induced by central angiotensin II in conscious rats. *Exp Physiol* 92:903–11.

Reis, W. L., W. A. Saad, L. A. A. Camargo, L. L. K. Elias, and J. Antunes-Rodrigues. 2010. Central nitrergic system regulation of neuroendocrine secretion, fluid intake and blood pressure induced by angiotensin-II. *Behav Brain Funct* 6:64 (online).

Reis, W. L., V. C. Biancardi, S. Son, J. Antunes-Rodrigues, and J. E. Stern. 2012. Enhanced expression of heme oxygenase-1 and carbon monoxide excitatory effects in oxytocin and vasopressin neurones during water deprivation. *J Neuroendocrinol* 24:653–63.

Reppert, S. M., D. R. Weaver, J. H. Stehle, and S. A. Rivkees. 1991. Molecular cloning and characterization of the rat A1-adenosine receptor that is widely expressed in brain and spinal cord. *Mol Endocrinol* 5:1037–48.

Reul, J. M., and E. R. de Kloet. 1985. Two receptor systems for corticosterone in rat brain: Microdistribution and differential occupation. *Endocrinology* 117:2505–11.

Ribeiro, J. A. 1978. ATP related nucleotides and adenosine on neurotransmission. *Life Sci* 22:1373–80.

Richmon, J. D., K. Fukuda, K., N. Maida et al. 1998. Induction of heme oxygenase-1 after hyperosmotic opening of the blood-brain barrier. *Brain Res* 780:108–18.

Rivier, C. 1998. Role of nitric oxide and carbon monoxide in modulating the ACTH response to immune and nonimmune signals. *Neuroimmunomodulation* 5:203–13.

Robichaud, M., and G. Debonnel. 2005. Oestrogen and testosterone modulate the firing activity of dorsal raphe nucleus serotonergic neurones in both male and female rats. *J Neuroendocrinol* 17:179–85.

Rubinow, D. R., P. J. Schmidt, and C. A. Roca. 1998. Oestrogen–serotonin interactions: Implications for affective regulation. *Biol Psychiatry* 44:839–50.

Ruginsk, S. G., F. R. T. Oliveira, L. O. Margatho, L. Vivas, L. L. K. Elias, and J. Antunes-Rodrigues. 2007. Glucocorticoid modulation of neuronal activation and hormone secretion induced by blood volume expansion. *Exp Neurol* 206:192–200.

Ruginsk, S. G., E. T. Uchoa, L. L. K. Elias, and J. Antunes-Rodrigues. 2010. CB (1) modulation of hormone secretion, neuronal activation and mRNA expression following extracellular volume expansion. *Exp Neurol* 224:114–22.

Ruginsk, S. G., E. T. Uchoa, L. L. K. Elias, and J. Antunes-Rodrigues. 2012. Cannabinoid CB(1) receptor mediates glucocorticoid effects on hormone secretion induced by volume and osmotic changes. *Clin Exp Pharmacol Physiol* 39:151–4.

Saad, W. A., L. A. A. Camargo, C. R. Silva-Netto, C. G. Gentil, J. Antunes-Rodrigues, and M. R. Covian. 1975. Natriuresis, kaliuresis and diuresis in the rat following microinjections of carbachol into the septal area. *Pharmacol Biochem Behav* 3:985–92.

Saad, W. A., L. A. A. Camargo, F. G. Graeff, C. R. Silva-Netto, J. Antunes-Rodrigues, and M. R. Covian. 1976. The role of central muscarinic and nicotinic receptors in the regulation of sodium and potassium renal excretion. *Gen Pharmacol* 7:145–58.

Salter, M., R. G. Knowles, and S. Moncada. 1991. Widespread tissue distribution, species distribution and changes in activity of Ca(2+)-dependent and Ca(2+)-independent nitric oxide synthases. *FEBS Lett* 291:145–9.

Sanchez-Ferrer, C. F., J. C. Burnett Jr., R. R. Lorenz, and P. M. Vanhoutte. 1990. Possible modulation of release of atrial natriuretic factor by endothelium-derived relaxing factor. *Am J Physiol Heart Circ Physiol* 259:H982–6.

Sandi, C., C. Venero, and C. Guaza. 1996. Novelty-related rapid locomotor effects of corticosterone in rats. *Eur J Neurosci* 8:794–800.

Schnackenberg, C. G., B. L. Tabor, M. H. Strong, and J. P. Granger. 1997. Inhibition of intrarenal NO stimulates renin secretion through a macula densa-mediated mechanism. *Am J Physiol Regul Integ Comp Physiol* 272:R879–86.

Serino, R., Y. Ueta, M. Hanamiya et al. 2001. Increased levels of hypothalamic neuronal nitric oxide synthase and vasopressin in salt-loaded Dahl rat. *Auton Neurosci* 87:225–35.

Shade, R. E., and L. Share. 1975. Volume control of plasma antidiuretic hormone concentration following acute blood volume expansion in the anesthetized dog. *Endocrinology* 97:1048–57.

Share, L. 1988. Role of vasopressin in cardiovascular regulation. *Physiol Rev* 68:1248–84.

Shibahara, S., R. M. Muller, and H. Taguchi. 1987. Transcriptional control of rat hemeoxygenase by heat shock. *J Biol Chem* 262:12889–92.

Shinomura, T., S. Nakao, and K. Mori. 1994. Reduction of depolarization-induced glutamate release by hemeoxygenase inhibitor: Possible role of carbon monoxide in synaptic transmission. *Neurosci Lett* 166:131–4.

Shughrue, P. J., M. V. Lane, and I. Merchenthaler. 1997. Comparative distribution of oestrogen receptor-alpha and -beta mRNA in the rat central nervous system. *J Comp Neurol* 388:507–25.

Sierra, A., A. Gottfried-Blackmore, T. A. Milner, B. S. McEwen, and K. Bulloch. 2008. Steroid hormone receptor expression and function in microglia. *Glia* 56:659–74.

Skowsky, W. R., L. Swan, and P. Smith. 1979. Effects of sex steroid hormones on arginine vasopressin in intact and castrated male and female rats. *Endocrinology* 104:105–8.

Skvorak, J. P., and J. R. Dietz. 1997. Endothelin and nitric oxide interact to regulate stretch-induced ANP secretion. *Am J Physiol Regul Integ Comp Physiol* 273:R301–6.

Smith, A. M. Z., and A. K. Johnson. 1995. Chemical topography of efferent projections from the median preoptic nucleus to pontine monoaminergic cell groups in the rat. *Neurosci Lett* 199:215–9.

Snyder, S. H., S. R. Jaffrey, and R. Zakhary. 1998. Nitric oxide and carbon monoxide: Parallel roles as neural messengers. *Brain Res Brain Res Rev* 26:167–75.

Somponpun, S. J., A. K. Johnson, T. Beltz, and C. D. Sladek. 2004. Oestrogen receptor-alpha expression in osmosensitive elements of the lamina terminalis: Regulation by hypertonicity. *Am J Physiol Regul Integr Comp Physiol* 287:R661–9.

Srisawat, R., V. R. Bishop, P. M. Bull et al. 2004. Regulation of neuronal nitric oxide synthase mRNA expression in the rat magnocellular neurosecretory system. *Neurosci Lett* 369:191–6.

Stern, J. E., and M. Ludwig. 2001. NO inhibits supraoptic oxytocin and vasopressin neurons via activation of GABAergic synaptic inputs. *Am J Physiol Regul Integr Comp Physiol* 280:R1815–22.

Stocker, R., Y. Yamamoto, A. F. McDonagh, A. N. Glazer, and B. N. Ames. 1987. Bilirubin is an antioxidant of possible physiological importance. *Science* 235:1043–6.

Stricker, E. M., E. Thiels, and J. G. Verbalis. 1991. Sodium appetite in rats after prolonged dietary sodium deprivation: A sexually dimorphic phenomenon. *Am J Physiol Regul Integr Comp Physiol* 260:R1082–8.

Suematsu, M. 2003. Quartet signal transducers in gas biology. *Antioxid Redox Signal* 5:435–7.

Sumner, B. E., K. E. Grant, R. Rosie, C. Hegele-Hartung, K. H. Fritzemeier, and G. Fink. 1999. Effects of tamoxifen on serotonin transporter and 5-hydroxytryptamine (2A) receptor binding sites and mRNA levels in the brain of ovariectomized rats with or without acute estradiol replacement. *Brain Res Mol Brain Res* 73:119–8.

Sun, Y., M. O. Rotenberg, and M. D. Maines. 1990. Developmental expression of hemeoxygenase isozymes in rat brain. Two HO-2 mRNAs are detected. *J Biol Chem* 265:8212–7.

Suyemitsu, T., and H. Terayama. 1975. Specific binding sites for natural glucocorticoids in plasma membranes of rat liver. *Endocrinology* 96:1499–508.

Swenson, K. L., and C. D. Sladek. 1997. Gonadal steroid modulation of vasopressin secretion in response to osmotic stimulation. *Endocrinology* 138:2089–97.

Tarttelin, M. F., and R. A. Gorski. 1971. Variations in food and water intake in the normal and acyclic female rat. *Physiol Behav* 7:847–52.

Tenhunen, R., H. S. Marver, and R. Schmid. 1968. The enzymatic conversion of heme to bilirubin by microsomal hemeoxygenase. *Proc Natl Acad Sci USA* 61:748–55.

Tenhunen, R., H. S. Marver, and R. Schmid. 1969. Microsomal hemeoxygenase. Characterization of the enzyme. *J Biol Chem* 244:6388–94.

Thirion, S., J. D. Troadec, and G. Nicaise. 1996. Cytochemical localization of ecto-ATPases in rat neurohypophysis. *J Histochem Cytochem* 44:103–11.

Thrasher, T. N. 1994. Baroreceptor regulation of vasopressin and renin secretion: Low-pressure versus high-pressure receptors. *Front Neuroendocrinol* 15:157–96.

Tóth, A., J. Boczán, N. Kedei et al. 2005. Expression and distribution of vanilloid receptor 1 (TRPV1) in the adult rat brain. *Brain Res Mol Brain Res* 135:162–8.

Troadec, J. D., S. Thirion, G. Nicaise, J. R. Lemos, and G. Dayanithi. 1998. ATP-evoked increases in $[Ca^{2+}]i$ and peptide release from rat isolated neurohypophysial terminals via a P2X2 purinoceptor. *J Physiol* 511:89–103.

Turner, C. P., M. Bergeron, P. Matz et al. 1998. Heme oxygenase-1 is induced in glia throughout brain by subarachnoid hemoglobin. *J Cereb Blood Flow Metab* 18:257–73.

Ueta, Y., A. Levy, and S. L. Lightman. 2002. Gene expression in the supraoptic nucleus. *Microsc Res Technol* 56:158–63.

Vacher, C. M., H. Hardin-Pouzet, H. W. Steinbusch, A. Calas, and J. De Vente. 2003. The effects of nitric oxide on magnocellular neurons could involve multiple indirect cyclic GMP-dependent pathways. *Eur J Neurosci* 17:455–66.

Ventura, R. R., D. A. Gomes, W. L. Reis et al. 2002. Nitrergic modulation of vasopressin, oxytocin and atrial natriuretic peptide secretion in response to sodium intake and hypertonic blood volume expansion. *Braz J Med Biol Res* 35:1101–9.

Ventura, R. R., A. Giusti-Paiva, D. A. Gomes, L. L. K. Elias, and J. Antunes-Rodrigues. 2005. Neuronal nitric oxide synthase inhibition differentially affects oxytocin and vasopressin secretion in salt loaded rats. *Neurosci Lett* 379:75–80.

Ventura, R. R., J. F. Aguiar, J. Antunes-Rodrigues, and W. A. Varanda. 2008. Nitric oxide modulates the firing rate of the rat supraoptic magnocellular neurons. *Neurosci* 155:359–65.

Verma, A., D. J. Hirsch, C. E. Glatt, G. V. Ronnett, and S. H. Snyder. 1993. Carbon monoxide: A putative neural messenger. *Science* 259:381–4.

Verney, E. B. 1947. The antidiuretic hormone and the factors which determine its release. *Proc Soc Lond B Biol* 135:25–105.

Vijande, M., M. Costales, and B. Marín. 1978. Sex difference in polyethylenglycol-induced thirst. *Experientia* 34:742–3.

Vilhena-Franco, T., A. S. Mecawi, L. L. K. Elias, and J. Antunes-Rodrigues. 2011. Oestradiol potentiates hormone secretion and neuronal activation in response to hypertonic extracellular volume expansion in ovariectomised rats. *J Neuroendocrinol* 23:481–9.

Vincent, S. R., S. Das, and M. D. Maines. 1994. Brain heme-oxygenase isoenzymes and nitric oxide synthase are co-localized in select neurons. *Neurosci* 63:223–31.

Voisin, D. L., S. X. Simonian, and A. E. Herbison. 1997. Identification of oestrogen receptor-containing neurons projecting to the rat supraoptic nucleus. *Neurosci* 78:215–28.

Weber, C. M., B. C. Eke, and M. D. Maines. 1994. Corticosterone regulates heme oxygenase-2 and NO synthase transcription and protein expression in rat brain. *J Neurochem* 63:953–62.

Weisinger, R. S., P. Considine, D. A. Denton et al. 1982. Role of sodium concentration of the cerebrospinal fluid in the salt appetite of sheep. *Am J Physiol Regul Integr Comp Physiol* 242:R51–63.

Weisinger, R. S., D. A. Denton, M. J. McKinley, A. F. Muller, and E. Tarjan. 1985. Cerebrospinal fluid sodium concentration and salt appetite. *Brain Res* 326:95–105.

White, K. A., and M. A. Marletta. 1992. Nitric oxide synthase is a cytochrome P-450 type hemoprotein. *Biochem* 31:6627–31.

White, T. D. 1988. Role of adenine compounds in autonomic neurotransmission. *Pharmacol Ther* 38:129–68.

Wilcox, C. S., W. J. Welch, F. Murad et al. 1992. Nitric oxide synthase in macula densa regulates glomerular capillary pressure. *Proc Natl Acad Sci USA* 89:11993–7.

Windle, R. J., and M. L. Forsling. 1993. Variations in oxytocin secretion during the 4-day oestrous cycle of the rat. *J Endocrinol* 136:305–11.

Windle, R. J., S. A. Wood, N. Shanks, S. L. Lightman, and C. D. Ingram. 1998. Ultradian rhythm of basal corticosterone release in the female rat: Dynamic interaction with the response to acute stress. *Endocrinology* 139:443–50.

Wittmann, G., L. Deli, I. Kalló et al. 2007. Distribution of type 1 cannabinoid receptor (CB1)-immunoreactive axons in the mouse hypothalamus. *J Comp Neurol* 503:270–9.

Xu, X., J. C. Xiao, L. F. Luo et al. 2008. Effects of ovariectomy and 17beta-oestradiol treatment on renin–angiotensin system, blood pressure, and endothelial ultrastructure. *Int J Cardiol* 130:196–204.

Zakhary, R., S. P. Gaine, J. L. Dinerman, M. Ruat, N. A. Flavahan, and S. H. Snyder. 1996. Hemeoxygenase 2: Endothelial and neuronal localization and role in endothelium-dependent relaxation. *Proc Natl Acad Sci USA* 93:795–8.

Zhuo, M., S. A. Small, E. R. Kandel, and R. D. Hawkins. 1993. Nitric oxide and carbon monoxide produce activity-dependent long-term synaptic enhancement in hippocampus. *Science* 260:1946–50.

Zou, A. P., and A. W. Cowley Jr. 1999. Role of nitric oxide in the control of renal function and salt sensitivity. *Curr Hypertens Rep* 1:178–86.

4 Preoptic–Periventricular Integrative Mechanisms Involved in Behavior, Fluid–Electrolyte Balance, and Pressor Responses

José Vanderlei Menani, Alexandre A. Vieira, Débora S. A. Colombari, Patrícia M. De Paula, Eduardo Colombari, and Laurival A. De Luca Jr.

CONTENTS

4.1 INTRODUCTION

Body fluid homeostasis and arterial pressure are intimately related to the point that their control share many common mechanisms. The diagram shown in Figure 4.1 illustrates an interactive network (antidehydration network) activated by the dehydration of the two major body fluid compartments, extracellular [represented by the production of angiotensin II (ANG II)] and intracellular (represented by hyperosmolarity). The operation of the network involves redundancy and reciprocity and results in effector mechanisms that counteract dehydration. Although highly simplified (many important factors, e.g., aldosterone, are omitted), the network diagram

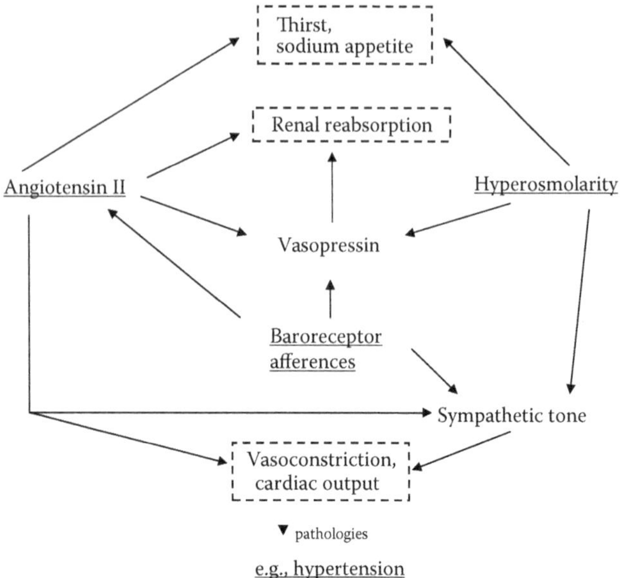

FIGURE 4.1 Interactive network of signals generated by extracellular (angiotensin II) and intracellular (hyperosmolarity) dehydration and effector mechanisms (dotted boxes) that counteract dehydration (antidehydration network). Arrows only indicate the flow of information; their functions are described in the text.

suggests complex control systems orchestrated by the brain. When deranged, the operation of such systems may lead to pathologies, for example, hypertension.

The brain circuit that counteracts dehydration has two main entrances or input paths for sensory information arising from the periphery (blood and viscera). They are located at opposite poles of an axis of multiple connections formed between hindbrain and forebrain. The preoptic periventricular tissue surrounding the anteroventral third ventricle (AV3V) forms a key region in the forebrain pole that integrates mechanisms to control the antidehydration network. The AV3V and the lamina terminalis share the organum vasculosum (OVLT) and the ventral median preoptic nucleus (MnPO). Also belonging to the lamina terminalis is the subfornical organ (SFO). The AV3V extends from the OVLT to the periventricular preoptic tissue until the rostral limits of the anterior hypothalamic area (Figure 4.2) (Brody and Johnson 1980; Menani et al. 1988b). The OVLT, along with the SFO, functions as a primary sensory station of the forebrain that monitors humoral factors such as circulating ANG II and osmolarity (Johnson 2007; McKinley et al. 2001; Chapter 2). The other entrance to the antidehydration brain circuit is located in the hindbrain and involves primary visceral sensory inputs in the nucleus of the solitary tract (NTS) and another circumventricular organ devoid of blood–brain barrier, such as the OVLT and SFO, the area postrema (AP).

The AV3V functions as a nodal structure that integrates and redistributes signals originated in visceral sensory organs to pattern generators of neuroendocrine, autonomic and somatic effector actions against dehydration and reduction in blood volume. It has an intimate connection with the lamina terminalis and connects with

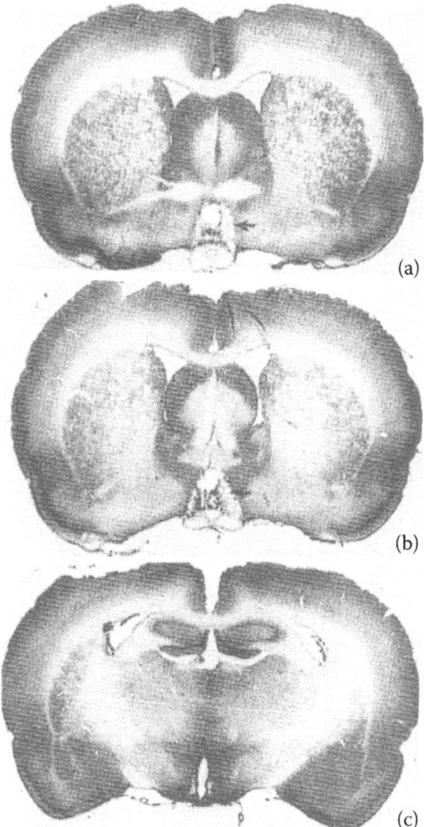

FIGURE 4.2 Photomicrographs showing (arrows) typical AV3V lesion. Extension of the lesion also defines the region. (a) Anterior portion of AV3V lesion at the level of OVLT, (b) lesion at level of preoptic periventricular area, (c) anterior hypothalamus, no lesion. (Reprinted with permission from Menani, J.V. et al., *Brain Res*, 446, 295–302, 1988.)

many other areas in the fore and hindbrain. As suggested in Figure 4.3, signals generated in the OVLT and SFO make their way out of the lamina terminalis through projections to forebrain structures such as the paraventricular and supraoptic nuclei of the hypothalamus (PVN and SON, respectively), the lateral hypothalamus and the medial septal area (MSA) (Brody and Johnson 1980). The AV3V region also direct or indirectly connects with areas of the hindbrain that control blood pressure, including the NTS, AP and the rostral ventrolateral medulla (RVLM) (Johnson 2007; Ricardo and Koh 1978; Saper et al. 1983; Whalen et al. 1999).

The NTS and RVLM are the main areas of the medullary circuitry involved in cardiovascular control. The NTS located dorsally in the hindbrain is the site of the first synapse of baroreceptor, chemoreceptor, and cardiopulmonary receptor afferent fibers in the central nervous system, whereas the RVLM located in the ventral surface of the hindbrain is the main premotor sympathetic nuclei, projecting directly to the intermediolateral (IML) column in the spinal cord and responsible for the generation and

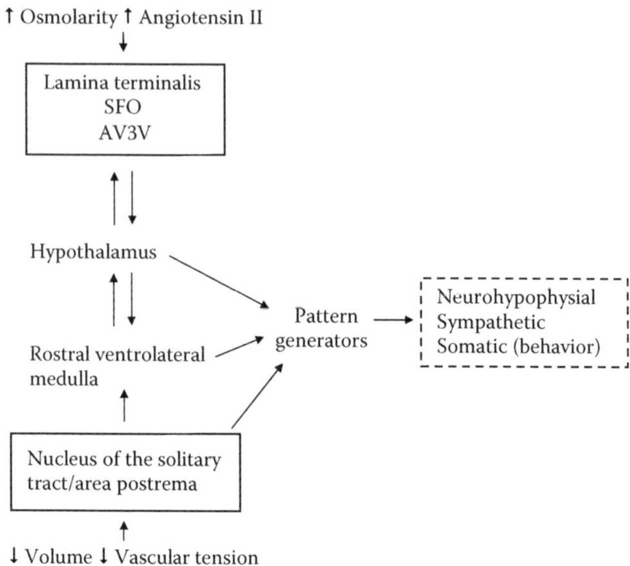

FIGURE 4.3 Summary diagram for a neural circuit that controls the antidehydration network illustrated in Figure 4.1. Two main inputs to the circuit occur in hindbrain and forebrain (see text for details). Box with continuous line: input. Box with dotted line: output.

maintenance of sympathetic vasomotor tone (Guyenet 2006). The reciprocal direct or indirect connections of the AV3V distributed along the forebrain–hindbrain axis form the neuroanatomical basis for the AV3V as an integrative region.

An important indirect connection of AV3V with hindbrain is made through the PVN, which mono- or polysynaptically connects with sympathetic neurons in IML (Westerhaus and Loewy 1999) or indirectly affects sympathetic activity through connections with the RVLM (Yang and Coote 1998). Moreover, the AV3V also has connections with salivatory nuclei in the pons (Hübschle et al. 2001).

The role of the lamina terminalis for the sensory integration of thirst and sodium appetite has deserved a special review in a relatively recent past (Johnson 2007). In this chapter, we first briefly refresh data on the general role of AV3V on the control of body fluid homeostasis and its role for the secretion of the atrial natriuretic peptide (ANP). Then, we review more recent data emphasizing the involvement of AV3V with salivary secretion, hindbrain control of cardiovascular function, and brain plasticity, in this order. Pilocarpine, a useful cholinergic agonist for therapeutics and experimental investigation about salivation, also affects arterial pressure and fluid balance. Early evidence for a central action of pilocarpine-induced salivation derived from studies with damage to the AV3V, but now we see that the same damage also interferes with the cardiovascular effects of pilocarpine. A role for AV3V on salivation linked to thermoregulation is also discussed. Then we show how AV3V influences the control of arterial blood pressure, first by presenting its role to sustain arterial pressure taking hemorrhage as a model, and second, by discussing compelling evidence for its role in the modulation of hindbrain mechanisms involved with short-term and long-term control of arterial pressure. Finally, we recall early evidence for brain recovery from AV3V damage and its implication for brain plasticity associated with sensitization of sodium intake before leading to conclusions.

4.2 ELECTROLYTIC LESIONS OF AV3V: GENERAL EFFECTS ON CONTROL OF BODY FLUID HOMEOSTASIS

Damage to the AV3V produces adipsia, inhibits sodium appetite, and reduces arterial pressure and renal sodium excretion. Much of these effects may have their source in the removal of sensory and membrane protein receptors. The AV3V, especially the OVLT, but also the MnPO, is rich in receptors for ANG II. In addition, OVLT and SFO have osmoreceptors. The MnPO receives important projections from the SFO, another area rich in ANG II receptors. Several other receptors for neurotransmitters—acetylcholine, glutamate, GABA, and oxytocin—and sexual hormone receptors are located in the AV3V (Cotman and Iversen 1987; Lenkey et al. 1995; Morris et al. 1977; Simerly et al. 1990; Yamagushi and Watanabe 2005; Yamaguchi and Yamada 2008; Yoshimura et al. 1993; Chapter 2).

One of the most remarkable effects of the AV3V lesions in rats is the adipsia that occurs in the first 4 to 7 days after the AV3V lesions, without significant change in food intake (Brody and Johnson 1980; Buggy and Johnson 1977a,b, 1978; Johnson and Buggy 1978). Giving to the AV3V lesioned rats access to palatable sweet solution is an efficient method to keep the animals hydrated during the acute critical phase of adipsia. After this initial period of adipsia, rats with AV3V lesions recover most of their *ad libitum* daily water intake, are usually alert, move easily, and continue to ingest food and palatable solutions (Buggy and Johnson 1977a,b), thereby sustaining a regular life in the laboratory. However, although rats with AV3V lesions may also overdrink, particularly when the ratio of water to food intake is taken into account (Lind and Johnson 1983), they respond badly to challenges that affect fluid–electrolyte balance. For example, the water intake induced by different dipsogenic stimuli, such as ANG II, increased plasma osmolarity or central cholinergic stimulation is still attenuated (Buggy and Johnson 1977a,b; Menani et al. 1990). The same happens with angiotensin II–induced sodium appetite (De Luca et al. 1992).

Acute AV3V lesions also reduce the secretion of vasopressin and ANP, two hormones that mediate the control of body fluid homeostasis. They reduce vasopressin secretion and produce acute increase in urinary volume which, combined with adipsia, results in up to 25% loss in body weight, and increased plasma sodium and osmolality (Bealer et al. 1979; Brody and Johnson 1980; Buggy and Johnson 1977a). Acute AV3V lesions also reduce basal secretion of the ANP and abolish the remarkable increase in plasma ANP concentration induced by central cholinergic activation or by isotonic blood volume expansion (Antunes-Rodrigues et al. 1991, 1997; Baldissera et al. 1989). The AV3V lesion also strongly reduces ANP concentration in several forebrain areas (e.g., medial basal hypothalamus, median eminence, neurohypophysis); in the atrium, it induces a trend to decrease when acute and to increase when chronic (Baldissera et al. 1989). Central cholinergic-induced natriuresis strongly correlated to the increase in plasma ANP produced by the activation of central cholinergic mechanisms, and the impaired central cholinergic-induced natriuresis by AV3V lesions is correlated with the reduction of ANP in these animals. It is possible that their acute increase in plasma sodium concentration and osmolality results from combined reduction of renal sodium excretion and increased diuresis (Bealer et al. 1979; Brody and Johnson 1980; Buggy and Johnson 1977a). However, the reduced natriuresis of animals with AV3V lesions is unlikely a result of the adipsia because they still show impaired central cholinergic-induced natriuresis when pair-hydrated to sham-lesioned animals (De Luca et al. 1991).

Acute or chronic AV3V lesions block the pressor and dipsogenic responses induced by intracerebroventricular (icv) injections of ANG II and reduce or block

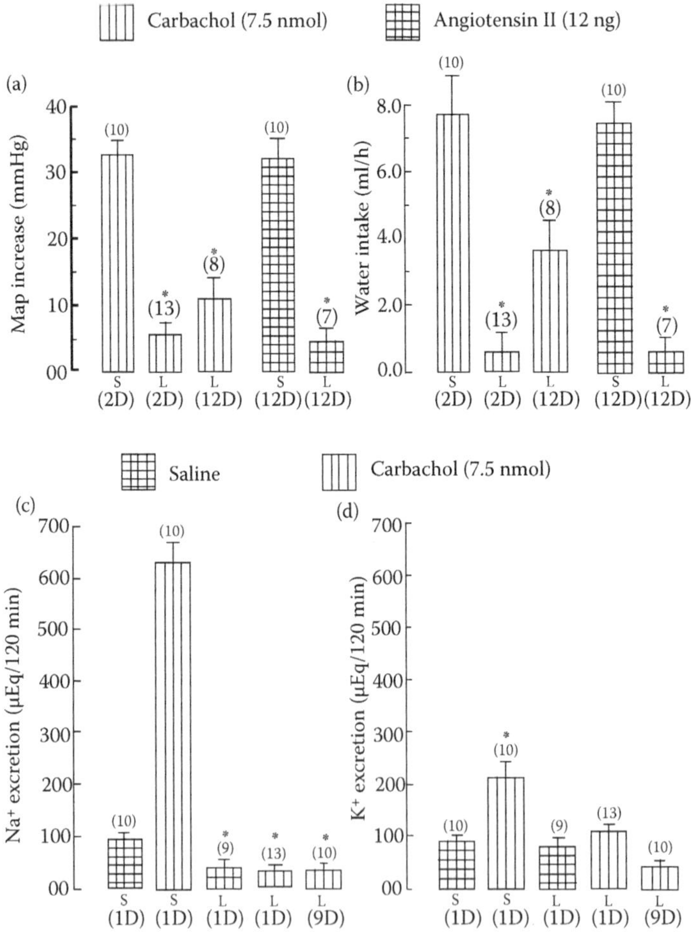

FIGURE 4.4 (a) Increase in mean arterial pressure (MAP) and (b) water intake in sham (S) or AV3V lesioned rats (L, 2 or 12 days) treated with icv injection carbachol (7.5 nmol/1 μL) or angiotensin II (12 ng/1 1 μL). Urinary (c) sodium and (d) potassium excretion in sham or AV3V lesioned rats (1 or 9 days) treated with icv injection of saline (control) or carbachol (7.5 nmol/1 μL). Results are presented as means ± standard error of means. Number of rats is indicated above each bar. (Reprinted with permission from Menani, J.V. et al., *Neurosci Lett*, 113, 339–344, 1990.)

the pressor, dipsogenic, natriuretic, and kaliuretic responses to central cholinergic activation with icv injection of the cholinergic agonist carbachol (Figure 4.4) or injection of carbachol into the SFO, MSA, ventromedial hypothalamus (VMH), and locus coeruleus (LC) (Brody and Johnson 1980; Colombari et al. 1992a,b; De Luca et al. 1991; De Luca and Menani 1996; Valladão et al. 1992). The AV3V lesions do not produce a consistent effect on the antidiuresis produced by central cholinergic activation.

4.3 AV3V REGION: CONTROL OF SALIVARY GLAND FUNCTION AND THERMOREGULATION

Saliva spreading over the body surface when grooming is a thermoregulatory behavior of utmost importance for water balance in the rat (Ritter and Epstein 1974). Damage to AV3V reduces salivation induced by pilocarpine (Renzi et al. 1993) and heat stress (Whyte and Johnson 2002). Rats with chronic (3 weeks) AV3V lesion have higher basal core temperature and increased peritoneal temperature when ambient temperature is elevated to 37°C (Whyte and Johnson 2007). This difficulty to dissipate heat is accompanied by attenuation of cardiovascular responses—increase in arterial pressure, heart rate, and mesenteric artery resistance—to increased ambient temperature. Such impaired cardiovascular responses are not accompanied by alterations in grooming, but by reduction in both heat-defensive behavior and salivation.

The reduced salivation to heat stress shown by animals with damage to the AV3V is likely a result of combined alterations in salivary glands and disruption of brain pathways. Acute or chronic AV3V lesions (2 to 30 days) cause atrophy of the acini, fibrosis of the connective tissue and reduced diameter of blood vessels, possibly resulting in ischemia, of the submandibular gland (Renzi et al. 1990). Those lesions also disrupt neural pathways linking all components of the lamina terminalis to hindbrain nuclei that project to the submandibular and sublingual glands (Hübschle et al. 2001). It is also possible that AV3V lesion disrupts central cholinergic pathways involved with thermoregulation (Dilsaver and Alessi 1988). Acute (first 7 days) AV3V lesions strongly impair the salivation induced by the intraperitoneal (ip) administration of pilocarpine (Renzi et al. 1993). However, in spite of the persistent morphological changes in the salivary glands, chronic AV3V lesions (15 days) produce no change in pilocarpine-induced salivation, suggesting the recruitment of alternative central pathways activated by pilocarpine to produce salivation (Renzi et al. 1993).

Pilocarpine injected systemically acts both in the brain and salivary glands to produce salivation (Takakura et al. 2003), and it is possible that damage to the AV3V alters these actions. Pilocarpine injected ip increases arterial pressure and superior mesenteric artery vascular resistance paralleled by increased blood flow and reduction in vascular resistance of the submandibular/sublingual salivary gland complex (SSG) (Takakura et al. 2005). In addition to reduced cardiovascular responses to heat (Whyte and Johnson 2007), AV3V-lesioned rats have reduced increase in arterial pressure, reduced superior mesenteric artery and hindlimb vascular resistances, and reduced SSG blood flow, in response to pilocarpine injected ip in 1 h or 2 days after the AV3V lesion (Takakura et al. 2005). The SSG vasodilation is not altered by the lesion, but it is possible that the reduced SSG blood flow contributes to the reduced salivation.

Systemic adrenoceptor antagonism and sympathetic ganglionectomy partially reduces the salivation to central pilocarpine (Cecanho et al. 1999). Therefore, disruption of sympathetic action might be a cause of the reduced pilocarpine-induced salivation in AV3V lesioned rats; however, this remains to be demonstrated.

4.4 IMPORTANCE OF AV3V REGION FOR RECOVERY FROM HEMORRHAGE

Intravenous infusion of hypertonic saline has been shown to successfully restore cardiovascular function in hypotensive hemorrhage (Velasco et al. 1990). The intravenous infusion of hypertonic saline rapidly shifts fluid from the intracellular compartment to the intravascular space, producing an increase in plasma volume, which seems to influence venous capacitance, improves myocardial contractility, and promotes precapillary dilation that may facilitate the recovery and survival of the animal in a hypovolemic hemorrhage. However, the increase in plasma osmolarity produced by hypertonic saline infusion may activate central mechanisms to produce vasoconstriction, which together with the peripheral mechanisms are essential for the recovery of the arterial pressure to almost normal levels.

The AV3V is involved in the central osmoregulatory control and cardiovascular responses related to ANG II, hyperosmolarity or reflex in origin (Brody and Johnson 1980; Menani et al. 1988a). It may impair the responses to central osmoreceptor activation by the hypertonic saline or the pulmonary reflexes (Lopes et al. 1981) or the central action of angiotensinergic mechanisms (Velasco et al. 1990), also suggested as being part of the mechanisms activated by hypertonic saline to recover the arterial pressure during a hypovolemic hemorrhage (see Chapter 8 for additional information about hypertonic saline and hemorrhage).

The AV3V lesion reduces recovery and survival to hypovolemic shock produced by hemorrhage owing to lack of ability to redistribute blood flow to essential organs (Feuerstein et al. 1984). Such susceptibility seems independent from alterations in vasopressin, renin–angiotensin, or catecholamines, but another study suggests attenuated activation of α_1 adrenoceptors (Schaumloffel at al. 1990). In rats that have been bled to reach a stable arterial pressure around 60 mm Hg for at least 20 min, intravenous infusion of 7.5% NaCl (4 mL/kg of body weight) increases arterial pressure to about 100 mm Hg. As shown in Figure 4.5, the AV3V lesions (4 h to 20 days) abolish these beneficial effects of hypertonic saline on hemorrhagic shock in rats (Barbosa et al. 1990, 1992). The bleeding volume to reach the stable arterial pressure of about 60 mm Hg was similar in sham and AV3V-lesioned rats (4 h or 4 days); however, in 20-day AV3V-lesioned rats, the bleeding volume was a little smaller than that of sham rats (Barbosa et al. 1992). The reduced bleeding in 20-day AV3V-lesioned is similar to the results reported in another study in 14-day AV3V-lesioned rats (Schaumloffel at al. 1990), which suggests that differences may exist when comparing the compensatory responses during hemorrhagic shock in acute and chronic AV3V-lesioned rats.

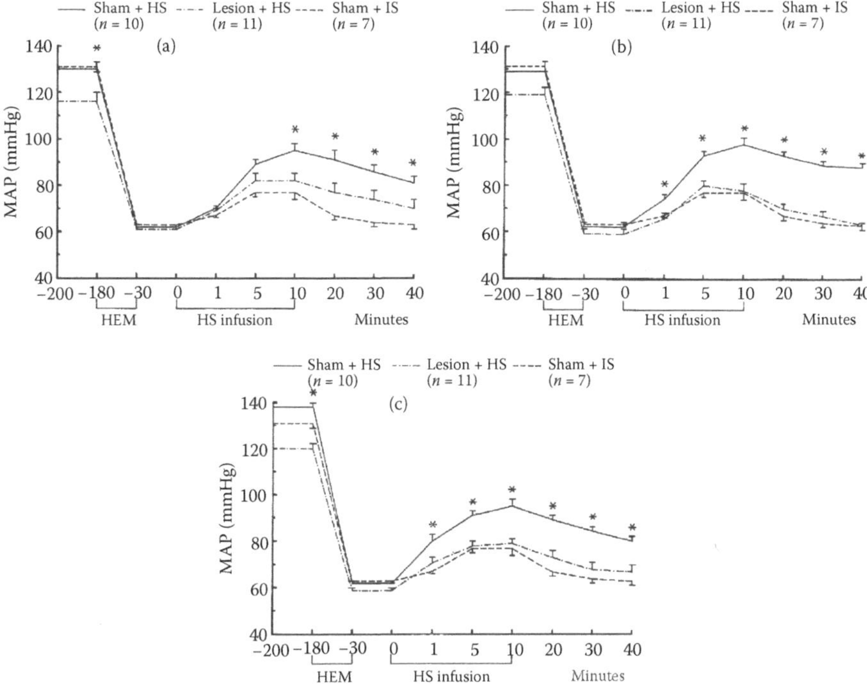

FIGURE 4.5 Temporal evolution of mean arterial pressure (MAP) in sham- or AV3V-lesioned rats (a, 1 h; b, 4 days; c, 20 days after lesion) submitted to hemorrhagic shock by bleeding that received an intravenous infusion of isotonic (IS) or hypertonic saline (HS, 7.5%) NaCl. Results are present as means ± standard error of means. *Different from AV3V-lesioned rats infused with HS. (Reprinted with permission from Barbosa, S.P. et al., *Brain Res*, 587, 109–114, 1992.)

At the end of the bleeding, plasma Na^+ concentration was reduced in sham rats, but not in AV3V-lesioned rats. The infusion of hypertonic NaCl increased plasma Na^+ concentration in sham and AV3V-lesioned rats; however, plasma Na^+ concentration increased more in AV3V-lesioned rats (Barbosa et al. 1992). These results suggest that AV3V lesion impairs the normal shift of fluid/solute between tissue and plasma.

4.5 IMPORTANCE OF AV3V FOR CARDIOVASCULAR RESPONSES TO ACTIVATION OF HINDBRAIN AREAS

The AV3V participates in the control of arterial pressure by affecting cardiovascular function and body fluid homeostasis as well. One of the main acute effects of the AV3V lesion is the intense tachycardia that may persist for more than 2 weeks. Most of the time, the AV3V lesions produce no significant change in the mean arterial pressure (MAP); however, a small increase in MAP may occur in some rats in the first 2 days after the lesions (Menani et al. 1988b; Vieira et al. 2004, 2006). Electrical stimulation of the AV3V produces renal and mesenteric vasoconstriction and hindquarters vasodilation associated with depressor responses and bradycardia. The vascular responses are dependent on direct sympathetic innervations and, in part, on the production of cathecolamines from the adrenal gland (Fink et al. 1978; Knuepfer et al. 1984). The renal vasodilation to body fluid expansion is also modified by the AV3V lesion (see Chapter 8 for details). It is possible that these effects of AV3V lesion influence short- and long-term control of arterial pressure. Here, we focus on how AV3V influences such controls when dependent on hindbrain mechanisms.

4.5.1 Short-Term Control

The pressor response produced by central angiotensinergic or cholinergic activation depends on vasopressin secretion and sympathetic activation (Hoffman et al. 1977; Imai et al. 1989), and AV3V lesions reduce vasopressin secretion and sympathetic activation produced by different stimuli (Bealer et al. 1979; Vieira et al. 2004, 2006). Moreover, cholinergic responses produced by the activation of more caudal areas such as the VMH or LC or even responses to glutamate into hindbrain areas are reduced or abolished by the AV3V lesions (De Luca et al. 1991; Valladão et al. 1992; Vieira et al. 2004, 2006). As we will show in this section, it is also possible that baseline AV3V activity releases facilitatory signals to caudal areas facilitating sympathetic activation and the pressor responses produced by the stimulation of these areas (Vieira et al. 2004, 2006).

Initial indications that the AV3V is involved with facilitation of cardiovascular reflexes integrated in the hindbrain were shown by the effect of AV3V lesions on the pressor response produced by bilateral common carotid occlusion (de Castro et al. 1993; Menani et al. 1988a). The pressor responses produced by bilateral common carotid occlusion in rats is a consequence of the deactivation of carotid baroreceptors combined with the activation of carotid and central chemoreceptors and depend on sympathetic activation and vasopressin secretion; the AV3V lesions impaired the pressor responses to common carotid occlusion by reducing sympathetic activation and vasopressin secretion activated by carotid occlusion (de Castro et al. 1993; Menani et al. 1988a).

More recently, we have shown that the AV3V also controls or modulates the effects of neurotransmission in the hindbrain with consequences to arterial pressure. The excitatory amino acid L-glutamate is one of the main neurotransmitters released by the afferent projections from peripheral baroreceptors and chemoreceptors in the NTS (Colombari et al. 1994; Haibara et al. 1995; Talman 1980). In unanesthetized rats, the injection of L-glutamate into the NTS or RVLM increases sympathetic activity and produces pressor responses (Colombari et al. 1994; Vieira et al. 2004, 2006). Acute (1 day) or chronic (15 days) AV3V lesions abolish the pressor response to L-glutamate injected into the NTS and attenuate the pressor response to L-glutamate injected into the RVLM as shown in Figure 4.6 (Vieira et al. 2004, 2006). In contrast to what they do to the more complex responses to carotid occlusion, AV3V lesions do not reduce baroreflex, chemoreflex, or vascular reactivity when evaluated separately (Bealer 1995; Vieira et al. 2004, 2006; Whalen et al. 1999). These results suggest the involvement of mechanisms dependent on the forebrain areas, particularly, the

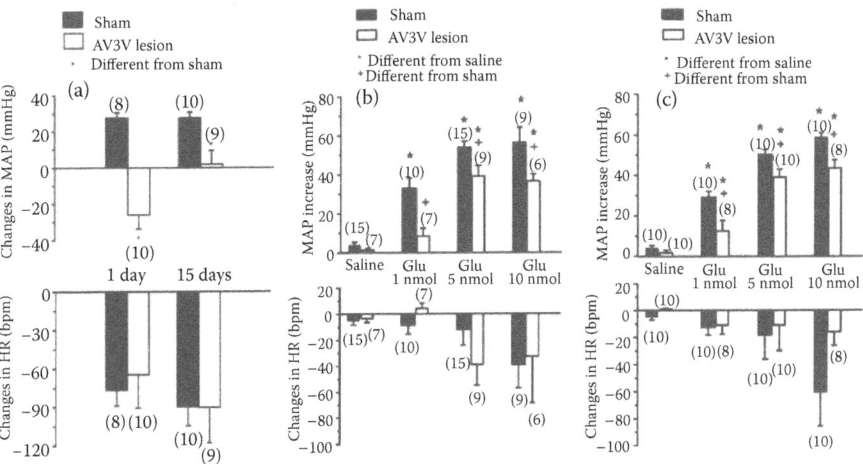

FIGURE 4.6 (a) Changes in mean arterial pressure (MAP) and heart rate (HR) produced by injections of L-glutamate (5 nmol/100 nL) into NTS in acute (1 day) or chronic (15 days) sham- or AV3V-lesioned rats. Changes in MAP and HR produced by injections of L-glutamate (1, 5 or 10 nmol/100 nL) into the RVLM in (b) acute (1 day) or (c) chronic (15 days) sham- or AV3V-lesioned rats. Results are presented as means ± standard error of means. Number of rats is indicated above and below each bar. (Reprinted with permission from Vieira, A.A. et al., *Brain Res*, 1025, 106–112, 2004; Vieira, A.A. et al., *Brain Res*, 1086, 160–167, 2006.)

AV3V region, to facilitate the pressor responses induced by glutamatergic activation in the NTS or RVLM.

The reduction of the pressor response to glutamate injected into the RVLM by the AV3V lesions suggests that the lesions may reduce both the glutamate-induced excitability of RVLM neurons and the resulting sympathetic activation by a yet unknown mechanism. A possible reduction in neuronal excitability of the RVLM may result from changes in basal activity or in the mechanisms activated by L-glutamate. Reduction in RVLM neuron excitability may attenuate the activation of RVLM neurons produced by L-glutamate, that is, L-glutamate does not easily activate RVLM neurons in the absence of facilitatory mechanisms dependent on the AV3V region that might increase the excitability of RVLM neurons (Vieira et al. 2006). The RVLM receives inhibitory and excitatory influences from different hindbrain and forebrain areas (Agarwal and Calaresu 1993; Dampney et al. 2003; Guyenet 2006). The AV3V lesions may produce an imbalance in those influences, by either reducing facilitatory or increasing inhibitory signals, or both, to the RVLM, thus reducing the excitability of RVLM neurons involved in sympathetic activation. In addition, besides the decrease in RVLM excitability, it is necessary to consider that AV3V lesions might also reduce the pressor responses to L-glutamate into the RVLM by directly changing the excitability of sympathetic neurons of the IML (Vieira et al. 2006). The disruption of the signals that facilitate the effects of L-glutamate into the RVLM by the AV3V lesions might be an additional mechanism that may account for the antihypertensive effects of AV3V lesions in some models of hypertension.

Anatomical connections between the NTS and the forebrain, including nuclei in the AV3V region have been described (Dampney et al. 2003; Ricardo and Koh 1978). Therefore, signals produced by L-glutamate injected into the NTS could ascend to the AV3V region, which in turn activates the sympathetic system. Similar to the effects on RVLM glutamatergic pressor responses, another possibility is that the AV3V region may send descending signals to the NTS that tonically facilitate the pressor mechanisms activated by L-glutamate into the NTS (Vieira et al. 2004). In addition to these possibilities, the reduction in the effects of L-glutamate injected into the RVLM might also be a mechanism that reduces the pressor response to L-glutamate into the NTS in AV3V lesioned rats.

Central cholinergic or angiotensinergic blockade with icv injections of the muscarinic antagonist atropine or the AT1 angiotensinergic antagonists losartan or ZD 7155 reduces the pressor response to L-glutamate injected into the RVLM similar to AV3V lesions (Vieira et al. 2007, 2010), which suggests that the effects of the AV3V lesions reducing the pressor response to L-glutamate injected into the RVLM might be a consequence of the blockade of signals produced by the baseline activation of forebrain angiotensinergic or cholinergic mechanisms.

4.5.2 Long-Term Control: Hypertension

The AV3V lesions prevents the development of experimental hypertension in different models—most of them resulting from the central action of ANG II and sympathetic activation—such as renal artery clip, aortic ligation, sinoaortic denervation, and Dahl salt-sensitive rats (Buggy et al. 1977; Goto et al. 1982; Haywood et al. 1983; Johnson 1980; Menani et al. 1988b).

The hypertension of spontaneously hypertensive rats (SHR) or the one produced by massive destruction of the NTS seems refractory to AV3V lesions (Catelli and Sved 1988; Gordon et al. 1982). Nevertheless, recent data suggest that such resistance is not absolute.

Lesions of the commissural portion of the NTS (commNTS) alone produce a transitory 7-day reduction of the arterial pressure in SHR (Moreira et al. 2009; Sato et al. 2001, 2003). Yet, as shown in Figure 4.7, lesions of both the AV3V and commNTS produced a long-term reduction in arterial pressure that lasted until the end of the experimental period (40 days) in adult SHR (Moreira et al. 2009). This finding suggests that forebrain and brainstem mechanisms, particularly those related to AV3V region and commNTS, act together to maintain hypertension in SHR.

The treatment with intravenous AT1 receptor or ganglionic blocker produced smaller reduction in MAP in AV3V + commNTS-lesioned SHR compared to sham-lesioned SHR, which suggests that the antihypertensive effects of the combined lesions are probably related to the reduction in ANG II– and sympathetic-mediated pressor mechanisms (Moreira et al. 2009). Changes in fluid–electrolyte balance produced by the AV3V lesions or in food intake produced by commNTS lesions are not important for the antihypertensive effects of the lesions (Moreira et al. 2009). In normotensive rats, combined commNTS + AV3V lesions produced only a small and transitory reduction of MAP (less than 20 mm Hg only in the first day after lesions), which provides evidence that these lesions affect only the mechanisms

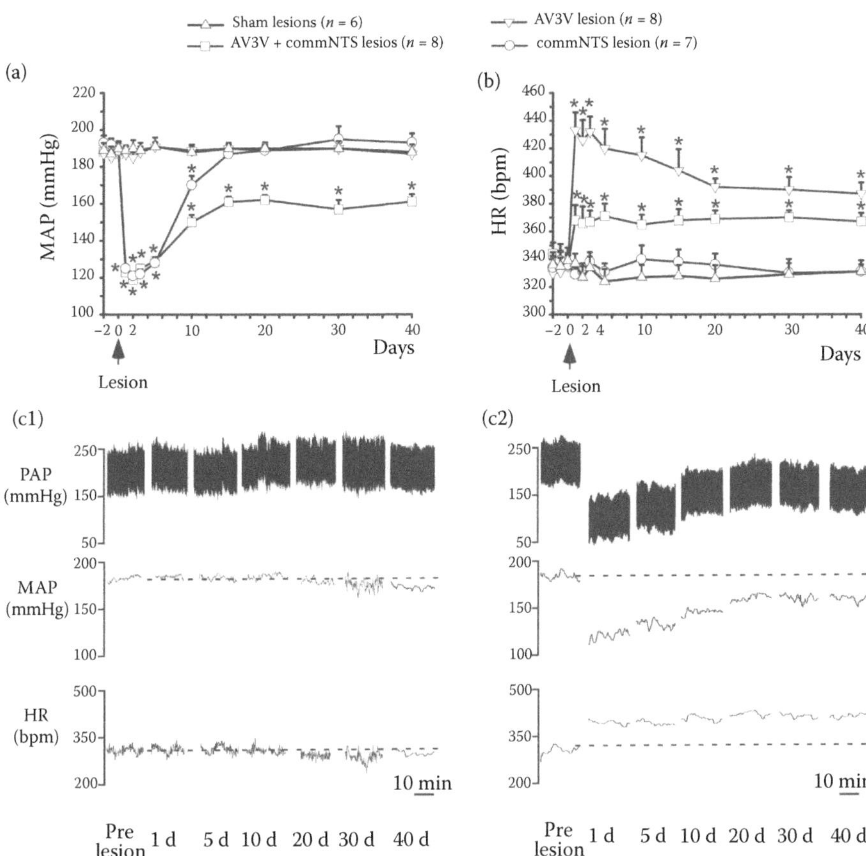

FIGURE 4.7 Baseline (a) mean arterial pressure (MAP) and (b) heart rate (HR) on control days (pre-lesion) and 1, 2, 3, 5, 10, 15, 20, 30, and 40 days after sham, AV3V, commNTS or AV3V + commNTS lesions in SHR. Representative recordings of pulsatile arterial pressure (PAP), MAP and HR in (cl) one SHR with sham lesions and (c2) one SHR with combined commNTS + AV3V lesions. Results in panels a and b are presented as means ± standard errors of means. *Different from sham-lesioned group. Dashed lines in cl and c2 show pre-lesion levels of MAP and HR. (Reprinted with permission from Moreira, T.S. et al., *Am J Physiol Regul Integr Comp Physiol*, 296, R1797–1806, 2009.)

activated to produce hypertension and not those that maintain normotensive levels of arterial pressure.

The AV3V lesions may affect the activity of hindbrain areas by disrupting the control the PVN exerts on sympathetic activity through direct connections with the IML or RVLM. The NTS receives important cardiovascular afferent signals and sends projections to hindbrain areas such as caudal ventrolateral medulla (CVLM), RVLM, parabrachial nucleus, and medullary regions containing cathecolamine cell bodies and to forebrain areas such as specific hypothalamic nuclei (Ricardo and Koh 1978; Saper et al. 1983). Therefore, the commNTS lesion may impair important signals that control the activity of CVLM and RVLM, two main areas of the medullary circuitry involved in cardiovascular control and, in addition, also signals that might reach directly or indirectly the AV3V region. Besides the inhibitory signals that reach the RVLM through CVLM, the NTS and PVN are also important sources of excitatory inputs to RVLM and sympathetic system. Lesion of the commNTS disrupts mainly excitatory inputs from the NTS to RVLM without changing the inhibitory inputs, and AV3V lesions may impair excitatory inputs from the PVN. It seems that the blockade of both mechanisms is necessary to reduce MAP in SHR.

It seems that hypertension in SHR is strongly dependent on central complex mechanisms involving multiple central areas such as the AV3V region and the commNTS and probably different mechanisms. One mechanism may compensate for the absence of the other, which increases the probability of hypertension in these animals.

4.6 AV3V LESION: BRAIN PLASTICITY AND RESILIENCE

As already mentioned, animals with AV3V lesions are capable of recovering their daily water intake and salivary response to pilocarpine, which is likely a result of plasticity mechanisms.

It is possible that the function to control water intake is transferred to the more lateral areas of the brain as time goes by after surgery (Gonçalves et al. 1992). Cholinergic-induced thirst by carbachol injected into medial areas such as SFO, MSA, and VMH (Colombari et al. 1992a,b; Valladão et al. 1992), but not the lateral preoptic area (LPOA), is persistently impaired by AV3V lesions. The thirst induced by carbachol injection into the LPOA is only transiently affected by AV3V lesions, suggesting that this area recovers the function to control water intake (Gonçalves et al. 1992).

It is not known what exactly is recovered through lateral areas. One possibility to check is if LPOA is linked, at least partially, to recovery of central cathecolaminergic neurotransmission. Central cholinergic-induced thirst is inhibited by central injection of a beta-blocker (Saad et al. 1985), and improvement of central noradrenergic function restores water intake induced by ANG II in animals with AV3V lesions (Cunningham and Johnson 1991).

In spite of the striking impaired response to the dipsogenic and natriorexigenic effects of exogenously generated ANG II, animals with AV3V lesions still express behaviors naturally mediated by this peptide in response to water deprivation and sodium depletion (De Luca et al. 1992; Lind and Johnson 1983). Their response, however, is not as fast as that of sham-lesioned animals, and their thirst or sodium appetite is initially blunted at the beginning of sodium appetite tests. This is similar to what happens to animals with lesion of the SFO (Thunhorst et al. 1990; Weisinger et al. 1990).

The importance of AV3V and SFO is reinforced by results about cell activation, as revealed by expression of the gene c-*fos*. Angiotensin II and sodium depletion produces c-*fos* expression in all components of AV3V and SFO (Rowland et al. 1996). c-*fos* expression is also highly correlated with production of thirst and sodium appetite in water-deprived animals in the lamina terminalis, but particularly in the

SFO of animals with hyperactive central renin–angiotensin system such as the SHR (Pereira-Derderian et al. 2010).

The importance of c-*fos* expression is also patent in the sensitization of sodium appetite produced by repeated episodes of sodium depletion. Angiotensin II and aldosterone are important mechanisms that subserve the sensitization of sodium appetite (Pereira et al. 2010; Sakai et al. 1987), and enhanced c-*fos* expression in the SFO is correlated with the enhanced sodium appetite of animals with history of sodium depletion (Na et al. 2007). Enhanced activity in an important area to control sodium intake and sensitive to ANG II, such as the SFO, plus morphological alterations in spiny neurons in the nucleus accumbens (Na et al. 2007; Roitman et al. 2002) suggest a neural basis to the enhanced sodium appetite. However, AV3V lesioned animals submitted to multisodium depletions are capable of enhancing both their sodium appetite to restore sodium deficit acutely as well as their need-free daily hypertonic sodium intake, as sham-lesioned animals do (De Luca et al. 1992).

It is also possible that in spite of the strong link between the effects of ANG II and the AV3V, redundant mechanisms may take hold to control water and sodium intake when this region is damaged. For example, in addition to the LPOA mechanism (Gonçalves et al. 1992), the AV3V lesioned animals remain responsive to the natriorexigenic effect of deoxycorticosterone, which mimics the effect of aldosterone (De Luca et al. 1992; Fitts et al. 1990). Moreover, similar to what happens when combined lesions of AV3V and commissural NTS reduce the arterial pressure of SHR (Moreira et al. 2009), combined lesions of the OVLT and SFO are more effective than single lesions to reduce sodium appetite (Fitts et al. 2004). This suggests neural redundancy outside and within the lamina terminalis itself. Finally, ANG II belonging to renin–angiotensin systems localized in other parts of the brain (Wright et al. 2008) might also contribute to redundancy, particularly to long-term mechanisms such as those related to behavioral sensitization; however, remains to be checked.

4.7 CONCLUSIONS

The AV3V broadly participates in the control of behavioral and physiological responses related to fluid–electrolyte balance, thermoregulation, and cardiovascular function. Several of such responses depend on sensory information that enters the brain through the lamina terminalis. Several others result from the activation of different forebrain areas. More specifically, in the forebrain, the AV3V region processes sensory information that arises from circulating humoral factors, and its damage makes animals insensitive to the effects of ANG II and increased osmolarity. The AV3V is also necessary for responses to forebrain cholinergic activation such as thirst, natriuresis, ANP release, and increase in arterial pressure and salivation.

In addition to its importance as an integrative area for forebrain mechanisms, it is interesting to note the strong connections between the AV3V and the hindbrain circuitry involved with the control of sympathetic activity. Although baseline sympathetic activity and arterial pressure in normotensive rats do not seem to depend on AV3V signals, sympathetic activation produced by different stimuli acting at different levels of the central nervous system are strongly dependent on the AV3V, including the sympathetic activation produced by the glutamatergic stimulation of

TABLE 4.1
Effects of Acute and Chronic AV3V Lesions in Cardiovascular System and Fluid–Electrolyte Balance

Effects	Acute	Chronic
Adipsia	Yes	No
Decrease of vasopressin release	Yes	Yes
Hypernatremia	Yes	Yes
Impaired hypertension and thirst induced by central angiotensin II	Yes	Yes
Reduction of sodium intakc dependent on angiotensin II	Yes	Yes
Impaired hypertension, thirst intake and natriuresis central cholinergic activation	Yes	Yes
Blockade of pilocarpine-induced pressor response and salivation	Yes	No
Attenuated pressor response to bilateral carotid occlusion	Yes	Yes
Suppressed pressor response to L-glutamate into NTS (conscious rat)	Yes	Yes
Attenuated pressor response to L-glutamate into RVLM (conscious rat)	Yes	Yes
Reduced ANP secretion	Yes	?
Reduced MAP recovery: hemorrhagic shock	Yes	Yes

Note: Acute lesion, less than 7 days; chronic lesion, more than 7 days.
ANP, atrial natriuretic peptide; ip, intraperitoneal; MAP, mean arterial pressure; NTS, nucleus of the solitary tract; RVLM, rostral ventrolateral medulla.

the NTS–RVLM circuitry that directly activates the sympathetic neurons. Another good evidence of the connections between the AV3V and the hindbrain pressor mechanisms is the chronically persistent reduction of the hypertension in SHR by the combined AV3V and commNTS lesions, an effect not produced by the lesion of each area alone. In addition, the AV3V is also essential for the activation of hypertensive mechanisms in many models of experimental hypertension and is part of the central circuitry that sustains arterial pressure in response to body fluid challenges such as hemorrhage, all indications of the importance of the AV3V for the activation of pressor mechanisms.

The importance of the AV3V for signals originating at different poles of the brain suggests it works as an integrator of autonomic mechanisms by processing information along an axis between the forebrain and the hindbrain. The impact of AV3V lesions on several aspects of body fluid balance and cardiovascular function are summarized in Table 4.1. The recovery and resilience of some of those aspects suggest plastic and redundant phenomena waiting for further experimental scrutiny.

REFERENCES

Agarwal, S. K., and F. R. Calaresu. 1993. Supramedullary inputs to cardiovascular neurons of rostral ventrolateral medulla in rats. *Am J Physiol Regul Integr Comp Physiol* 265:R111–6.

Antunes-Rodrigues, J., M. J. Ramalho, L. C. Reis et al. 1991. Lesions of the hypothalamus and pituitary inhibit volume-expansion-induced release of atrial natriuretic peptide. *Pro Natl Acad Sci USA* 88:2956–60.

Antunes-Rodrigues, J., A. L. Favaretto, J. Gutkowska, and S. M. McCann. 1997. The neuroendocrine control of atrial natriuretic peptide release. *Mol Psychiatry* 2:359–67.

Baldissera, S., J. V. Menani, L. F. dos Santos et al. 1989. Role of the hypothalamus in the control of atrial natriuretic peptide release. *Proc Natl Acad Sci USA* 86:9621–5.

Barbosa, S. P., W. A. Saad, L. A. A. Camargo et al. 1990. Lesion of the anteroventral third ventricle region abolishes the beneficial effects of hypertonic saline on hemorrhagic shock in rats. *Brain Res* 530:342–4.

Barbosa, S. P., L. A. A. Camargo, W. A. Saad, A. Renzi, L. A. De Luca Jr., and J. V. Menani. 1992. Lesion of the anteroventral third ventricle region impairs the recovery of arterial pressure induced by hypertonic saline in rats submitted to hemorrhagic shock. *Brain Res* 587:109–14.

Bealer, S. L. 1995. Preoptic recess ablation selectively increases baroreflex sensitivity to angiotensin II in conscious rats. *Peptides* 16:1197–201.

Bealer, S. L., M. I. Phillips, A. K. Johnson, and P. G. Schmid. 1979. Anteroventral third ventricle lesions reduce antidiuretic responses to angiotensin II. *Am J Physiol Endocrinol Metab* 236:E610–5.

Brody, M. J., and A. K. Johnson. 1980. Role of the anteroventral third ventricle region in fluid and electrolyte balance, arterial pressure regulation and hypertension. In: *Frontiers in Neuroendocrinology* ed. L. Martini, and W. F. Ganong, 249–92. New York: Raven Press.

Buggy, J., and A. K. Johnson. 1977a. Preoptic–hypothalamic periventricular lesions: Thirst deficits and hypernatremia. *Am J Physiol Regul Integr Comp Physiol* 233:R44–52.

Buggy, J., and A. K. Johnson. 1977b. Anteroventral third ventricular ablation: Temporary adipsia and persisting thirst deficits. *Neurosci Lett* 5:177–82.

Buggy, J., G. D. Fink, A. K. Johnson, and M. J. Brody. 1977. Prevention of the development of renal hypertension by anteroventral third ventricular tissue lesions. *Circ Res* 40:I110–7.

Buggy, J., and A. K. Johnson. 1978. Angiotensin-induced thirst: Effects of third ventricle obstruction and periventricular ablation. *Brain Res* 149:117–28.

Catelli, J. M., and A. F. Sved. 1988. Lesions of the AV3V region attenuate sympathetic activation but not the hypertension elicited by destruction of the nucleus tractus solitarius. *Brain Res* 439:330–6.

Cecanho, R., M. Anaya, A. Renzi, J. V. Menani, and L. A. De Luca Jr. 1999. Sympathetic mediation of salivation induced by intracerebroventricular pilocarpine in rats. *J Auton Nerv Syst* 76:9–14.

Colombari, E., W. A. Saad, L. A. A. Camargo, A. Renzi, L. A. De Luca Jr., and J. V. Menani. 1992a. AV3V lesion suppresses the pressor, dipsogenic and natriuretic responses to cholinergic activation of the septal area in rats. *Brain Res* 572:172–5.

Colombari, D. S. A., W. A. Saad, L. A. A. Camargo et al. 1992b. AV3V lesion impairs responses induced by cholinergic activation of SFO in rats. *Am J Physiol Regul Integr Comp Physiol* 263:R1277–83.

Colombari, E., L. G. H. Bonagamba, and B. H. Machado. 1994. Mechanisms of pressor and bradycardic responses to L-glutamate microinjected into the NTS of conscious rats. *Am J Physiol Regul Integr Comp Physiol* 266:R730–8.

Cotman, C. W., and L. L. Iversen. 1987. Excitatory amino acids in the brain—focus on NMDA receptors. *Trends Neurosci* 10:263–5.

Cunningham, J. T., and A. K. Johnson. 1991. The effects of central norepinephrine infusions on drinking behavior induced by angiotensin after 6-hydroxydopamine injections into the anteroventral region of the third ventricle (AV3V). *Brain Res* 558:112–6.

Dampney, R. A., J. Horiuchi, T. Tagawa, M. A. Fontes, P. D. Potts, and J. W. Polson. 2003. Medullary and supramedullary mechanisms regulating sympathetic vasomotor tone. *Acta Physiol Scand* 177:209–18.

de Castro, M. T., J. C. de Castro, and J. V. Menani. 1993. The effects of forebrain multiple lesions on the pressor response induced by bilateral carotid occlusion in conscious rats. *Brain Res* 612:243–46.

De Luca Jr., L. A., C. R. Franci, W. A. Saad, L. A. A. Camargo, and J. Antunes-Rodrigues. 1991. Natriuresis, not seizures, induced by cholinergic stimulation of the locus coeruleus is affected by forebrain lesions and water deprivation. *Brain Res Bull* 26:203–10.

De Luca L. A. Jr., O. Galaverna, J. Schulkin et al. 1992. The anteroventral wall of the third ventricle and the angiotensinergic component of need-induced sodium intake in the rat. *Brain Res Bull* 28:73–87.

De Luca Jr., L. A., and J. V. Menani. 1996. Preoptic–periventricular tissue (AV3V): Central cholinergic-induced hydromineral and cardiovascular responses, and salt intake. *Rev Bras Biol* 56:233–8.

Dilsaver, S. C., and N. E. Alessi. 1988. Temperature as a dependent variable in the study of cholinergic mechanisms. *Prog Neuropsychopharmacol Biol Psychiatry* 12:1–32.

Fink, G. D., J. Buggy, J. R. Haywood, A. K. Johnson, and M. J. Brody. 1978. Hemodynamic responses to electrical stimulation of areas of rat forebrain containing angiotensin on osmosensitive sites. *Am J Physiol Heart Circ Physiol* 235:H445–51.

Fitts, D. A., J. A. Freece, J. E. Van Bebber, D. K., and J. E. Basset. 2004. Effects of forebrain circumventricular organ ablation on drinking or salt appetite after sodium depletion or hypernatremia. *Am J Physiol Regul Integr Comp Physiol* 287:R1325–34.

Fitts, D. A., D. S. Tjepkes, and R. O. Bright. 1990. Salt appetite and lesions of the ventral part of the ventral median preoptic nucleus. *Behav Neurosci* 104:818–27.

Feuerstein, G., A. K. Johnson, R. L. Zerbe, R. Davis-Kramer, and A. I. Faden. 1984. Anteroventral hypothalamus and hemorrhagic shock: cardiovascular and neuroendocrine responses. *Am J Physiol Regul Integr Comp Physiol* 246:R551–7.

Gonçalves, P. C., M. B. Alves, J. E. Silveira et al. 1992. Effect of AV3V lesion on the cardiovascular, fluid, and electrolytic changes induced by activation of the lateral preoptic area. *Physiol Behav* 52:173–7.

Gordon, F. J., J. R. Haywood, M. J. Brody, and A. K. Johnson. 1982. Effect of lesions of the anteroventral third ventricle (AV3V) on the development of hypertension in spontaneously hypertensive rats. *Hypertension* 4:387–93.

Goto, A., M. Ganguli, L. Tobian, M. A. Johnson, and J. Iwai. 1982. Effect of anteroventral third ventricle lesion on NaCl hypertension in Dahl salt-sensitive rats. *Am J Physiol Heart Circ Physiol* 234:H614–8.

Guyenet, P. G. 2006. The sympathetic control of blood pressure. *Nat Rev Neurosci* 7: 335–46.

Haibara, A. S., E. Colombari, D. A. Chianca Jr., L. G. Bonagamba, and B. H. Machado. 1995. NMDA receptors in NTS are involved in bradycardic but not in pressor response of chemoreflex. *Am J Physiol Heart Circ Physiol* 269:H1421–7.

Haywood, J. R., G. D. Fink, S. Buggy, S. Boutelle, A. K. Johnson, and M. J. Brody. 1983. Prevention of two-kidney, one clip renal hypertension in rat by ablation of AV3V tissue. *Am J Physiol Heart Circ Physiol* 245:H683–9.

Hoffman, W. E., M. I. Philips, P. G. Schmid, J. Falcon, and J. F. Weet. 1977. Antidiuretic hormone release and the pressor response to central angiotensin II and cholinergic stimulation. *Neuropharmacology* 16:463–72.

Hübschle, T., M. L. Mathai, M. J. McKinley, and B. J. Oldfield. 2001. Multisynaptic neuronal pathways from the submandibular and sublingual glands to the lamina terminalis in the rat: A model for the role of the lamina terminalis in the control of osmo- and thermoregulatory behavior. *Clin Exp Pharmacol Physiol* 28:558–69.

Imai, Y., K. Abe, and S. Sasaki. 1989. Role of vasopressin in cardiovascular response to central cholinergic stimulation in rats. *Hypertension* 13:549–57.

Johnson, A. K., and J. Buggy. 1978. Periventricular preoptic–hypothalamus is vital for thirst and normal water economy. *Am J Physiol Regul Integr Comp Physiol* 234:R122–9.

Johnson, A. K. 1980. The sensory psychobiology of thirst and salt appetite. *Med Sci Sports Exerc* 39:1388–400.

Knuepfer, M. M., A. K. Johnson, and M. J. Brody. 1984. Vasomotor projections from the anteroventral third ventricle (AV3V) region. *Am J Physiol Heart Circ Physiol* 247:H139–45.

Lenkey, Z., P. Corvol, and C. Llorens-Cortes. 1995. The angiotensin receptor subtype AT1A predominates in rat forebrain areas involved in blood pressure, body fluid homeostasis and neuroendocrine control. *Brain Res Mol Brain Res* 30:53–60.

Lind, R. W., and A. K. Johnson. 1983. A further characterization of the effects of AV3V lesions on ingestive behavior. *Am J Physiol Regul Integr Comp Physiol* 245:R83–90.

Lopes, O. U., V. Pontieri, M. Rocha e Silva Jr., and I. T. Velasco. 1981. Hyperosmotic NaCl and severe hemorrhagic shock: Role of the innervated lung. *Am J Physiol Heart Circ Physiol* 241:H883–90.

McKinley, M. J., A. M. Allen, C. N. May et al. 2001. Neural pathways from the lamina terminalis influencing cardiovascular and body fluid homeostasis. *Clin Exp Pharmacol Physiol* 28:990–2.

Menani, J. V., M. T. Bedran de Castro, and E. M. Krieger. 1988a. Influence of the anteroventral third ventricle region and sinoaortic denervation on the pressor response to carotid occlusion. *Hypertension* 11:I-178–81.

Menani, J. V., B. H. Machado, E. M. Krieger, and H. C. Salgado. 1988b. Tachycardia during the onset of one kidney one-clip renal hypertension: Role of renin–angiotensin system and AV3V tissue. *Brain Res* 446:295–302.

Menani, J. V., W. A. Saad, L. A. A. Camargo, A. Renzi, L. A. De Luca Jr., and E. Colombari. 1990. The anteroventral third ventricle (AV3V) region is essential for pressor, dipsogenic and natriuretic responses to central carbachol. *Neurosci Lett* 113:339–44.

Moreira, T. S., A. C. Takakura, E. Colombari, and J. V. Menani. 2009. Anti-hypertensive effects of central ablations in spontaneously hypertensive rats. *Am J Physiol Regul Integr Comp Physiol* 296:R1797–806.

Morris, M., S. M. McCann, and R. Orias. 1977. Role of transmitters in mediating hypothalamic control of electrolyte excretion. *Can J Physiol Pharmacol* 55:1143–54.

Na, E. S., M. J. Morris, R. F. Johnson, T. G. Beltz and A. K. Johnson. 2007. The neural substrates of enhanced salt appetite after repeated sodium depletions. *Brain Res.* 1171:104–10.

Pereira, D. T. B., J. V. Menani, and L. A. De Luca Jr. 2010. FURO/CAP: A protocol for sodium intake sensitization. *Physiol Behav* 99:472–81.

Pereira-Derderian, D. T. B., R. C. Vendramini, J. V. Menani, and L. A. De Luca Jr. 2010. Water deprivation-induced sodium appetite and differential expression of encephalic c-Fos immunoreactivity in the spontaneously hypertensive rat. *Am J Physiol Regul Integr Comp Physiol* 298:R1298–309.

Renzi, A., R. A. Lopes, M. A. Sala et al. 1990. Morphological, morphometric and stereological study of submandibular glands in rats with lesion of the anteroventral region of the third ventricle (AV3V). *Exp Pathol* 38:177–87.

Renzi, A., E. Colombari, T. R. Mattos Filho et al. 1993. Involvement of the central nervous system in the salivary secretion induced by pilocarpine in rats. *J Dent Res* 72:1481–4.

Ricardo, J. A., and E. T. Koh. 1978. Anatomical evidence of direct projections from the nucleus of the solitary tract to the hypothalamus, amygdala, and other forebrain structures in the rat. *Brain Res* 153:1–26.

Ritter, R. C., and A. N. Epstein. 1974. Saliva lost by grooming: A major item in the rat's water economy. *Behav Biol* 11:581–5.

Roitman, M. F., E. Na, G. Anderson, T. A. Jones, and I. L. Bernstein. 2002. Induction of a salt appetite alters dendritic morphology in nucleus accumbens and sensitizes rats to amphetamine. *J Neurosci* 22:RC225.

Rowland, N. E., M. J. Fregly, L. Han, and G. Smith. 1996. Expression of Fos in rat brain in relation to sodium appetite: Furosemide and cerebroventricular renin. *Brain Res* 728:90–6.

Saad, W. A., J. V. Menani, L. A. Camargo, and W. Abrão-Saad. 1985. Interaction between cholinergic and adrenergic synapses of the rat subfornical organ and the thirst-inducing effect of angiotensin II. *Braz J Med Biol Res* 18:37–46.

Sakai, R. R., W. B. Fine, A. N. Epstein, and S. P. Frankmann. 1987. Salt appetite is enhanced by one prior episode of sodium depletion in the rat. *Behav Neurosci* 101:724–31.

Saper, C. B., D. J. Reis, and T. Joh. 1983. Medullary catecholamine inputs to the anteroventral third ventricular cardiovascular regulatory region in the rat. *Neurosci Lett* 42:285–91.

Sato, M. A., J. V. Menani, O. U. Lopes, and E. Colombari. 2001. Lesions of the commissural nucleus of the solitary tract reduce arterial pressure in spontaneously hypertensive rats. *Hypertension* 38:560–4.

Sato, M. A., G. H. Schoorlemmer, J. V. Menani, O. U. Lopes, and E. Colombari. 2003. Recovery of high blood pressure after chronic lesions of the commissural NTS in SHR. *Hypertension* 42:713–8.

Schaumloffel, V., V. Pugh, and S. L. Bealer. 1990. Preoptic hypothalamic lesions reduce adrenergic vascular compensation during hemorrhagic shock. *Circ Shock* 31:19–202.

Simerly, R. B., C. Chang, M. Muramatsu, and L. W. Swanson. 1990. Distribution of androgen and estrogen receptor mRNA-containing cells in the rat brain: An in situ hybridization study. *J Comp Neurol* 294:76–95.

Takakura, A. C., T. S. Moreira, S. C. Laitano, L. A. De Luca Jr., A. Renzi, and J. V. Menani. 2003. Central muscarinic receptors signal pilocarpine-induced salivation. *J Dent Res* 82:993–7.

Takakura, A. C., T. S. Moreira, L. A. De Luca Jr., A. Renzi, J. V. Menani, and E. Colombari. 2005. Effects of AV3V lesion on pilocarpine-induced pressor response and salivary gland vasodilation. *Brain Res* 1055:111–21.

Talman, W. T., M. H. Perrone, and D. J. Reis. 1980. Evidence for L-glutamate as the neurotransmitter of baroreceptor afferent nerve fibers. *Science* 209:813–5.

Thunhorst, R. L., K. J. Ehrlich, and J. B. Simpson. 1990. Subfornical organ participates in salt appetite. *Behav Neurosci* 104:637–42.

Valladão, A. S., W. A. Saad, L. A. A. Camargo, A. Renzi, L. A. De Luca Jr., and J. V. Menani. 1992. AV3V lesion reduces the pressor, dipsogenic, and natriuretic responses to ventromedial hypothalamus activation. *Brain Res Bull* 28:909–14.

Velasco, I. T., R. C. Baena, M. Rocha e Silva, and M. I. Loureiro. 1990. Central angiotensinergic system and hypertonic resuscitation from severe hemorrhage. *Am J Physiol Heart Circ Physiol* 259:H1752–8.

Vieira, A. A., E. Colombari, L. A. De Luca Jr., D. S. A. Colombari, and J. V. Menani. 2004. Cardiovascular responses to microinjection of L-glutamate into the NTS in AV3V-lesioned rats. *Brain Res* 1025:106–12.

Vieira, A. A., E. Colombari, L. A. De Luca Jr., D. S. A. Colombari, and J. V. Menani. 2006. AV3V lesions reduce the pressor response to L-glutamate into the RVLM. *Brain Res* 1086:160–7.

Vieira, A. A., E. Colombari, L. A. De Luca Jr., D. S. A. Colombari, and J. V. Menani. 2007. Central cholinergic blockade reduces the pressor response to L-glutamate into the rostral ventrolateral medullary pressor area. *Brain Res* 1155:100–7.

Vieira, A. A., E. Colombari, L. A. De Luca Jr., D. S. A. Colombari, P. M. de Paula, and J. V. Menani. 2010. Importance of angiotensinergic mechanisms for the pressor response to L-glutamate into the rostral ventrolateral medulla. *Brain Res* 1322:72–80.

Weisinger, R. S., D. A. Denton, R. Di Nicolantonio et al. 1990. Subfornical organ lesion decreases sodium appetite in the sodium-depleted rat. *Brain Res* 526:23–30.

Whalen, E. J., T. G. Beltz, S. J. Lewis, and A. K. Johnson. 1999. AV3V lesions attenuate the cardiovascular responses produced by blood-borne excitatory amino acid analogs. *Am J Physiol Heart Circ Physiol* 276:H1409–15.

Westerhaus, M. J., and A. D. Loewy. 1999. Sympathetic-related neurons in the preoptic region of the rat identified by viral transneuronal labeling. *J Comp Neurol* 414:361–78.

Whyte, D. G., and A. K. Johnson. 2002. Lesions of periventricular tissue surrounding the anteroventral third ventricle (AV3V) attenuate salivation and thermal tolerance in response to a heat stress. *Brain Res* 951:146–9.

Whyte, D. G., and A. K. Johnson. 2007. Lesions of the anteroventral third ventricle region exaggerate neuroendocrine and thermogenic but not behavioral responses to a novel environment. *Am J Physiol Regul Integr Comp Physiol* 292:R137–42.

Wright, J. W., B. J. Yamamoto, and J. W. Harding. 2008. Angiotensin receptor subtype mediated physiologies and behaviors: New discoveries and clinical targets. *Prog Neurobiol* 84:157–81.

Yamagushi, K., and K. Watanabe. 2005. Anteroventral third ventricular N-methyl-D-aspartate receptors, but not metabotropic glutamate receptors are involved in hemorrhagic AVP secretion. *Brain Res Bull* 66:59–69.

Yamaguchi, K., and T. Yamada. 2008. Roles of forebrain GABA receptors in controlling vasopressin secretion and related phenomena under basal and hyperosmotic circumstances in conscious rats. *Brain Res Bull* 77:61–9.

Yang, Z., and J. H. Coote. 1998. Influence of the hypothalamic paraventricular nucleus on cardiovascular neurons in the ventrolateral medulla of the rat. *J Physiol* 513:521–30.

Yoshimura, R., H. Kiyama, T. Kimura et al. 1993. Localization of oxytocin receptor messenger ribonucleic acid in the rat brain. *Endocrinology* 133:1239–46.

5 Diverse Roles of Angiotensin Receptor Intracellular Signaling Pathways in the Control of Water and Salt Intake

Derek Daniels

CONTENTS

5.1 INTRODUCTION

Angiotensin II (AngII) is a key component in the maintenance of body fluid homeostasis. It has potent dipsetic* and natriorexigenic effects when injected into the brain of rats. This ingestive response to AngII is so reliable that AngII is frequently used to verify proper placement of cannulae in forebrain ventricles, even by laboratories with little interest in fluid intake. The actions of AngII have been an important focus in the study of fluid intake, but have also been particularly informative in other fields of study. Much of our knowledge about the means by which peripherally derived

* Those familiar with the literature in the field of drinking behavior will likely find "dipsetic" less familiar than the alternate "dipsogenic." Both words arise from the Greek "dipsitikos," which means provoking thirst. Even though "dipsogenic" is more commonly used in the field, it is not found in current editions of many dictionaries, including *The American Heritage Dictionary*, *Merriam-Webster*, and the *Oxford English Dictionary*. The term "dipsetic," although less commonly used in science, is, however, found in the *Oxford English Dictionary*. Accordingly, our laboratory freely alternates between the two terms.

peptides act in the brain, for example, has been greatly influenced by earlier studies that identified central targets of AngII.

The central targets of AngII are numerous. Perhaps the most interesting among them are the forebrain circumventricular organs (CVOs), which reside on the anterior wall of the third ventricle. The subfornical organ (SFO) and the organum vasculosum of the lamina terminalis (OVLT) (for review, see Daniels and Fluharty 2004; Phillips 1987) are critical for many central actions of AngII. The hypothesis that AngII acts directly on CVOs is supported by a variety of experimental approaches. Lesion techniques have shown that ablation of the SFO or the periventricular area containing the OVLT dramatically reduces the water or NaCl intake stimulated by AngII (Buggy and Johnson 1978; Morris et al. 2002; Simpson et al. 1978). Small amounts of AngII injected directly into the SFO or OVLT, on the other hand, increase water and NaCl intake (Mangiapane and Simpson 1979, 1980). Experiments using receptor autoradiography demonstrated high levels of AngII receptor expression in CVO structures (for review, see Allen et al. 2000) and studies of brain activation using deoxyglucose (Kadekaro et al. 1989) or Fos immunohistochemistry (McKinley et al. 1995; Rowland et al. 1994a,b, 1995) highlight these areas as central AngII targets. Electrophysiological recordings have also been useful in the study of central responses to AngII. Early whole-cell patch clamp recordings from SFO neurons demonstrated AngII-induced excitation (Li and Ferguson 1993a,b) and suggested that the excitation occurred by inhibition of transient outward currents (Ferguson and Li 1996). More recent reports have supported these earlier findings (e.g., Ono et al. 2005), firmly establishing the SFO as a brain area that is sensitive to AngII.

Although many studies have confirmed the importance of the forebrain CVOs in the actions of AngII, these areas initially were an attractive focus primarily because they lack a blood–brain barrier and AngII does not cross the blood–brain barrier (Harding et al. 1988). AngII receptors have been found in numerous brain areas and, other than the CVOs, these brain areas do not appear accessible to peripherally derived AngII. The apparent paradox of AngII responsive brain areas without an endogenous source of AngII has been solved by the more recent discovery that the brain contains components of the renin–angiotensin system (Sakai and Sigmund 2005). In what appears to be a fascinating coincidence, cells in the same structure that responds to AngII made in the periphery, the SFO, use AngII as a peptide transmitter (Li and Ferguson 1993a,b).

The receptors for AngII fall into two main subtypes, type 1 (AT_1) and type 2 (AT_2) receptors. The intake effects of AngII appear to be exclusively driven by action at the AT_1 receptor. AT_1 receptors are expressed at relatively high levels in the CVOs and other CNS structures (Bunnemann et al. 1992; Rowe et al. 1992; Song et al. 1992), and experimental manipulation of these receptors has notable effects on fluid intake. Mice with genetic disruption of the gene for the AT_1 receptor have a severely impaired water intake response to AngII injection (Li et al. 2003), and AT_1-specific antagonists or antisense oligonucleotides attenuate the intake responses to AngII (Beresford and Fitzsimons 1992; Sakai et al. 1994, 1995; Weisinger et al. 1997). Although most commonly referred to as the AT_1 receptor, this nomenclature actually comprises two receptor isoforms, the AT_{1a} and AT_{1b} receptors. These isoforms are identical in the number of amino acids (359 in both the rat and mouse), but have

17 residues that differ in the rat and 22 residues that differ in the mouse. Although there are several reports of anatomical localization of the two AT_1 receptor subtypes, inconsistencies in the reports and the potential for species differences remain to be resolved. For example, an early study in rats used polymerase chain reaction to demonstrate expression of AT_{1a} and AT_{1b} receptors in CVO structures and hypothalamus (Kakar et al. 1992), but subsequent *in situ* hybridization studies indicate that the AT_{1a} receptor is the predominant or the only receptor type in rat CVO structures and in the hypothalamus, whereas the AT_{1b} receptor is expressed in the cerebral cortex and the hippocampus (Johren et al. 1995). Studies conducted predominantly in mice, however, have revealed interesting differences in the role of these AT_1 receptor isoforms and their regulation by perturbations in fluid balance. Specifically, selective gene targeting found that the AT_{1b} receptor is critical for the drinking response to central injections of AngII in mouse (Davisson et al. 2000), and other studies found that dehydration increased AT_{1a} receptor expression in forebrain CVOs, but AT_{1b} receptor expression was unaffected (Chen and Morris 2001). The latter of these studies, however, reported expression of both AT_{1a} and AT_{1b} receptors in CVO structures, suggesting that the distribution in the mouse and the rat may be different. A different anatomical distribution and different roles of the receptor isoforms may explain the differences between mice and rats in their responsiveness to AngII. Specifically, lateral ventricle injections of ~5 ng of AngII are sufficient to stimulate water intake in rats, but far greater amounts (~100–200 ng) are needed to stimulate water intake in mice and AngII is surprisingly not dipsetic when injected peripherally in mice (Rowland et al. 2003). Accordingly, it is appropriate to be cautious when attempting to apply findings generated in one species to another species.

5.2 ANGIOTENSIN RECEPTOR STRUCTURE AND SIGNALING

The structure and function of AT_1 receptors has been well studied using *in vitro* approaches (de Gasparo et al. 2000). The receptor is a prototypical seven-transmembrane receptor (7TMR; also commonly referred to as G protein–coupled receptors) that, like other members of the 7TMR superfamily, engages a variety of intracellular pathways when activated. In the case of the AT_1 receptor, agonist binding leads to stimulation of G_q, which activates phospholipase C, causing the resultant formation of inositol trisphosphate (IP_3) and diacylglycerol (DAG). IP_3 binds intracellular receptors to liberate stores of intracellular Ca^{2+}, while DAG activates protein kinase C (PKC). In addition to this more "traditional" signal transduction pathway, AT_1 agonists stimulate the activation (phosphorylation) of mitogen-activated protein kinase (MAP kinase) family members through actions of β-arrestin (Wei et al. 2003). Although AngII-induced MAP kinase activation has been shown to require the G_q/IP_3/PKC pathway in some settings (Chiu et al. 2003), several lines of research indicate that the G protein– and β-arrestin–mediated responses are separable. AT_1 receptor point mutants revealed that MAP kinase activation can occur independent of G protein activation and IP_3 formation (Hines et al. 2003). Specifically, point-mutated AT_1 receptors that failed to stimulate the IP_3 pathway retained their ability to activate MAP kinase when treated with AngII. Likewise, β-arrestin–biased ligands have been an important tool in the discovery of AT_1 receptor signaling

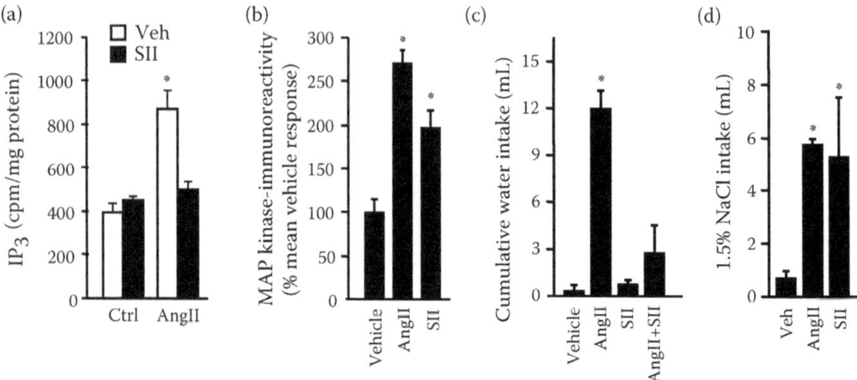

FIGURE 5.1 Response to treatment with biased agonist, SII. Using measures of IP_3 formation in cultured cells transiently transfected with AT_1 receptor, (a) SII appears to function as a classic competitive antagonist; (b) however, measures of activated MAP kinase reveal the agonist properties of SII. (c) Experiments with rats show that brain injections of SII do not increase water intake and block water intake in response to AngII, (d) but SII increases 1.5% saline intake with a magnitude similar to that after AngII injection. (Reproduced with permission from Daniels, D. et al., *Endocrinology*, 146, 12, 5552–5560, Copyright 2005, The Endocrine Society. Reprinted from *Physiol Behav*, 100, 5, Daniels, D., Alan N. Epstein award: Intracellular signaling and ingestive behaviors, 496–502, Copyright 2010, with permission from Elsevier.)

cascades (Violin and Lefkowitz 2007). Arguably, the most critical step forward in this line of research was the development of $Sar^1Ile^4Ile^8$-AngII (SII) (Holloway et al. 2002; Miura and Karnik 1999). In the original report, Miura and Karnik (1999) demonstrated that SII bound the AT_1 receptor with high affinity, but the authors considered it "inactive" because it completely failed to stimulate IP_3 production in a transient transfection preparation. In subsequent studies, however, Holloway and her colleagues (2002) showed that, despite the inability to stimulate IP_3 formation, SII activated MAP kinase in AT_1 receptor-transfected cells. Building on these findings, we replicated the earlier studies and modestly extended them to show that SII competitively antagonizes the IP_3 response to AngII, but increases MAP kinase activation, in a transfected cell model (Figure 5.1a,b) (Daniels et al. 2005). Thus, there is a wealth of information indicating that activation of MAP kinase can occur independent of G protein–mediated signaling cascades.

5.3 DIVERGENT BEHAVIORAL RELEVANCE OF ANGIOTENSIN RECEPTOR SIGNALING PATHWAYS

Although many studies had focused on the intracellular signaling properties of AngII receptors, little attention had been focused on the behavioral relevance of these pathways. The studies of Fleegal and Sumners (2003) show that the water intake response to AngII requires PKC, which is a downstream effector of Gq, but nothing was known about the requirement of MAP kinase or the effect of these pathways on saline intake. This open question was initially addressed using the biased agonist, SII, introduced above. These initial studies (Daniels et al. 2005) found that lateral ventricle injections of SII produced no water intake and blocked water intake stimulated by AngII in single-bottle tests (Figure 5.1c). Two-bottle testing, however, found an effect of SII on saline intake that was statistically indistinguishable from the effect of AngII (Figure 5.1d). Taken together, these findings were used to generate a model—initially described by Daniels et al. (2007) and shown in Figure 5.2—that proposes separable gating of neural function by G protein– and MAP kinase–mediated pathways. Specifically, it was proposed that there is some neural event or set of events gated by the G protein–dependent responses to AngII (likely mediated by calcium or by PKC) that is predominantly involved in the water intake that occurs after AngII administration and some other event, gated by MAP kinase or something downstream of MAP kinase, that plays a role in the saline intake stimulated by AngII. Although the illustration of the model suggests that these events occur in separate cells, whether or not this is the case requires empirical testing. Indeed, it seems likely that there is at least some overlap in the structures and cells that mediate the responses.

The first test of this model, which we now refer to as the "divergent signaling hypothesis of angiotensin-induced water and saline intake" (Daniels 2010), measured intake of water and saline in a two-bottle test after injection of AngII in the presence or absence of a PKC or a MAP kinase inhibitor (Daniels et al. 2009). In the first set of experiments, the PKC inhibitor chelerythrine, attenuated AngII-induced water intake, but had no effect on saline intake in the same test (Figure 5.3a). This finding confirms the earlier study of Fleegal and Sumners (2003) and extends it to show that the effect of chelerythrine does not generalize to saline intake under the same conditions. In a second set of experiments, pretreatment with U0126, a MAP

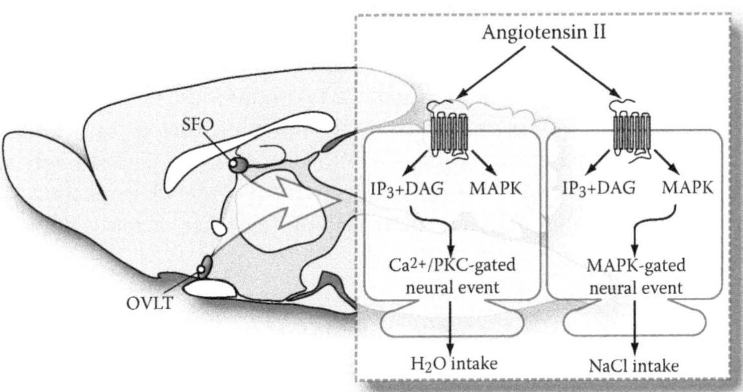

FIGURE 5.2 An illustration of divergent signaling hypothesis of AngII stimulation described by Daniels et al. (2007) and Daniels (2010). (Daniels, D., D. K. Yee, and S. J. Fluharty: Angiotensin II receptor signalling. *Exp Physiol.* 2007. 92(3). 523–527. Copyright Wiley-VCH Verlag GmbH & Co. KGaA. Reproduced with permission.)

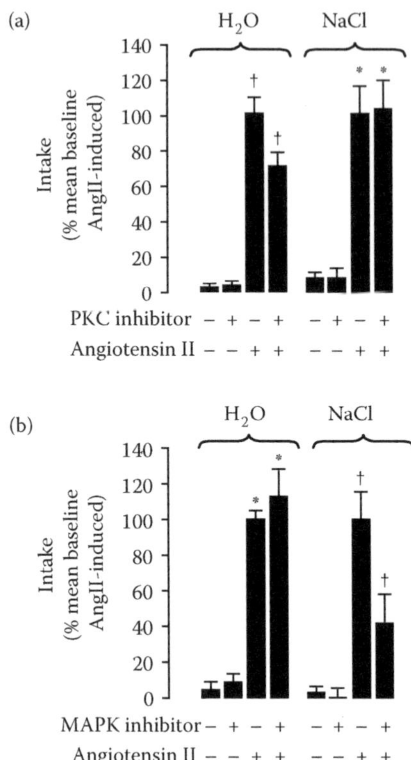

FIGURE 5.3 Inhibition of (a) PKC or (b) MAP kinase selectively affect water or saline intake, respectively, after injection of AngII. (Based on Daniels, D. et al., *Exp Physiol*, 94, 130–137, 2009.)

kinase inhibitor, was used to test the requirement of MAP kinase in the fluid intake stimulated by AngII (Figure 5.3b). Again, a two-bottle test was used, but in this experiment water intake was unaffected by the inhibitor whereas saline intake was markedly reduced in the group pretreated with U0126. These data strongly support the divergent signaling hypothesis and collectively provide what we found to be a fascinating demonstration of divergent behavioral relevance of intracellular signaling pathways stimulated by a single ligand.

5.4 DIVERSE ROLES OF ANGIOTENSIN RECEPTOR SIGNALING PATHWAYS IN THE CONTROL OF WATER INTAKE

These data clearly demonstrate divergent roles of AngII-induced signaling pathways in the control of water and saline intake, but as suggested by the title of this chapter, there appears to be additional diversity in the roles of these pathways in fluid intake. Experiments in our laboratory on this diversity have used a model of behavioral desensitization (i.e., tachyphylaxis or tolerance) that occurs after repeated central injections of AngII. The AT_1 receptor is known to desensitize *in vitro* (Guo et al. 2001; Thomas 1999; Thomas et al. 1996), and earlier studies suggest that this also happens *in vivo*. Pretreatment with AngII reduces the response to an AngII challenge in rats allowed to drink after each of the repeated injections (Quirk et al. 1988; Torsoni et al. 2004). The authors of these studies concluded that it was the repeated AngII that caused the difference in intake, and they controlled for water load as a factor by giving gastric infusions of water in a control experiment. These infusions did not reduce drinking in response to subsequent AngII, but it is possible that the very act of drinking, regardless of the water that actually enters or stays in the system, affects subsequent behavior. Indeed, when we attempted to replicate the earlier studies, using the same AngII injection paradigm, but did not allow access to water during the first injections or injections (which we refer to as the "treatment regimen") and only let the rats drink after the final injection of AngII (which we call the "test injection"), we found that much higher doses and more injections of AngII were needed to show a difference from controls (Vento and Daniels 2010). In these experiments, we found that the optimal procedure for producing behavioral desensitization involved a treatment regimen of three injections of large amounts of AngII (300 ng each, approximately 60 times the dose needed to stimulate drinking), each separated by 20 min, before a final test injection of 100 ng AngII (Figure 5.4a). These experiments showed that rats in the AngII treatment regimen group drank less water than controls after both groups were injected with 100 ng AngII (Figure 5.4b) even though rats in the experimental group received 900 ng more AngII than the controls. A variety of control experiments were conducted, and we were able to show that the response required AT_1 receptors, did not affect saline intake, was not a result of some generalized motor impairment, and was not an effect of differences in blood pressure at the time of the intake test (Vento and Daniels 2010; Vento et al. 2012).

Our original report on the effect of repeated injections of AngII contained an additional finding that is guiding current research in the laboratory. Specifically, when we repeated the experiment, but used a two-bottle test (water and saline) to measure intake, we found that water intake was different between the groups, but saline intake was unaffected by repeated injections of AngII (Figure 5.4c) (Vento and Daniels 2010). This is a potentially important finding for at least two reasons. First, it suggests that the process of desensitization ascertained from *in vitro* studies does not always occur in the brain. More specifically, the current understanding of desensitization from *in vitro* studies indicates that the process occurs through internalization of the receptors, thereby reducing the number of available surface receptors and reducing the response to AngII (Hunyady et al. 2000; Thomas et al. 1996). If this internalization occurred after repeated injections of AngII in our studies, all effects of AngII, including saline intake, would be affected. The fact that saline intake was, however, unaffected by repeated exposure to AngII suggests that there is a mechanism other than internalization by which the response to AngII is mediated. Second, the separability of water and saline intake, taken together with the divergent signaling hypothesis, suggests the possibility of differences in the desensitization of some, but not all, of the signaling pathways engaged by AT_1 receptor activation. Indeed, experiments attempting to understand the roles of intracellular signaling cascades in this phenomenon led us to uncover the diverse nature of these pathways.

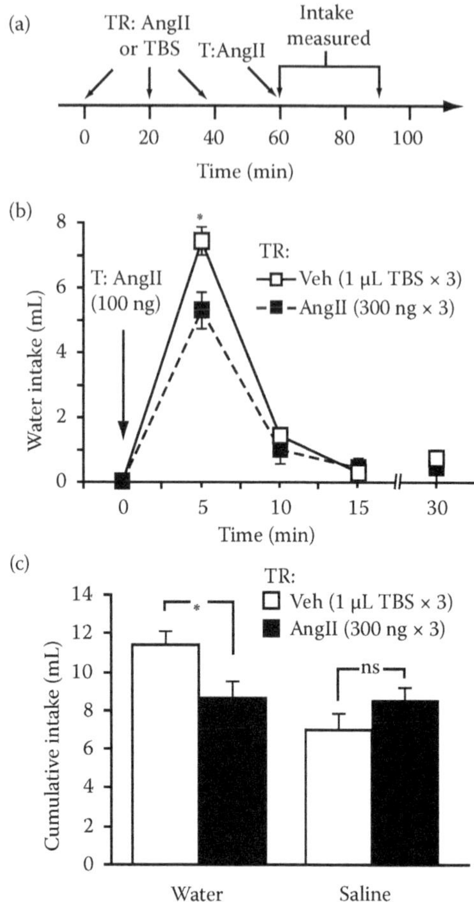

FIGURE 5.4 Repeated injections of AngII reduce water intake response to a subsequent injection of AngII. (a) Timeline used for this type of experiment in our laboratory. Fluid intake responses in (b) a typical one-bottle water intake test or (c) a two-bottle water and saline intake test are shown. (Based on Vento, P.J., Daniels, D., *Exp Physiol*, 95, 736–745, 2010.)

Preliminary studies representing our first foray into the role of intracellular signaling pathways in the effect of repeated injections of AngII focused on the activation of MAP kinase by AT_1 receptors. To test the requirement of G protein–mediated pathways in the differences in water intake observed in our repeated injection paradigm, we repeated the initial experiment comparing repeated injections of AngII to repeated injections of vehicle, but we added a third group that received three injections of a relatively large amount of the biased agonist, SII, before all three groups were injected with a test injection of AngII. In response to this test injection, rats receiving a treatment regimen of either AngII or SII drank less water than was consumed by controls. Subsequent experiments demonstrated that the reduced drinking by SII-treated rats was not the effect of simple receptor antagonism because the blockade of AngII-induced water intake by SII was short lasting and likely did not account for the differences observed 20 min after the last injection of SII. These studies suggested that G protein–mediated pathways are not required for the desensitization, but whether MAP kinase—which does not seem to play an important role in the *stimulation* of AngII-induced water intake—was important for desensitization remained an open question.

To determine the requirement of MAP kinase in the effect of repeated AngII injections, we gave repeated injections of AngII to rats, but did so with or without pretreatment with the MAP kinase inhibitor, U0126. These preliminary experiments found a complete reversal of the effect of repeated injections of AngII by the MAP kinase inhibitor. More specifically, rats given U0126 drank the same amount as controls and more than rats given repeated injections of AngII without U0126 after all rats were given a test injection of AngII. These data suggest that although MAP kinase appears unimportant for the stimulation of water intake by AngII, it is necessary for the desensitization that occurs after repeated AngII injections. Given that our earlier experiments suggested an important role for MAP kinase in the stimulation of saline intake, and our experiments showing that normal AngII-induced saline intake persists in animals given repeated injections of AngII, we used Western blotting to measure active MAP kinase in tissue punches containing the SFO and OVLT from rats sacrificed 5 min after a test injection of AngII given after a treatment regimen of either AngII or vehicle. We found no differences in the amount of active (phosphorylated) MAP kinase in any of the groups, suggesting that, like saline intake, the activation of MAP kinase is unaffected by repeated injections of AngII. Indeed, it is tempting to speculate that this persistent activation of MAP kinase underlies the ability of the test injection to stimulate saline intake, even though water intake is reduced.

5.5 CONCLUSIONS

Taken together, these studies support and extend the model proposed earlier and build on the divergent signaling hypothesis to add more information about the means by which intracellular signaling pathways can affect both saline and water intake (Figure 5.5). A summary of our findings is given below.

AngII stimulates both G protein–dependent and G protein–independent events. The current model proposes that the G protein–dependent events are critical for the full expression of water intake, but are irrelevant for saline intake. This is supported by the empirical findings that SII, which activates AT_1 receptors without engaging G protein–dependent pathways, fails to stimulate water intake and blocks water intake stimulated by AngII (Daniels et al. 2005). Moreover, pretreatment with the PKC inhibitor, chelerythrine, reduces water intake, without affecting saline intake, after central injection of AngII (Daniels et al. 2009). The G protein–independent activation of MAP kinase, however, has a more prominent role in the generation of saline intake by AngII because SII, which activates MAP kinase, stimulates saline intake when given alone and the MAP kinase inhibitor U0126 attenuates saline intake, but not water intake, when given before central application of AngII (Daniels et al. 2005, 2009). Although we previously believed, with

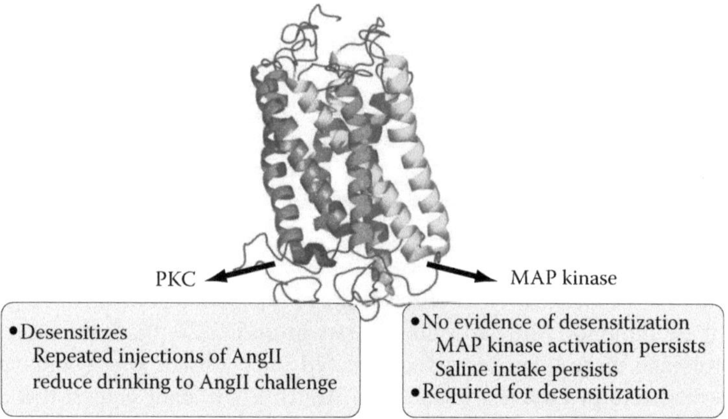

FIGURE 5.5 Summary of current understanding of diverse roles of AT_1 receptor intracellular signaling pathways in fluid intake. Although there is evidence that the role of MAP kinase in facilitation of drinking is specific for saline intake, additional studies indicate that it is a critical part of behavioral desensitization that occurs after repeated injections of MAP kinase. (Figure includes an illustration of AT_1 receptor based on the work of Baleanu-Gogonea, C., Karnik, S., *J Mol Model*, 12, 325–337, 2006.)

respect to the acute actions of AngII on behavior, that the role of MAP kinase was limited to the generation of saline intake, more recent, preliminary experiments in the laboratory suggest a more diverse role for this cascade that also provides some control of water intake. Specifically, the behavioral desensitization that develops after repeated injections of AngII (Vento and Daniels 2010) appears to require MAP kinase because preliminary experiments found that treatment with U0126, given before and concurrently with the desensitizing treatment regimen of repeated AngII injections, blocked the desensitizing properties of the repeated injections of AngII. We also now believe that whatever mechanism accounts for the behavioral desensitization, it likely affects G protein–dependent pathways more than the pathways that activate MAP kinase. This is because repeated injections of AngII affect AngII-induced water intake, without affecting saline intake (Vento and Daniels 2010) and because preliminary studies found persistent MAP kinase activation in the brains of rats treated with the desensitizing treatment regimen. These findings will guide future research on the behavioral relevance of the diverse functioning of intracellular signaling pathways. Such research has the potential to not only influence our thinking about the control of fluid intake, but could become relevant to a variety of systems as the role of intracellular signaling pathways in behavior becomes better understood.

ACKNOWLEDGMENTS

The research described was supported, in part, by grants from the National Institute of Diabetes and Digestive and Kidney Diseases (DK073800) and from the National Heart Lung and Blood Institute (HL091911).

REFERENCES

Allen, A. M., J. Zhuo, and F. A. Mendelsohn. 2000. Localization and function of angiotensin AT1 receptors. *Am J Hypertens* 13:31S–8.

Baleanu-Gogonea, C., and S. Karnik. 2006. Model of the whole rat AT1 receptor and the ligand-binding site. *J Mol Model* 12:325–37.

Beresford, M. J., and J. T. Fitzsimons. 1992. Intracerebroventricular angiotensin II-induced thirst and sodium appetite in rat are blocked by the AT1 receptor antagonist, losartan (DuP 753), but not by the AT2 antagonist, CGP 42112B. *Exp Physiol* 77:761–4.

Buggy, J., and A. K. Johnson. 1978. Angiotensin-induced thirst: Effects of third ventricle obstruction and periventricular ablation. *Brain Res* 149:117–28.

Bunnemann, B., N. Iwai, R. Metzger, K. Fuxe, T. Inagami, and D. Ganten. 1992. The distribution of angiotensin II AT1 receptor subtype mRNA in the rat brain. *Neurosci Lett* 142:155–8.

Chen, Y., and M. Morris. 2001. Differentiation of brain angiotensin type 1a and 1b receptor mRNAs: A specific effect of dehydration. *Hypertension* 37:692–7.

Chiu, T., C. Santiskulvong, and E. Rozengurt. 2003. ANG II stimulates PKC-dependent ERK activation, DNA synthesis, and cell division in intestinal epithelial cells. *Am J Physiol Gastrointest Liver Physiol* 285:G1–11.

Daniels, D. 2010. Alan N. Epstein award: Intracellular signaling and ingestive behaviors. *Physiol Behav* 100:496–502.

Daniels, D., and S. J. Fluharty. 2004. Salt appetite: A neurohormonal viewpoint. *Physiol Behav* 81:319–37.

Daniels, D., E. G. Mietlicki, E. L. Nowak, and S. J. Fluharty. 2009. Angiotensin II stimulates water and NaCl intake through separate cell signalling pathways in rats. *Exp Physiol* 94:130–7.

Daniels, D., D. K. Yee, L. F. Faulconbridge, and S. J. Fluharty. 2005. Divergent behavioral roles of angiotensin receptor intracellular signaling cascades. *Endocrinology* 146:5552–60.

Daniels, D., D. K. Yee, and S. J. Fluharty. 2007. Angiotensin II receptor signalling. *Exp Physiol* 92:523–7.

Davisson, R. L., M. I. Oliverio, T. M. Coffman, and C. D. Sigmund. 2000. Divergent functions of angiotensin II receptor isoforms in the brain. *J Clin Invest* 106:103–6.

de Gasparo, M., K. J. Catt, T. Inagami, J. W. Wright, and T. Unger. 2000. International union of pharmacology: XXIII. The angiotensin II receptors. *Pharmacol Rev* 52:415–72.

Ferguson, A. V., and Z. Li. 1996. Whole cell patch recordings from forebrain slices demonstrate angiotensin II inhibits potassium currents in subfornical organ neurons. *Regul Pept* 66:55–8.

Fleegal, M. A., and C. Sumners. 2003. Drinking behavior elicited by central injection of angiotensin II: Roles for protein kinase C and Ca^{2+}/calmodulin-dependent protein kinase II. *Am J Physiol Regul Integr Comp Physiol* 285:R632–40.

Guo, D. F., Y. L. Sun, P. Hamet, and T. Inagami. 2001. The angiotensin II type 1 receptor and receptor-associated proteins. *Cell Res* 11:165–80.

Harding, J. W., M. J. Sullivan, J. M. Hanesworth, L. L. Cushing, and J. W. Wright. 1988. Inability of [^{125}I]Sar1, Ile8-angiotensin II to move between the blood and cerebrospinal fluid compartments. *J Neurochem* 50:554–7.

Hines, J., S. J. Fluharty, and D. K. Yee. 2003. Structural determinants for the activation mechanism of the angiotensin II type 1 receptor differ for phosphoinositide hydrolysis and mitogen-activated protein kinase pathways. *Biochem Pharmacol* 66:251–62.

Holloway, A. C., H. Qian, L. Pipolo et al. 2002. Side-chain substitutions within angiotensin II reveal different requirements for signaling, internalization, and phosphorylation of type 1A angiotensin receptors. *Mol Pharmacol* 61:768–77.

Hunyady, L., K. J. Catt, A. J. Clark, and Z. Gaborik. 2000. Mechanisms and functions of AT(1) angiotensin receptor internalization. *Regul Pept* 91:29–44.

Johren, O., T. Inagami, and J. M. Saavedra. 1995. AT1A, AT1B, and AT2 angiotensin II receptor subtype gene expression in rat brain. *Neuroreport* 6:2549–52.

Kadekaro, M., S. Cohen, M. L. Terrell, H. Lekan, H. Gary, Jr., and H. M. Eisenberg. 1989. Independent activation of subfornical organ and hypothalamo–neurohypophysial system during administration of angiotensin II. *Peptides* 10:423–9.

Kakar, S. S., K. K. Riel, and J. D. Neill. 1992. Differential expression of angiotensin II receptor subtype mRNAs (AT-1A and AT-1B) in the brain. *Biochem Biophys Res Commun* 185:688–92.

Li, Z., and A. V. Ferguson. 1993a. Angiotensin II responsiveness of rat paraventricular and subfornical organ neurons in vitro. *Neuroscience* 55:197–207.

Li, Z., and A. V. Ferguson. 1993b. Subfornical organ efferents to paraventricular nucleus utilize angiotensin as a neurotransmitter. *Am J Physiol Regul Integr Comp Physiol* 265:R302–9.

Li, Z., M. Iwai, L. Wu, T. Shiuchi, T. Jinno, T. X. Cui, and M. Horiuchi. 2003. Role of AT2 receptor in the brain in regulation of blood pressure and water intake. *Am J Physiol Heart Circ Physiol* 284:H116–21.

Mangiapane, M. L., and J. B. Simpson. 1979. Pharmacologic independence of subfornical organ receptors mediating drinking. *Brain Res* 178:507–17.

Mangiapane, M. L., and J. B. Simpson. 1980. Subfornical organ: Forebrain site of pressor and dipsogenic action of angiotensin II. *Am J Physiol Regul Integr Comp Physiol* 239:R382–9.

McKinley, M. J., E. Badoer, L. Vivas, and B. J. Oldfield. 1995. Comparison of c-*fos* expression in the lamina terminalis of conscious rats after intravenous or intracerebroventricular angiotensin. *Brain Res Bull* 37:131–7.

Miura, S., and S. S. Karnik. 1999. Angiotensin II type 1 and type 2 receptors bind angiotensin II through different types of epitope recognition. *J Hypertens* 17:397–404.

Morris, M. J., W. L. Wilson, E. M. Starbuck, and D. A. Fitts. 2002. Forebrain circumventricular organs mediate salt appetite induced by intravenous angiotensin II in rats. *Brain Res* 949:42–50.

Ono, K., T. Toyono, E. Honda, and K. Inenaga. 2005. Transient outward K^+ currents in rat dissociated subfornical organ neurones and angiotensin II effects. *J Physiol* 568:979–91.

Phillips, M. I. 1987. Functions of angiotensin in the central nervous system. *Annu Rev Physiol* 49:413–35.

Quirk, W. S., J. W. Wright, and J. W. Harding. 1988. Tachyphylaxis of dipsogenic activity to intracerebroventricular administration of angiotensins. *Brain Res* 452:73–8.

Rowe, B. P., D. L. Saylor, and R. C. Speth. 1992. Analysis of angiotensin II receptor subtypes in individual rat brain nuclei. *Neuroendocrinology* 55:563–73.

Rowland, N. E., B. E. Goldstein, and K. L. Robertson. 2003. Role of angiotensin in body fluid homeostasis of mice: Fluid intake, plasma hormones, and brain Fos. *Am J Physiol Regul Integr Comp Physiol* 284:R1586–94.

Rowland, N. E., B. H. Li, M. J. Fregly, and G. C. Smith. 1995. Fos induced in brain of spontaneously hypertensive rats by angiotensin II and co-localization with AT-1 receptors. *Brain Res* 675:127–34.

Rowland, N. E., B. H. Li, A. K. Rozelle, M. J. Fregly, M. Garcia, and G. C. Smith. 1994a. Localization of changes in immediate early genes in brain in relation to hydromineral balance: Intravenous angiotensin II. *Brain Res Bull* 33:427–36.

Rowland, N. E., B. H. Li, A. K. Rozelle, and G. C. Smith. 1994b. Comparison of fos-like immunoreactivity induced in rat brain by central injection of angiotensin II and carbachol. *Am J Physiol* 267:R792–8.

Sakai, K., and C. D. Sigmund. 2005. Molecular evidence of tissue renin–angiotensin systems: A focus on the brain. *Curr Hypertens Rep* 7:135–40.

Sakai, R. R., P. F. He, X. D. Yang et al. 1994. Intracerebroventricular administration of AT1 receptor antisense oligonucleotides inhibits the behavioral actions of angiotensin II. *J Neurochem* 62:2053–6.

Sakai, R. R., L. Y. Ma, P. F. He, and S. J. Fluharty. 1995. Intracerebroventricular administration of angiotensin type 1 (AT1) receptor antisense oligonucleotides attenuate thirst in the rat. *Regul Pept* 59:183–92.

Simpson, J. B., A. N. Epstein, and J. S. Camardo, Jr. 1978. Localization of receptors for the dipsogenic action of angiotensin II in the subfornical organ of rat. *J Comp Physiol Psychol* 92:581–601.

Song, K., A. M. Allen, G. Paxinos, and F. A. Mendelsohn. 1992. Mapping of angiotensin II receptor subtype heterogeneity in rat brain. *J Comp Neurol* 316:467–84.

Thomas, W. G. 1999. Regulation of angiotensin II type 1 (AT1) receptor function. *Regul Pept* 79:9–23.

Thomas, W. G., T. J. Thekkumkara, and K. M. Baker. 1996. Cardiac effects of AII. AT1A receptor signaling, desensitization, and internalization. *Adv Exp Med Biol* 396:59–69.

Torsoni, M. A., J. B. Carvalheira, V. C. Calegari et al. 2004. Angiotensin II (AngII) induces the expression of suppressor of cytokine signaling (SOCS)-3 in rat hypothalamus—a mechanism for desensitization of AngII signaling. *J Endocrinol* 181:117–28.

Vento, P. J., and D. Daniels. 2010. Repeated administration of angiotensin II reduces its dipsogenic effect without affecting saline intake. *Exp Physiol* 95:736–45.

Vento, P. J., K. P. Myers, and D. Daniels. 2012. Investigation into the specificity of angiotensin II-induced behavioral desensitization. *Physiol Behav* 105:1076–81.

Violin, J. D., and R. J. Lefkowitz. 2007. Beta-arrestin-biased ligands at seven-transmembrane receptors. *Trends Pharmacol Sci* 28:416–22.

Wei, H., S. Ahn, S. K. Shenoy et al. 2003. Independent beta-arrestin 2 and G protein-mediated pathways for angiotensin II activation of extracellular signal-regulated kinases 1 and 2. *Proc Natl Acad Sci USA* 100:10782–7.

Weisinger, R. S., J. R. Blair-West, P. Burns, D. A. Denton, and E. Tarjan. 1997. Role of brain angiotensin in thirst and sodium appetite of rats. *Peptides* 18:977–84.

6 Separating Thirst from Hunger

Gina L. C. Yosten and Willis K. Samson

CONTENTS

6.1 INTRODUCTION

Classical thirst stimuli can be modified by other ongoing physiologic conditions. In addition, although food intake and water drinking are traditionally seen as contemporaneous behaviors, stimuli for food intake can be separated from water drinking and vice versa. In this chapter, we address two issues: independence of the two behaviors and the impact of non-thirst–related stimuli on water drinking behaviors.

Do all stimuli for food intake induce concomitant water drinking? Although pattern analysis of *ad libitum* food and water consumption definitively demonstrated the coordination of those two behaviors (de Castro 1988; Fitzsimons and LeMagnen 1969; Kissileff 1969), it is clear that water drinking can occur under physiologic conditions in the absence of food intake (Zorilla et al. 2005), and even under conditions of water restriction some food intake does occur (Fitzsimons and LeMagnen 1969; Zorilla et al. 2005). However, a more physiologic approach to the concordance of the two behaviors is the examination of the intervals of water drinking during 24 h, *ad libitum* periods, and under these conditions, significant amounts of water are consumed independent of food intake (Zorilla et al. 2005). This may seem obvious when one considers the importance of osmotic and volemic stimuli for drinking; however, when one considers water drinking in the presence of food intake, a certain necessity for the coincident behaviors overrides their independence. This is in no small part due to the lubricative function of fluid taken with solid food, and the resulting osmotic stimuli following solute absorption in the stomach and intestines. But do all stimuli for food intake also stimulate drinking? Evidence from the Daniels laboratory (Mietlicki et al. 2009) provides a clear, negative answer.

6.2 DRINKING SEPARATED FROM FEEDING

One of the most powerful stimuli for food intake is ghrelin, acting in hypothalamic centers to stimulate food intake (Wren et al. 2000). However, when studied under conditions more suitable for the examination of drinking behavior, such as pharmacologic (angiotensin II) or physiologic (hyperosmolar) challenges, quite the opposite is observed. Indeed, ghrelin administration, while stimulating food intake, inhibits angiotensin II–induced drinking, as well as drinking in response to hyperosmolar challenge (Mietlicki et al. 2009). Similarly, when drinking behavior is the primary focus, food intake can be dissociated. Indeed, dipsogenic doses of angiotensin II inhibit rather than stimulate food intake (Porter and Potratz 2004).

6.3 FEEDING AS A CONSEQUENCE OF DRINKING

Can decreased drinking behavior be the cause of decreased food intake? We believe so. Obestatin was originally identified as a posttranslational product of the ghrelin preprohromone (Zhang et al. 2005). It was named on the basis of its apparent ability to inhibit feeding, an activity that was almost immediately challenged by other researchers (Seone et al. 2006). In our hands, central administration of obestatin did reduce food intake to a slight but not significant degree; however, more impressive was the effect to inhibit water drinking in those same animals for up to 24 h after central administration before the onset of the dark phase (Samson et al. 2006). This inhibition of water, but not food, intake resulted in a significant loss of body weight over the 24-h period. The fact that neither food intake nor open field locomotor behaviors were altered by these same doses of obestatin suggested to us that the action of the peptide was selective for thirst. To test this possibility more directly, we examined the action of obestatin on pharmacologically driven thirst and observed a significant inhibitory effect of obestatin on angiotensin II–induced water drinking (Samson et al. 2006). Water drinking in response to a hypovolemic challenge also was inhibited by intracerebroventicularly (icv) administered obestatin (Samson et al. 2008a). Finally, we identified the subfornical organ as the potential site of the action of obestatin (Samson et al. 2006), again suggesting a unique action of the peptide on thirst mechanisms, independent of any significant action on food intake.

Is the selective action of obestatin on thirst physiologically relevant? Because obestatin is processed from the same preprohormone as ghrelin, gene knockout (embryonic gene ablation) or translation compromise (antisense oligonucleotides, small interfering RNA, ribozymes) approaches would abrogate the production of both peptides. How then would you interpret any behavioral phenotype that resulted? Would it be caused uniquely by loss of obestatin or by loss of ghrelin, or even loss of both? Additionally, no consensus exists on the identity of the obestatin receptor (Lauwers et al. 2006; Moechars et al. 2006), and no selective obestatin receptor antagonist has been developed. Therefore, we turned to passive immunoneutralization as an approach to examine the question of physiological relevance. Indeed, pretreatment with a selective obestatin antiserum resulted in a highly significant increase in water drinking in *ad libitum* fed and watered rats (Samson et al. 2008a). Food intake was also elevated in the antiserum-treated animals, although that increase failed to reach statistical significance. These results together with our observation of direct actions of the peptide on subfornical organ neurons pointed to a physiologically relevant action of obestatin on fluid and electrolyte homeostasis. This became even more evident when we demonstrated that in addition to inhibiting water intake, obestatin, administered icv, inhibited vasopressin, but not oxytocin, secretion in overnight water–restricted animals and vasopressin secretion in response to central angiotensin II administration (Samson et al. 2008a). This action of obestatin also appeared to have physiologic relevance because central administration of anti-obestatin antiserum resulted in exaggerated vasopressin, but not oxytocin, secretion in response to overnight water restriction (Samson et al. 2008a). Here, then, is an example of an endogenous neuropeptide that controls fluid and electrolyte homeostasis by acting primarily to inhibit water drinking, and if any effects on food intake are observed, those are in all likelihood secondary to its antidipsogenic actions.

6.4 THIRST INDEPENDENT OF HUNGER: ROLE OF BAROREFLEX

Is drinking behavior affected by non-thirst–related cues, in addition to those associated with food intake? Certainly, the answer is yes. Most obvious would be the effect of nauseogenic stimuli on drinking behavior. A second would be anxiety-related drinking, also known as psychogenic polydipsia. However, more subtle influences can also modify drinking behavior, not the least of which is ambient circulatory pressure. Several research groups have demonstrated the ability of changes in mean arterial pressures to modulate pharmacologically driven water drinking (Evered et al. 1988; Hosutt et al. 1978; Robinson and Evered 1987; Thunhorst and Johnson 1993; Thunhorst et al. 1993). It should make sense then that increased arterial pressure, via high- and low-pressure baroreceptive mechanisms, might not only reduce vasopressin secretion, but also the drive to consume fluids. In the opposite direction, hypotensive stimuli, in particular hypovolemia, stimulate both vasopressin release and water drinking, as well as increase autonomic outflow, resulting in restoration of the normovolemia. But can perfusion pressures drop even lower, past some threshold for drinking? Certainly work from Stricker's group established that possibility (Hosutt et al. 1978). Is the effect of hypotension robust enough to alter *ad libitum* drinking behavior, such as that entrained to the onset of the dark phase?

6.4.1 Can Baroreflex Also Present Confusing Signals for Thirst and Hunger?

In 2008, our group discovered a novel neuropeptide encoded in the somatostatin preprohormone that when administered icv, unlike somatostatin, inhibited food and water intake in a significant and dose-dependent fashion (Samson et al. 2008b). The antidipsogenic and anorexigenic action of this peptide, named neuronostatin, could be blocked by pretreatment with a melanocortin receptor antagonist,

SHU9119. This suggested an interaction of neuronostatin with pro-opiomelanocortin (POMC) producing neurons (Yosten and Samson 2010), something we established in collaboration with Alastair Ferguson's assistance by electrophysiologic approaches (Samson et al. 2008b). At that point, it appeared that the effects of neuronostatin on water intake might be mainly a reflection of its anorexigenic activity because not only were POMC neurons activated by the peptide, NPY producing neurons were also inhibited (Samson et al. 2008b). But are those anorexigenic and antidipsogenic actions physiologically relevant? Because neuronostatin is a product of posttranslational processing of the somatostatin preprohormone, any attempt to compromise the peptide's production would also compromise somatostatin production and thus the effects observed would be uninterpretable. In fact, the absence of an altered growth phenotype of, or growth hormone secretion in, somatostatin gene knockout animals proves this point (Low et al. 2001).

The neuronostatin receptor has not been identified and thus we could not compromise the peptide's action by deletion of its receptor. Similarly, selective neuronostatin antagonists currently are not available; therefore, we sought to selectively compromise the action of neuronostatin *in vivo* by passive immunoneutralization. Adult male rats bearing lateral cerebroventricle cannulas were habituated to metabolic cages as previously described (Yosten and Samson 2010). Daily *ad libitum* food and water intakes were monitored for minimally 3 days before the administration of 3 μL preimmune rabbit serum (NRS) or 3 μL rabbit antirat neuronostatin antiserum (Phoenix Pharmaceuticals, Burlingame, CA, USA) 1 h before lights were turned off (lights on 0600–1800 hours). Food and water intakes were then monitored hourly until 2100 hours and again 24 and 48 h later. We had hypothesized that compromise of neuronostatin action (i.e., neutralization with antiserum) would result in exaggerated eating and drinking because the pharmacologic effect of the peptide was an inhibition of both. Surprisingly, the anti-neuronostatin treated rats actually ate less food and drank less water than NRS-treated controls (Figure 6.1). Because they responded with an appropriate startle response, we did not consider the decreased behavior to be related to a deficit in general motor function. No differences were observed between the two groups in terms of food and water intakes or body weight gains after 24 h. Thus, the inhibitory effects observed on the day of treatment were reversible.

We then hypothesized that the failure of the animals to eat or drink was not primarily due to loss of the action of endogenous neuronostatin on feeding and drinking circuits in the brain, but was instead secondary to a loss of the action of endogenous peptide on autonomic centers in the medulla. We have demonstrated that central administration of exogenous neuronostatin elevates mean arterial pressure by two mechanisms: increased sympathetic efferent tone and increased vasopressin release (Yosten et al. 2011). Was it possible then that in the absence of endogenous neuronostatin, arterial pressures were low enough to compromise baroreflex function and thus the rats were experiencing orthostatic hypotension when they rose to access the food and water? To test this hypothesis, we monitored mean arterial pressure in conscious, unrestrained rats before and after central administration of NRS or anti-neuronostatin antiserum. Indeed, arterial pressures fell significantly in the anti-neuronostatin treated rats, were unstable, and fell even more when the animals rose to move around the cages.

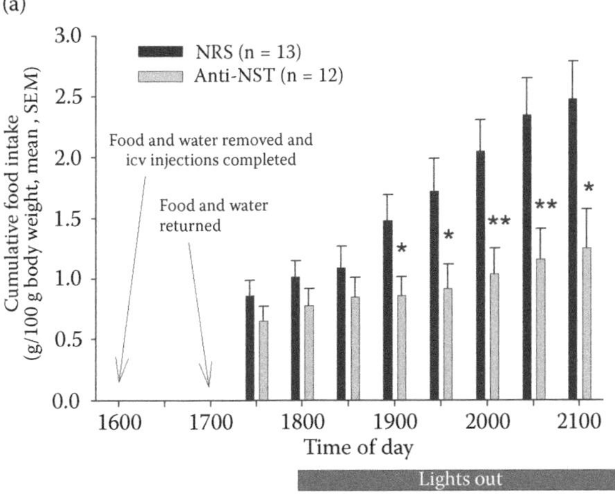

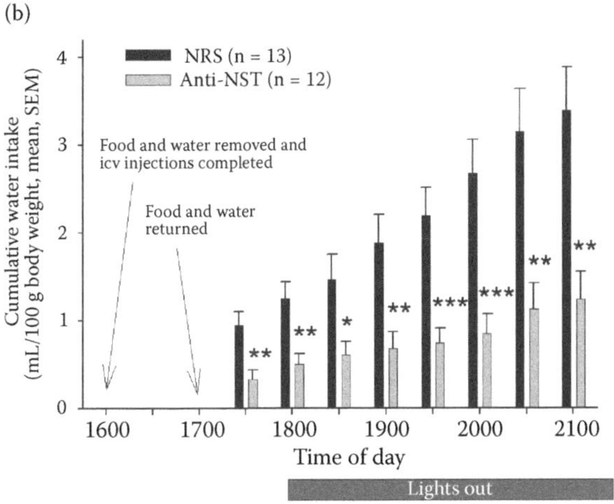

FIGURE 6.1 *Ad libitum* food (a) and water (b) intakes in male rats following icv administration of normal rabbit serum (NRS) or anti-neuronostatin (NST) antiserum. Intakes in NRS-treated rats did not differ significantly from those administered saline previously published with the same experimental design (Samson et al. 2008). $^*p < 0.05$, $^{**}p < 0.01$, $^{***}p < 0.001$ versus intakes in NRS-treated animals (independent Student's t-test).

6.4.2 Does Loss of Baroreflex Alter Thirst and Hunger?

To test the hypothesis that orthostasis was the potential cause of the failure of the antiserum treated rats to eat and drink at the onset of the dark phase, we instrumented animals with lateral cerebroventricular cannulas, and 5 days later catheters were placed in the carotid artery and jugular vein, as previously described (Yosten and Samson 2010). On the following day, animals were habituated to the testing

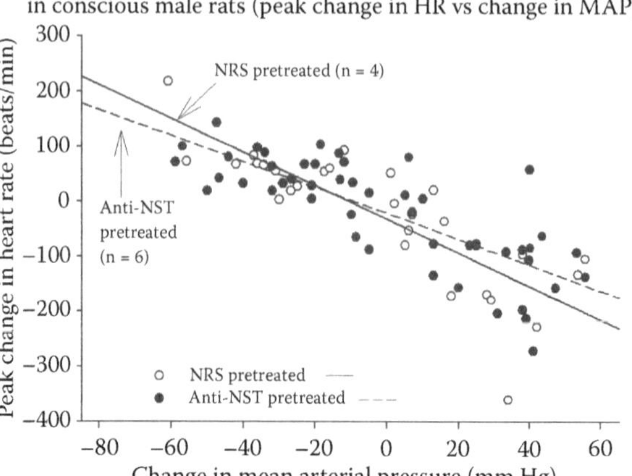

FIGURE 6.2 Intracerebroventricular administration of anti-neuronostatin (NST) antiserum alters baroreflex function in conscious, unrestrained male rats (slopes compared by linear regression analysis). Pressures were lowered or raised by intra-arterial administration of sodium nitroprusside or phenylephrine, respectively, and peak heart rate response plotted against change in mean arterial pressure).

room, the carotid catheter was connected to a pressure transducer, and the jugular catheter was connected to an extension tube to facilitate intravenous administration of either sodium nitroprusside (2.5, 7.5, and 15 μg/kg body weight), to lower arterial pressures, or phenylephrine (5.0, 25, and 50 μg/kg body weight), to raise pressures. Heart rates and arterial pressures were monitored for at least 20 min before and continuously during testing. As can be seen in Figure 6.2, pretreatment with anti-neuronostatin antiserum resulted in a significant change in baroreflex function (comparison of slopes of the regression lines, $t_{\alpha}(2),73 = 29.2$, $p < 0.001$) such that those animals responded with less of a heart response to either a drop or a rise in mean arterial pressure. Thus, we conclude that the inhibitory effect of anti-neuronostatin pretreatment on food and water intake was in all likelihood secondary to a compromise in autonomic function that resulted in orthostasis and therefore a deficit in locomotor activity. These data, as well as current studies in our laboratory using molecular approaches to compromise neuronostatin function by abrogating production of the G protein-coupled receptor we are proposing to be the cognate receptor for the peptide, point to a physiologically relevant role for endogenous neuronostatin in the central control of autonomic function, but not a direct role in thirst or hunger.

6.5 SUMMARY AND CONCLUSIONS

In summary, elevations in mean arterial pressure can buffer drinking responses to thirst stimuli, probably through baroreflex activation, and hypotension can similarly alter drinking behavior, when the thirst centers in the brain are not alerted to the drop in pressure because of a compromised baroreflex. Investigators fluent in the thirst literature are well aware of these potential pitfalls when data from feeding studies are reported, but can the same be said about those less well versed in that literature? Clearly, the literature on obestatin tells a cautionary tale (Zhang et al. 2007). Additionally, when novel peptides are first tested for potential effects on food and water intakes, experiments should be designed that potentially separate the two behaviors, and the possibility that either of those two behaviors are a result of changes in autonomic function must be examined. Our experiences certainly have taught us that valuable lesson.

ACKNOWLEDGMENTS

Drs. Yosten and Samson are supported by the National Institutes of Health (HL66023) and the American Heart Association (10GRNT4470043).

REFERENCES

de Castro, J. 1988. A microregulatory analysis of spontaneous fluid intake in humans: Evidence that the amount of liquid ingested and its timing is mainly governed by feeding. *Physiol Behav* 3:705–14.

Evered, M.D., M.M. Robinson, and P.A. Rose. 1988. Effect of arterial pressure on drinking and urinary responses to angiotensin II. *Am J Physiol Regul Integr Comp Physiol* 254:R69–74.

Fitzsimons, J.T., and J. Le Magnen. 1969. Eating as a regulatory control of drinking in the rat. *J Comp Physiol Psychol* 67:273–83.

Hosutt, J.A., N. Rowland, and E.M. Stricker. 1978. Hypotension and thirst in rats after isoproterenol treatment. *Physiol Behav* 21:593–8.

Kissileff, H.R. 1969. Food-associated drinking in the rat. *J Comp Physiol Psychol* 67:284–300.

Lauwers, E., B. Landuyt, L. Arckens, L. Schoof, and W. Luyten. 2006. Obestatin does not activate orphan G protein-coupled receptor GPR39. *BiochemBiophys Res Commun* 351:21–25.

Low, M.J., V. Otero-Corchon, A.F. Parlow et al. 2001. Somatostatin is required for masculinization of growth hormone – regulated hepatic gene expression but not growth. *J Clin Invest* 107:1571–80.

Mietlicki, E.G., E.L. Nowak, and D. Daniels. 2009. The effect of ghrelin on water intake during dipsogenic conditions. *Physiol Behav* 96:37–43.

Moechars, D., I. Depoortere, B. Moreaux et al. 2006. Altered gastrointestinal and metabolic function in the GPR39-obestatin receptor knockout-mouse. *Gastroenterology* 131:1131–41.

Porter, J.P., and K.R. Potratz. 2004. Effect of intracerebroventricular angiotensin II on body weight and food intake in adult rats. *Am J Physiol Regul Integr Comp Physiol* 287:R422–8.

Robinson, M.M., and M.D. Evered. 1987. Pressor action of intravenous angiotensin II reduces drinking response in rats. *Am J Physiol Regul Integr Comp Physiol* 252:R754–9.

Samson, W.K., M.M. White, C. Price, and A.V. Ferguson. 2006. Obestatin acts in brain to inhibit thirst. *Am J Physiol Regul Integr Comp Physiol* 292:R637–43.

Samson, W.K., G.L.C. Yosten, J.K. Chang, A.V. Ferguson, and M.M. White. 2008a. Obestatin inhibits vasopressin secretion: Evidence for a physiological action in the control of fluid homeostasis. *J Endocrinol* 196:1–7.

Samson, W.K., J.V. Zhang, O. Avsian-Kretchmer et al. 2008b. Neuronostatin encoded by the somatostatin gene regulates neuronal, endocrine and metabolic functions. *J Biol Chem* 283:31949–59.

Seone, L.M., O. Al-Massadi, Y. Pazoz, U. Pagotto, and F.F. Casaneuva. 2006. Central obestatin administration does not modify either spontaneous or ghrelin-induced food intake in rats. *J Endocrinol Invest* 29:RC13–15.

Thunhorst, R.L., and A.K. Johnson. 1993. Effects of arterial pressure on drinking and urinary responses to intracerebroventricular angiotensin II. *Am J Physiol* 264:R211–7.

Thunhorst, R.L., S.J. Lewis, and A.K. Johnson. 1993. Role of arterial baroreceptor input on thirst and urinary responses to intracerebroventricular angiotensin II. *Am J Physiol* 265:R591–95.

Wren, A.M., C.J. Small, H.L. Ward et al. 2000. The novel hypothalamic peptide ghrelin stimulates food intake and growth hormone secretion. *Endocrinology* 141:4325–8.

Yosten, G.L., and W.K. Samson. 2010. The melanocortins, not oxytocin, mediate the anorexigenic and antidipsogenic actions of neuronostatin. *Peptides* 31:1711–4.

Yosten, G.L., A.T. Pate, and W.K. Samson. 2011. Neuronostatin acts in brain to biphasically increase mean arterial pressure through sympatho-activation, followed by vasopressin secretion: The role of melanocortin receptors. *Am J Physiol* 300:R1194–9.

Zhang, J.V., P.-G. Ren, O. Avsian-Kretchmer et al. 2005. Obestatin, a peptide encoded by the ghrelin gene, opposes ghrelin's effect on food intake. *Science* 310:996–9.

Zhang, J.V., C. Klein, P.-G. Ren et al. 2007. Response to comment on 'Obestatin, a peptide encoded by the Ghrelin gene, opposes effects on food intake.' *Science* 315:766.

Zorilla, E.P., K. Inoue, E.M. Fekete, A. Tabarin, G.R. Valdez, and G.F. Koob. 2005. Measuring meals: Structure of prandial food and water intake of rats. *Am J Physiol* 288:R1450–67.

7 Interdependent Preoptic Osmoregulatory and Thermoregulatory Mechanisms Influencing Body Fluid Balance and Heat Defense

Michael J. McKinley and Michael L. Mathai

CONTENTS

7.1 THERMOREGULATION AND BODY FLUID HOMEOSTASIS

Maintenance of core body temperature within narrow bounds and the constancy of composition and volume of body fluids in mammals are probably the most iconic manifestations of homeostasis in mammals. The ability to keep the internal environment of their bodies as unchanged as possible in the face of hostile environments is one of the reasons for mammals being able to populate and survive in many regions of the planet. Whereas mammals that inhabit the cooler and temperate regions of the earth are less likely to experience extremes of high ambient temperatures or insufficient supply of water for maintaining bodily hydration, this is not the case for those species that exist in warmer and drier parts of the earth. The evolutionary emergence and survival of many mammalian species has resulted from adaptations that permit such animals to minimize and correct changes in core body temperature as well as deficits in body fluid volume and composition arising in hot, dry environments.

7.2 THERMOREGULATORY RESPONSES TO INCREASING CORE BODY TEMPERATURE

Included in the homeostatic mechanisms that mammals use to prevent core temperature increasing to dangerous levels are behavioral responses such as reduced physical activity, lethargy, and avoidance of a hot environment. Physiological responses include sweating, panting, reduced thermogenesis, and increased blood supply to the skin. This latter adjustment usually requires dilatation of cutaneous blood vessels with compensatory reduction of blood flow to some internal organs and increase in cardiac output that enable the supply of blood to the skin to be increased (Johnson and Proppe 1996). The most effective cooling occurs when the heat being transferred from core to skin is dissipated by the evaporation of sweat, saliva, or moisture from respiratory membranes.

Evaporative cooling mechanisms such as those occurring with sweating (humans, horses), panting (ungulates, canines, felines), and the grooming of saliva on skin (rodents), although effective at reducing core temperature, incur obligatory losses of fluid from the body. If these losses of fluid are not replaced quickly (by ingestion of water), dehydration of the body ensues and increases further as sweating, panting, and salivary grooming continue. Not only is body temperature homeostasis perturbed, but a number of physiological mechanisms now come into operation to reduce the impact of fluid loss on the animal. As the physiological response to bodily heating almost always involves dissipation of body heat by evaporation of body fluids, thermoregulatory and fluid homeostasis are inexorably linked. In this chapter, these links and the neural mechanisms within the brain that enable them will be explored.

7.3 HOMEOSTATIC RESPONSES TO DEHYDRATION RESULTING FROM SWEATING, PANTING, OR SALIVA LOSS

7.3.1 Vasopressin Secretion, Thirst, and Natriuresis

When water is lost from the body as a result of sweating, panting, or salivation, hypovolemia will result unless the losses are compensated for by ingestion of fluid. The fluid lost from the body as sweat, saliva, or from respiratory surfaces is hypotonic relative to plasma. Consequently, the osmolality and tonicity of the extracellular fluid increase as hypovolemia develops. The osmotic pressures that develop across cell membranes cause movement of intracellular fluid to the extracellular compartment by osmosis, thereby causing intracellular dehydration as well as hypovolemia of the extracellular compartment. It is the reduction in cell volume of specific osmoreceptors within the lamina terminalis of the brain in response to systemic hypertonicity that is the *modus operandi* by which these sensors detect hypertonicity resulting from bodily dehydration (Ciura et al. 2011; Bourque 2008). Osmoreceptors located in the organum vasculosum of the lamina terminalis (OVLT) and subfornical organ initiate several physiological and behavioral responses that include increased vasopressin secretion and thirst (McKinley et al. 1978, 1983, 2004; Thrasher et al. 1982). An anorexic response is also observed that results in reduced food intake in dehydrated animals (Schloorlemmer and Evered 1993). The action of vasopressin on the kidney to concentrate urine and the increased renal sodium excretion that occurs in response to dehydration have the effect of minimizing both the hypovolemia and the increase in plasma tonicity that occurs in animals if they do not have access to water to replenish fluid losses. As mentioned above, if water-deprived animals are also subjected to a warm environment so that skin and core body temperature increase, the thermoregulatory responses of sweating, salivation, and panting cause further loss of body fluids. A question arises: are these thermoregulatory evaporative cooling responses sustained in dehydrated animals?

7.3.2 Thermoregulatory Inhibition of Sweating and Panting

Environmental conditions may arise where the homeostatic control of body temperature opposes body fluid homeostasis. In hot environments, body water may be depleted because of sweating or panting—two major homeostatic cooling mechanisms. Evaporation of fluid from the skin surface in the case of sweating, or from moist membranes of the respiratory tract during panting, requires energy in the form of heat from the body. Such dissipation of bodily heat for evaporative purposes is an effective means of preventing dangerously high increases in core body temperature. On the other hand, if drinking water is not available, these thermoregulatory responses of sweating and panting will eventually result in mammals becoming dehydrated. More than 70 years ago, Lee and Mulder (1935) investigated the effect of withholding water from human subjects exposed to a hot environment. They observed that the rate of sweating of these heated subjects fell if water was not ingested with prolonged exposure to heat over 5 h and core temperature increased. Interestingly, as soon as water became available and was ingested, sweating increased again. This suggested that uncompensated loss of fluid from the body as sweat eventually leads to a reduction of sweating, and therefore impaired ability to dissipate heat from the body.

More recently, it was shown that sweating in response to increased core temperature was reduced if hypertonic saline was infused systemically into heated humans (Takamata et al. 1995, 2001). So to was the increased blood flow to the skin in response to increased core temperature (Takamata et al. 1997). Hypertonicity of the blood results in intracellular dehydration due to diffusion of intracellular water to the hypertonic extracellular compartment by osmosis. As a result of the inhibition of sweating, core body temperature increased more than it did in control subjects administered with isotonic saline systemically (Takamata et al. 2001). Likewise, infusions of hypertonic saline into the carotid artery of dogs exposed to high ambient temperatures caused a reduction of their panting, thereby disrupting heat defense mechanisms (Baker and Dawson 1985). These inhibitory effects of systemic hypertonicity on thermoregulatory cooling responses in humans and dogs show that these species defend plasma osmolality at the expense of increasing core body temperature when plasma osmolality increases. These data also show the feasibility of an osmoreceptor mechanism having an inhibitory influence on thermoregulatory dissipation of heat by sweating or panting. We have observed that when sheep are deprived of water for 2 days, core temperature increases, but there is no compensatory panting response accompanying this hyperthermia. Because plasma hypertonicity results from dehydration in these animals, it is likely that there is an osmoregulatory inhibition of panting in dehydrated sheep.

7.4 OSMORECEPTOR MEDIATION OF INHIBITION OF PANTING

We investigated the feasibility of cerebral osmoreceptors mediating the inhibition of thermoregulatory panting in sheep by making infusions of isotonic or hypertonic solutions into the carotid artery of conscious sheep that were exposed to high ambient temperatures (McKinley et al. 2008). Both common carotid arteries had been previously enclosed in loops of skin to facilitate the placement of cannulae into the carotid arteries of conscious animals. Sheep that were kept in a thermal chamber in which temperature was maintained at 39°C for 1–2 h exhibited panting with respiratory frequencies of more than 100 breaths/min. Intracarotid infusions of either hypertonic (1.5 mol/L) or isotonic (0.15 mol/L) saline were made for 10 min in four of these sheep maintained at 37°C, and the effect on respiratory rate was observed. Whereas the control infusion of isotonic saline had no effect on panting, intracarotid infusion of hypertonic saline caused a significant and pronounced reduction of respiratory rate within 2–4 min (Figure 7.1). This effect

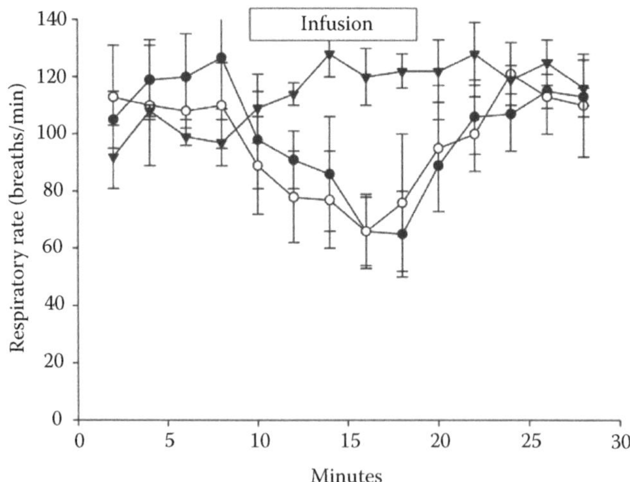

FIGURE 7.1 Effects of infusions of either isotonic (0.15 mol/L) or hypertonic NaCl (1.65 mol/L) solutions, or hypertonic sorbitol (3 mol/L in 0.15 mol/NaCl solution) into the carotid artery (at 1.6 mL/min for 10 min) on respiratory rate of four sheep maintained in an environmental temperature of 39°C. Panting was significantly reduced after 4 min of infusion of either hypertonic NaCl (closed circles) or sorbitol (open circles), then rapidly recovered after cessation of infusions. Intracarotid infusion of isotonic NaCl (closed triangles) was ineffective at reducing panting.

persisted throughout the 10 min of infusion; however, respiratory rate increased almost immediately on the termination of the infusion of hypertonic saline into the carotid artery, indicating it likely that the increased tonicity of blood supplying the head, and probably the brain, inhibited panting (McKinley et al. 2008). The rapid onset and termination of the effects make it unlikely that the inhibition of panting was due to an effect of hypertonicity of the remainder of the systemic circulation. Because it was possible that the reduced panting response caused by intracarotid hypertonic saline was a specific effect of NaCl rather than an osmotically mediated effect, we also investigated the effect on panting of infusion of another osmotic agent, hypertonic (3 mol/L sorbitol), into the carotid artery of these sheep. Intracarotid infusion of hypertonic 3 mol/L sorbitol also caused a significant reduction of respiratory rate in sheep at 37°C within 2–4 min (Figure 7.1). The effect was similar to that of intracarotid infusion of hypertonic NaCl, and it was reversed soon after ceasing the infusion. These results indicate that an osmoreceptor mechanism, probably within the brain, can have a strong inhibitory influence on thermoregulatory panting in sheep.

7.5 SITE OF CEREBRAL OSMORECEPTORS INFLUENCING THERMOREGULATORY PANTING

Cerebral osmoreceptors regulating vasopressin secretion, water drinking, and renal sodium excretion are located in circumventricular organs of the lamina terminalis—the OVLT and subfornical organ (Bourque 2008; Ciura et al. 2011; McKinley et al. 1982, 1992, 2004; Thrasher et al. 1982). A second question arises: do osmoreceptors in the lamina terminalis also have an inhibitory influence on thermoregulatory fluid losses such as occur during panting and sweating? We probed this question by investigating the effect of ablating the lamina terminalis on thermoregulatory panting responses in sheep exposed to high ambient temperature (39°C). In contrast to the rapid reduction in panting caused by intracarotid infusion of either hypertonic sodium chloride or sorbitol solution, intracarotid infusion of hypertonic saline did not inhibit panting in sheep in which the lamina terminalis had been ablated (Figure 7.1). This result shows that the lamina terminalis plays a crucial role in osmoregulatory inhibitory influences on thermoregulatory panting. The results from studies of lesions do not establish the exact functional role of the lamina terminalis in thermoregulatory panting. However, as osmoreceptors for thirst and vasopressin are located there, it seems likely that osmoreceptors exerting inhibitory influences on thermoregulatory panting in sheep may also be located in the lamina terminalis. By analogy, osmoreceptors in the lamina terminalis may also exert inhibitory influences on sweating in mammals that use this means for evaporative cooling.

It is of interest that the same general region of the brain, the medial part of the preoptic area, within which the osmoreceptors of the lamina terminalis are located (McKinley et al. 1996), is also a site of central thermoreceptors that detect increases in core body temperature to initiate sweating, skin vasodilatation, and panting. Studies that we have conducted on c-*fos* expression in rats in response to hypertonicity or dehydration, show that osmotically stimulated neurons in the preoptic region are sharply confined to the dorsal cap of the OVLT, the most midline part of the median preoptic nucleus and the outer shell of the subfornical organ (McKinley et al. 2004; Oldfield et al. 1993). By contrast, c-*fos* expression in rats in response to exposure to a hot environment (40°C for 1 h) was observed in neurons that in general were more laterally placed in the median preoptic nucleus, and ventral and lateral to the OVLT (Hubschle et al. 2001). These heat-activated neurons are likely to receive neural inputs from skin thermal sensors relayed via the spinal cord and lateral parabrachial nucleus. Some may also respond to increased core temperature. However, the neural connections that may exist between osmoregulatory units, and these thermoregulatory neurons are unknown and may be the focus of future research.

7.6 EFFECT OF DRINKING ON THERMOREGULATION

Although it is clear from the evidence presented in the previous paragraphs that systemic hypertonicity can exert a considerable inhibitory influence on thermoregulatory panting in dehydrated animals, we have observed that the act of drinking water *per se* can have the opposite effect. When sheep that had been deprived of water (and therefore had an elevated core temperature) were provided with water to drink, they immediately drank, as expected, a volume of water within 2–4 min that was equivalent to their loss of body weight. On ceasing the bout of drinking, the respiratory rate increased immediately, indicating that, upon rehydration, thermoregulatory panting had recommenced at a higher level (Figure 7.2). This recommencement of panting was accompanied by a sharp fall in core temperature that was not due to a cooling effect of the water that was ingested because it was provided at 39°C (the core temperature of the dehydrated sheep). The rapidity of the panting response also indicated that it was unlikely that the initial response was due to systemic absorption of water from the gastrointestinal tract. In agreement with this idea, such a rapid panting response to rehydration was not observed if the same volume of water was administered directly into the rumen via a nasoruminal tube to cause rehydration without the act of drinking (Figure 7.3). In this experiment, panting did eventually recommence, albeit after a delay of several minutes, and there was a slow decrease in core temperature after administration of the intraruminal water. It is these delayed responses that are probably due to systemic osmolality falling as the result of absorption of water from the gut, because when dehydrated sheep were administered saline by nasoruminal tube, no panting or fall in core temperature occurred at all (McKinley et al. 2009). Thus, it appears that the act of drinking water is a crucial aspect of initiating the panting response to rehydration in sheep, and the subsequent absorption of water from the gut that lowers plasma tonicity maintains the thermoregulatory panting until core temperature falls back to normal. Consistent with this explanation was the rapid, but transient, panting response and fall in core temperature that we observed when dehydrated sheep drank isotonic saline at 39°C (Figure 7.4). As no fall in plasma osmolality would have occurred with the drinking of saline solution, the panting response did not persist and the reduced core temperature was not maintained (McKinley et al. 2009). Other species (dogs, goats) have also been observed to rapidly increase panting after they drank water to restore fluid balance (Baker

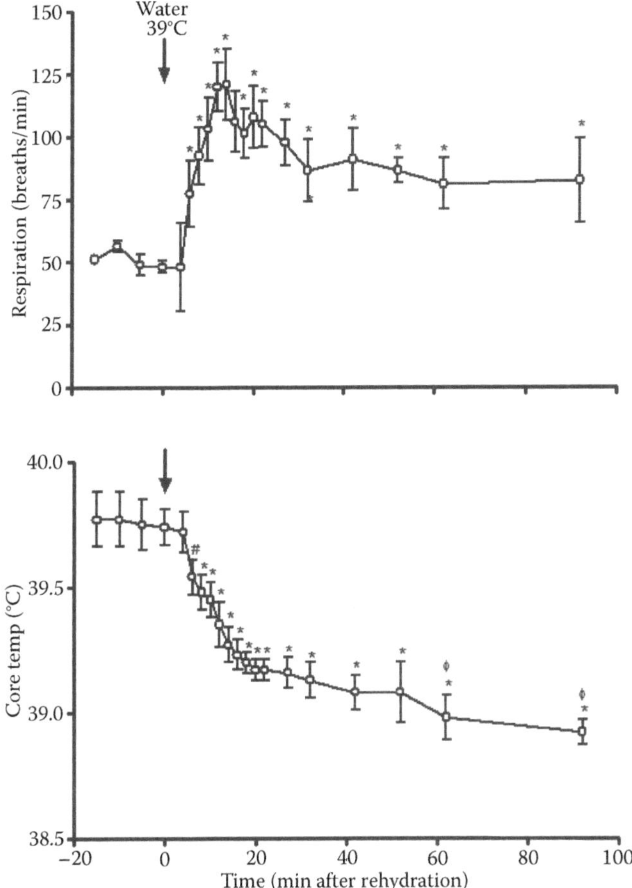

FIGURE 7.2 When water-deprived sheep are rehydrated by allowing them to drink water (water temperature was 39°C, similar to their core body temperature), a rapid increase in respiratory rate (panting) and fall in core body temperature ensues. Significant changes from prerehydrational levels are indicated by #$p < 0.05$, *$p < 0.01$; ϕ indicates significant difference ($p < 0.05$) from 20 min value ($n = 4$ sheep). (Reprinted with permission from McKinley, M.J. et al., *Am J Physiol Regul Integr Comp Physiol*, 296, R1881–R1888, 2009.)

and Turlejska 1989). As well, an analogous rapid increase in sweating occurs in warmed human subjects when they drink water to restore fluid balance (Lee and Mulder 1935; Takamata et al. 1995).

This rapid initiation of panting and sweating (Baker 1989) and fall in core temperature that results when dehydrated mammals drink water resembles the rapid decrease in vasopressin secretion that also occurs with rehydration by drinking (Blair-West et al. 1985; Thrasher et al. 1981). It seems likely that on drinking fluid, afferent nerves from the upper parts of the alimentary canal (oropharynx, esophagus,

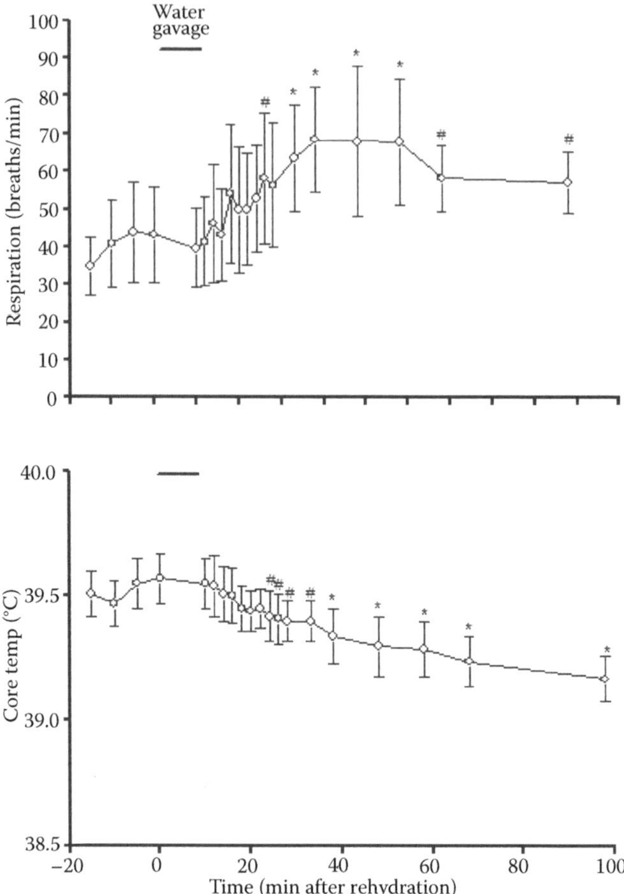

FIGURE 7.3 When water-deprived sheep are rehydrated by allowing administering water (water temperature was 39°C) into the rumen via a nasoruminal tube, respiratory rate (panting) increased and core temperature fell, but at a much slower onset and rate than if the sheep had drunk water as in Figure 7.2. Significant changes from prerehydrational levels are indicated by #$p < 0.05$, *$p < 0.01$ ($n = 4$ sheep). (Reprinted with permission from McKinley, M.J. et al., *Am J Physiol Regul Integr Comp Physiol*, 296, R1881–R1888, 2009.)

stomach) signal to the brain that rehydration has occurred, allowing the osmotic inhibition of thermoregulatory panting and stimulation of vasopressin secretion to be withdrawn. These responses to rehydration could be considered to be coordinated feed-forward mechanisms that contribute to a rapid shedding of an accumulated heat load as well as the return to normal neuroendocrine control of kidney function and water excretion when strong fluid conservation is no longer required. These mechanisms ensure that the homeostatic regulations of body temperature and body fluids are optimal after rehydration.

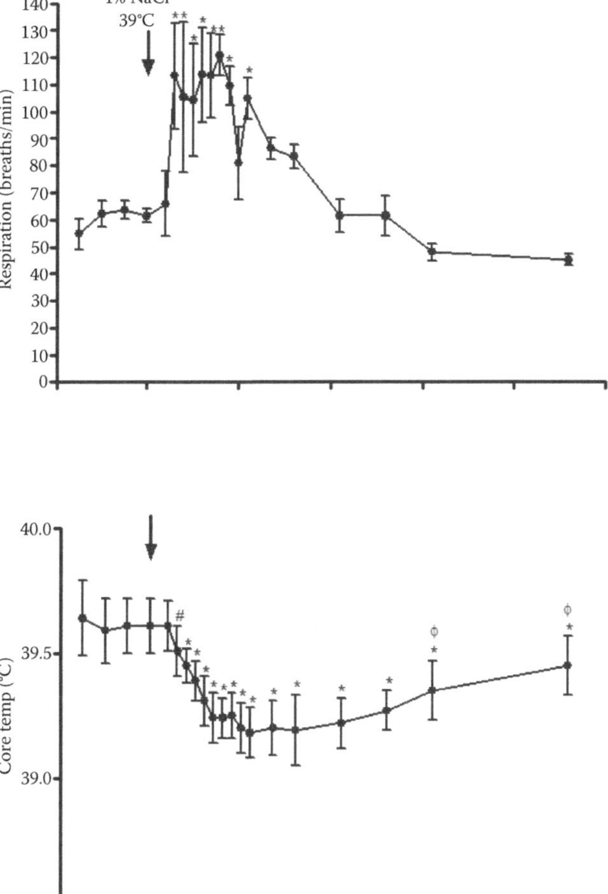

FIGURE 7.4 When water-deprived sheep are rehydrated by allowing them to drink saline solution (temperature 39°C), a rapid increase in respiratory rate (panting) and fall in core body temperature resulted, but was not as sustained as when they drank water as in Figure 7.2. Significant changes from prerehydrational levels are indicated by #$p < 0.05$, *$p < 0.01$; ϕ indicates significant difference ($p < 0.05$) from 20 min value ($n = 4$ sheep). (Reprinted with permission from McKinley, M.J. et al., *Am J Physiol Regul Integr Comp Physiol*, 296, R1881–R1888, 2009.)

7.7 CONCLUDING REMARKS

Data from several laboratories in a variety of mammals show that thermoregulation and body fluid regulation are interdependent especially when core temperature increases. Osmoregulatory neural pathways that arise within the preoptic and hypothalamic brain regions to regulate vasopressin secretion and thirst also appear to have inhibitory influences on thermoregulatory panting and sweating if thermoregulatory fluid losses are not replaced, thereby conserving body fluids at the expense of an increased core temperature. These inhibitory influences can be rapidly disengaged by the act of drinking. Immediately upon the ingestion of water by dehydrated animals, neural signals are generated within the upper parts of the gastrointestinal tract and are transmitted to and integrated within the brain to initiate panting, sweating, and reduced secretion of vasopressin. These effects are observed well before any absorption of fluid from the gastrointestinal tract could have occurred, and provide a rapid homeostatic feed-forward mechanism for optimal control of body temperature and fluid balance.

ACKNOWLEDGMENTS

This work was supported by the G. Harold and Leila Y. Mathers Charitable Trust, the Robert J. Jr. and Helen C. Kleberg Foundation and the National Health and Medical Research Council of Australia. MJMcK held an NHMRC Fellowship 454369.

REFERENCES

Baker, M. A. 1989. Effects of dehydration and rehydration on thermoregulatory sweating in goats. *J Physiol* 417:421–35.

Baker, M. A., and D. D. Dawson. 1985. Inhibition of thermal panting by intracarotid infusion of hypertonic saline in dogs. *Am J Physiol Regul Integr Comp Physiol* 249:R787–91.

Baker, M. A., and E. Turlejska. 1989. Thermal panting in dehydrated dogs: Effects of plasma volume expansion and drinking. *Pflügers Arch* 413:511–15.

Blair-West, J. R., A. P. Gibson, R. L. Woods, and A. H. Brook. 1985. Acute reduction of plasma vasopressin levels by rehydration in sheep. *Am J Physiol Regul Integr Comp Physiol* 248:R68–71.

Bourque, C. W. 2008. Central mechanisms of osmosensation and systemic osmoregulation. *Nature Rev Neurosci* 9:519–31.

Ciura, S., W. Liedtke, and C. W. Bourque. 2011. Hypertonicity sensing in organum vasculosum lamina terminalis neurons: A mechanical process involving TRPV1 but not TRPV4. *J Neurosci* 31:14669–76.

Hubschle, T., M. L. Mathai, M. J. McKinley et al. 2001. Multisynaptic neuronal pathways from the submandibular and sublingual glands to the lamina terminalis in the rat: A model for the lamina terminalis in the control of osmo- and thermo-regulatory behaviour. *Clin Exp Pharmacol Physiol* 28:558–69.

Johnson, J. M., and D. W. Proppe. 1996. Cardiovascular adjustments to heat stress. In *Handbook of Physiology. Environmental Physiology Section 4*, vol. 1, 215–43. Bethesda MD: American Physiological Society.

Lee, D. H. K., and A. G. Mulder. 1935. Some immediate effects of reduced cooling powers upon the water balance and related effects in the human subject. *J Physiol* 85:410–32.

McKinley, M. J., D. A. Denton, and R. S. Weisinger. 1978. Sensors for thirst and antidiuresis—osmoreceptors or CSF sodium detectors. *Brain Res* 141:89–103.

McKinley, M. J., D. A. Denton, L. G. Leksell et al. 1982. Osmoregulatory thirst in sheep is disrupted by ablation of the anterior wall of the optic recess. *Brain Res* 236:210–15.

McKinley, M. J., D. A. Denton, J. F. Nelson et al. 1983. Dehydration induces sodium depletion in rats, rabbits and sheep. *Am J Physiol Regul Integr Comp Physiol* 245: R287–92.

McKinley, M. J., B. Lichardus, J. G. McDougall et al. 1992. Periventricular lesions inhibit natriuresis to hypertonic but not isotonic NaCl loads. *Am J Physiol Renal Physiol* 262:F51–60.

McKinley, M. J., G. Pennington, and B. J. Oldfield. 1996. The anteroventral third ventricle and dorsal lamina terminalis: Headquarters for body fluid homeostasis. *Clin Exp Pharmacol Physiol* 23:271–81.

McKinley, M. J., M. L. Mathai, R. M. McAllen et al. 2004. Osmotic and hormonal regulation of vasopressin secretion by the lamina terminalis. *J Neuroendocrinol* 16:340–7.

McKinley, M. J., R. M. McAllen, D. Whyte et al. 2008. Central osmoregulatory influences on thermoregulation. *Clin Exp Pharmacol Physiol* 31:701–5.

McKinley, M.J., F. Weissenborn, and M. L. Mathai. 2009. Drinking induced thermoregulatory panting in rehydrated sheep: Influences of oropharyngeal/esophageal signals, core temperature, and thirst satiety. *Am J Physiol Regul Integr Comp Physiol* 296:R1881–88.

Oldfield, B. J., D. K. Hards, and M. J. McKinley. 1993. Fos production in retrogradely-labelled neurons of the lamina terminalis following intravenous infusion of either hypertonic saline or angiotensin II. *Neuroscience* 60:255–62.

Schloorlemmer, G. H., and M. D. Evered. 1993. Water and solute balance in rats during 10 h water deprivationand rehydration. *Can J Physiol Pharmacol* 71:379–86.

Takamata, A., G. W. Mack, C. M. Gillen et al. 1995. Osmoregulatory modulation of thermal sweating in humans: Reflex effects of drinking. *Am J Physiol Regul Integr Comp Physiol* 268:R414–22.

Takamata, A., K. Nagashima, H. Nose et al. 1997. Osmoregulatory inhibition of thermally induced cutaneous vasodilation in passively heated humans. *Am J Physiol Regul Integr Comp Physiol* 42:R197–204.

Takamata, A., T. Yoshida, N. Nishida et al. 2001. Relationship of osmotic inhibiton in thermoregulatory responses and sweat sodium concentration in humans. *Am J Physiol Regul Integr Comp Physiol* 280:R623–29.

Thrasher, T. N., L. C. Keil, and D. J. Ramsay. 1982. Lesions of the organum vasculosum of the lamina terminalis (OVLT) attenuate osmotically induced water drinking and vasopressin secretion in the dog. *Endocrinology* 110:1837–9.

Thrasher, T. N., J. F. Nistal-Herrera, L. C. Keil et al. 1981. Satiety and inhibition of vasopressin secretion after drinking in dehydrated dogs. *Am J Physiol Endocrinol Metab* 240:E394–401.

8 Catecholaminergic Medullary Pathways and Cardiovascular Responses to Expanded Circulating Volume and Increased Osmolarity

Gustavo R. Pedrino, Daniel A. Rosa, Oswaldo U. Lopes, and Sergio L. Cravo

CONTENTS

8.1 INTRODUCTION

Maintaining the composition and volume of the extracellular fluid (ECF) and intracellular fluid (ICF) within a restricted range of variation is critical for normal tissue perfusion and cellular function (Strange 1993). The ECF is primarily composed by the vascular and extravascular (interstitial) fluid, and has sodium as a determinant ion of its osmolarity and volume. The ECF/plasma volume ratio is of utmost importance for maintaining vascular capacitance and, consequently, venous return, cardiac output, and arterial pressure (Ramsay 1991). Maintaining the variability of the ECF volume within a strict range is a central goal of homeostatic mechanisms.

The central nervous system (CNS) is informed about alterations in volume and tonicity of the ECF through the activity of peripheral and central sensory receptors. Receptors that respond to alterations in the volume of ECF are sensitive to the degree of mechanical distension of blood vessels or cardiac chamber (reviewed by Bourque 2008). They are located in the wall of the atrium and pulmonary vessels (cardiopulmonary receptors), and in the adventitia of the aortic arch/carotid sinus (arterial baroreceptors), or renal afferent arterioles (renal baroreceptors). These receptors mediate central autonomic reflexes through primary connections to the nucleus of the solitary tract (NTS) or mediate renorenal reflexes. Receptors that respond to alterations in ICF are sensitive to ECF tonicity or effective osmotic concentration (Bourque 2008; Kuramochi and Kobayashi 2000; McKinley et al. 1992). They are located in renal, intestinal, and hepatic vessels, in the wall of the gastrointestinal tract and oropharyngeal cavity and in brain circumventricular organs, notably area postrema, subfornical organ (SFO), and organum vasculosum of the lamina terminalis (OVLT). Efferent responses to the activation of these receptors include renal sympathetic nerve activity, hormone secretion [atrial natriuretic peptide (ANP), oxytocin, vasopressin, renin–angiotensin II], and hemodynamic changes. Such responses culminate in controlled alterations of renal sodium and water excretion.

Among these several mechanisms and responses, we would like to call attention to those associated with ECF volume expansion. Dehydration is a menace to immediate survival, but volume expansion is associated with long-term changes that affect health, particularly hypertension and heart failure (reviewed by Antunes-Rodrigues et al. 2004; Toney and Stocker 2010; Toney et al. 2010). Conversely, hypertonic NaCl also has a potential therapeutic value to recovery from hemorrhagic shock (Pedrino et al. 2011; Rocha e Silva et al. 1986; Velasco et al. 1980). In this chapter, we will focus on cardiovascular responses to volume expansion and hypertonicity, their sensory afferences, and how these responses are subserved by a brainstem–hypothalamic–preoptic axis and its associated cathecolaminergic pathways.

8.2 HYPERTONIC VOLUME EXPANSION: NEUROENDOCRINE AND SYMPATHETIC REFLEXES

Hypertonic volume expansion produces combined neuroendocrine and sympathetic responses that ultimately act on the kidney, thereby facilitating the elimination of sodium and water.

Neuroendocrine responses include ANP and oxytocin release, which may act in concert by sharing common mechanisms of release and action. These peptides produce diuresis and natriuresis mediated by release of cGMP associated with renal vasodilation, increased blood flow, and glomerular filtration rate (Huang et al. 1995; Sjoquist et al. 1999; Soares et al. 1999; Verbalis et al. 1991). Mechanical distension of the atrial wall was originally shown to release ANP from myocytes (Lang et al. 1985), but later evidence showed that oxytocin also releases ANP in response to hypertonic blood volume expansion (Haanwinckel et al. 1995). This suggests that the neurohypophysis controls the relase of ANP through an autonomic neuroendocrine reflex.

Independent studies from the early 1990s (Antunes-Rodrigues et al. 1992; Morris and Alexander 1988) demonstrated that removal of aortic and carotid afferents by sinoaortic denervation drastically reduces ANP secretion, basal and induced by hypertonic expansion of the circulating volume. Thus, it is possible that ANP is released by activation of a reflex involving baroreceptors as the afferent loop and oxytocin released from neurohypophysis as the commanding efferent loop in response to volume expansion. Both oxytocin and ANP would then act in the kidneys producing vasodilation and all the events leading to the excretion of excess sodium and water.

Alterations in sympathetic activity may also influence the kidneys as the efferent loop of a similar reflex. Stimulation of the renal nerve promotes the release of renin and increases renal vascular resistance and tubular reabsorption of sodium (DiBona and Kopp 1997), which demonstrates the anti-natriuretic effect of sympathetic renal activation. The opposite occurs in response to hypertonicity. An increase in plasma sodium concentration promotes a reduction in sympathetic renal activity and consequent natriuresis (Badoer et al. 2003; Nishida et al. 1998; Pedrino et al. 2008; Weiss et al. 1996). Moreover, sinoaortic denervation abolishes renal vasodilation induced by expanded volume or hypernatremia, whereas bilateral vagotomy (removal of the cardiopulmonary receptors) does not modify this response (Colombari et al. 2000; Sera et al. 1999). These results suggest that the fibers associated with aortic and carotid baroreceptors not only play a role in the detection of changes in pressure and/or circulating volume but may also be involved in the detection of ECF composition. They also suggest that reduced sympathetic outflow to the kidney results in renal vasodilation and natriuresis in response to hypertonicity and volume expansion.

8.3 RECOVERY FROM HEMORRHAGIC SHOCK, HYPERTONICITY, AND CAROTID CHEMORECEPTION

In addition to their importance for cardiovascular and hormonal adjustments induced by increased volume or hypernatremia, the carotid and aortic afferents also maintain arterial pressure during the reduction of extracellular volume. The combined removal of peripheral baroreceptors and chemoreceptors enhances the hypotension induced by hypovolemic hemorrhaging in dogs (Thrasher and Keil 1998).

Hyperosmolarity induced by a hypertonic saline infusion has potential benefits in treating hypovolemia resulting from hemorrhage. The intravenous application of hypertonic solutions for the recovery of systemic arterial blood pressure during hemorrhagic shock dates back more than 85 years (Penfield 1919). However, a growing interest in the use of hypertonic solutions for treating hemorrhagic shock began with Velasco et al. (1980), who showed that the rapid intravenous administration of hypertonic saline (7.5% NaCl) at a volume equal to 10% of the volume removed to cause the hemorrhagic shock not only rapidly restored blood pressure and cardiac output, but also resulted in the survival of all of the dogs used in the study.

From that point on, various studies have sought to determine the cardiovascular effects of intravenous hypertonic saline infusions on experimental hemorrhagic shock (Lopes et al. 1981; Rocha e Silva et al. 1986; Younes et al. 1985). In their work with anesthetized dogs with hemorrhagic shock caused by controlled bleeding, these authors confirmed the findings of the pioneers' work and showed that it was possible to reverse cases of shock with infusions of reduced volumes (10% of the volume removed) of 7.5% NaCl. In these studies, the hypertonic saline infusion resulted in instantaneous elevation in average blood and pulse pressure, returned cardiac output to baseline values, and reestablished mesenteric, renal, and hindlimb circulation. In addition to the beneficial effects on circulation, hypertonic saline infusion reversed the metabolic acidosis produced by bleeding. These effects guaranteed the survival of all experimental animals, a dramatic effect compared to the 100% mortality of animals treated with an equal volume of isotonic saline (0.9% NaCl). Hypertonic solution also improved the hemodynamics of patients admitted to the intensive care unit with refractory to volume replacement shock (de Felippe et al. 1980).

Expansion of the plasmatic volume caused by hypertonic solutions is certainly part of the mechanism that enables the animal to recover from hypovolemic shock. In their study with 7.5% NaCl, Velasco et al. (1980) measured the plasma volume and variation of the hematocrit and demonstrated an increase in the plasma volume of 11 mL/kg, which then disappeared approximately 6 h later. The primary sources for the observed plasma volume expansion were red blood cells and endothelium, which lost nearly 8% of their volume to the intravascular compartment (Mazzoni et al. 1988, 1990; Moon et al. 1996). Cell shrinkage combined with hemodilution adds an important hemodynamic consequence to intravascular filling because it reduces viscosity and hydraulic resistance to microcirculation. Moreover, hypertonicity also directly relaxes the smooth vascular musculature, thereby producing a generalized precapillary vasodilation (Crystal et al. 1994; Gazitua et al. 1971; Kreimeier et al. 1990; Rocha e Silva et al. 1986; Velasco et al. 1980). The result is recovery of the microcirculatory blood flow in various vascular territories and generalized increased blood flow in peripheral circulation (Mazzoni et al. 1988, 1990).

Despite the abundance of evidence that supports the plasma volume expansion induced by infusion of a hypertonic solution in animals subjected to bleeding, relatively little is known regarding the involvement of the CNS in these responses. Over the past 40 years, several lines of evidence have demonstrated that survival from shock through hypertonic saline infusion depends on vagal integrity (Lopes et al. 1981; Velasco and Baena 2004; Younes et al. 1985). Section of the vagus nerve, its blockage through techniques such as freezing or anesthetic infiltration, or even the denervation of one of the lungs blocks the beneficial cardiovascular effects of the hypertonic saline solution in dogs that were bled. Reversible blockade of the nerve produced an irreversible impairment of the critical compensatory mechanisms, and the animals did not survive despite being treated with hypertonic solution. The authors concluded that hypertonic resuscitation depends on the activation of a possible neural reflex in addition to the direct hemodynamic effects of blood expansion and hypertonicity. It is possible that the sensory signals derived from cardiopulmonary receptor activation and conducted through the vagus (Goetz 1970) are impaired by vagus blockade; however, this remains to be demonstrated.

Peripheral chemoreception in the carotid corpuscle is another sensory mechanism possibly involved with the resuscitation produced by hypertonic saline (Pedrino et al. 2011). Carotid corpuscle chemoreceptors detect variations in arterial pO_2, arterial pCO_2, and pH (Gonzalez et al. 1994; Milsom and Burleson 2007), but they may also respond to hypertonicity. Hypertonic solutions promoted the depolarization of type I cells of the carotid corpuscle as shown by patch clamp studies (Gallego et al. 1979), and carotid deafferentation also impaired the recovery from shock, but the technique used possibly removed carotid baroreceptors and chemoreceptors (de Almeida Costa et al. 2009). Yet, when we did specific inactivation of the carotid chemoreceptors by ligation of the artery that irrigates the carotid corpuscle (Figure 8.1a), we abolished the recovery of arterial pressure induced by an infusion of hypertonic saline solution in rats subjected to hypovolemic hemorrhaging (Figure 8.1b; Pedrino et al. 2011). These results are compatible with the hypothesis that an increase in the plasma sodium concentration activates various reflex mechanisms that enable the rat to recover from hemorrhaging. Animals with inactive chemoreceptors lose these regulatory mechanisms. However, how the chemoreceptors located in the carotid corpuscle are affected by hypertonic solutions still requires clarification.

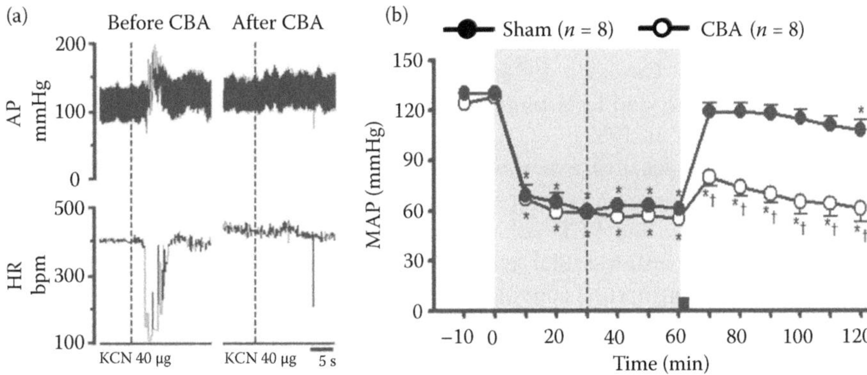

FIGURE 8.1 (a) Typical example of peripheral chemoreceptors inactivation induced by carotid body artery (CBA) ligation. CBA ligation abolished the cardiovascular adjustments induced by KCN injection (40 μg). (b) Effects of bilateral ligation of CBA on recovery of arterial pressure induced by hypertonic saline in rats subjected to hypovolemic hemorrhage. Shaded area indicates period of hemorrhage. Dashed line at $t = 30$ min indicates bilateral ligation of CBA or control treatment (no ligation). Blocks at $t = 60$ min indicate infusion of hypertonic saline. Mean arterial pressure (MAP). *Different from $t = -10$ min. †Different from time controls, both at $p < 0.05$. (Reprinted from *Auton Neurosci*, 160, Pedrino, G. R., M. V. Rossi, G. H. Schoorlemmer, O. U. Lopes, and S. L. Cravo, Cardiovascular adjustments induced by hypertonic saline in hemorrhagic rats: Involvement of carotid body chemoreceptors, 37–41, Copyright 2011, with permission from Elsevier.)

8.4 CENTRAL PROCESSING OF NEUROENDOCRINE AND SYMPATHETIC REFLEXES: THE BRAINSTEM–HYPOTHALAMIC/PREOPTIC AXIS

As already mentioned, the reflexes controlling kidney function in response to hypertonicity and ECF volume expansion may share similar sensory afferences (i.e., carotid afferents). Thus, they may also share similar central neural circuits that process the sensory signaling before distributing the appropriate commands to the hypothalamic neurosecretory nuclei and preganglionic sympathetic neurons.

Afferent fibers from the peripheral sensors to the brain form their first synapse in the NTS. Thus, information about the circulating fluid volume and composition are initially processed by the NTS. Studies by Hines et al. (1994) demonstrated that isotonic expansion of the right atrium causes an increase in the neural activity of the NTS. Additionally, Colombari et al. (1996) showed that renal vasodilation induced by volume expansion could be abolished by blocking the glutamatergic transmission of the NTS relay. From the NTS, the information from these sensors have a broad access to a number of regions in the CNS, including medullary structures, such as the motor nucleus of the vagus, the nucleus ambiguous, and the rostral (RVLM) and caudoventrolateral (CVLM) regions of the medulla oblongata; pontine structures, such as the parabrachial nucleus; and hypothalamic structures, such as the supraoptic nucleus (SON) and paraventricular nucleus of the hypothalamus (PVN) and the preoptic–periventricular tissue surrounding the anteroventral third ventricle (AV3V) (Andresen and Kunze 1994; Blessing et al. 1982; Ciriello and Caverson 1984; Cunningham and Sawchenko 1988; Day and Sibbald 1990; Saper and Levisohn 1983; Sawchenko and Swanson 1981; Tanaka et al. 1997; Tucker et al. 1987).

In the past few decades, experimental evidence has demonstrated that the AV3V region and PVN are important for integration, processing, and correction of variations in osmolarity and ECF volume (see also Chapter 4). Stimulation of these regions can elicit cardiovascular, endocrine, renal, and behavioral adjustments that aim at reestablishing normal conditions in this compartment. They maintain reciprocal connections with various CNS structures involved in the cardiovascular regulation and control of the hydroelectrolytic balance, such as the vasomotor nuclei of the ventral surface of the medulla, the RVLM and CVLM regions, the SFO, and the limbic system (Blessing et al. 1982; Sawchenko and Swanson 1981; Tanaka et al. 1997; Tucker et al. 1987; Zardetto-Smith and Johnson 1995).

The PVN can be functionally and anatomically subdivided into eight distinct subnuclei: three magnocellular subnuclei and five parvocellular subnuclei (Swanson and Sawchenko 1983). The magnocellular portions consist of the neurons that produce vasopressin (posterior magnocellular region) and oxytocin (anterior and medial magnocellular regions), which project to the posterior pituitary, where these hormones are released into the bloodstream. The parvocellular portion is composed of neurons that project to the median eminence and release factors that regulate the release of anterior pituitary hormones (anterior parvicellular, periventricular, and medial subdivisions). The dorsal and lateral parvicellular subdivisions of the

PVN have neuronal projections to numerous regions of the CNS, including various structures involved in the control of the sympathetic vascular tone, such as the RVLM and CVLM regions and the sympathetic preganglion neurons located in the intermediolateral column of the spinal cord (Hosoya et al. 1991). Thus, in addition to peptide secretion, the PVN can also control cardiovascular function and hydroelectrolytic balance through cardiac and renal sympathetic activity. For example, Coote et al. (1998) demonstrated that chemical stimulation of the PVN caused an increase in the arterial blood pressure and differential changes in sympathetic activity, with an observed increase in sympathetic cardiac, splanchnic, and adrenal activity and a decrease in sympathetic renal activity. Furthermore, studies have shown that lesion or inhibition (nanoinjection of GABAergic agonist) of the PVN reduces sympathoinhibition and renal vasodilation induced by hypertonic volume expansion (Badoer et al. 2003; Lovick et al. 1993).

The AV3V region also possesses a high concentration of angiotensin II receptors that respond by mediating an increase in arterial pressure (reviewed by Fitzsimons 1998). Through treatment with intracerebroventricular injections of saralasin, an angiotensin II receptor antagonist, Velasco et al. (1990) have suggested a possible interaction between hypertonic resuscitation in hemorrhage and angiotensinergic systems. Other studies have examined the protective effect of hypertonic solution on circulatory shock in rats with electrolytic lesions in the anteroventral region of the third cerebral ventricle, which is known as AV3V (Barbosa et al. 1992; Giraldelo et al. 1989; Chapter 4). Using different circulatory shock induction models, the authors demonstrated that the protective action of hypertonic solution (7.5% NaCl) was abolished in rats with lesions in AV3V, which indicates that the AV3V region plays a role in the arterial pressure response induced by hypertonic saline in rats subjected to circulatory shock.

Damage to the AV3V also affects sodium balance. It promotes initial acute natriuresis, followed by chronic sodium retention with plasma hypernatremia and impaired natriuretic response to isotonic ECF expansion (Bealer 1983). It also prevents increase in plasma ANP concentration in response to isotonic or hypertonic ECF expansion (Antunes-Rodrigues et al. 1991; Rauch et al. 1990) and reduces the sympathoinhibition induced by sodium overload (May et al. 2000). Furthermore, acute or chronic electrolytic lesion of the AV3V region (Figure 8.2a) abolishes the renal vasodilation induced by acute expansion of the circulating volume (Colombari and Cravo 1999) or hypernatremia (Pedrino et al. 2005; Figure 8.2b).

8.5 BRAIN NORADRENERGIC CONTROL OF RESPONSES TO VOLUME EXPANSION AND SODIUM LOAD

Noradrenergic neurotransmission in the AV3V region plays an important role in the responses induced by changes in the circulating volume. Isotonic expansion of the ECF releases noradrenaline in the MnPO nucleus (Bealer 2000), and nanoinjection of noradrenaline into the third ventricle and adjacent regions promotes ANP secretion and increases renal sodium excretion (Antunes-Rodrigues et al. 1993; Morris et al. 1977). In addition, alpha-adrenergic antagonists into the AV3V region reduce ANP and oxytocin secretions induced by sodium overload (Antunes-Rodrigues et al. 1993). Similarly, studies have recently shown that blocking the alpha-1 or alpha-2

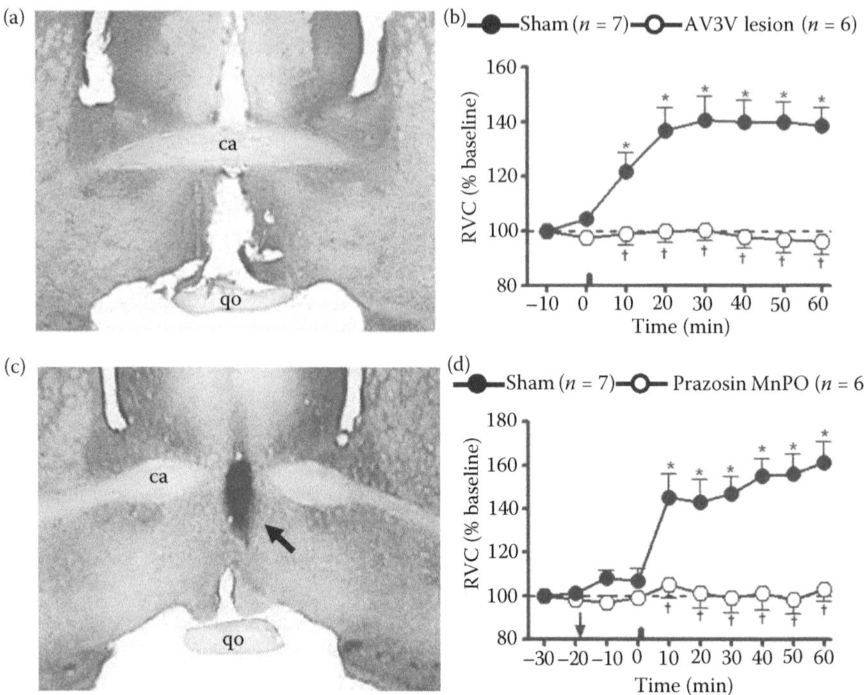

FIGURE 8.2 Effects of chronic lesion of anteroventral third ventricle region (AV3V; (a) or pretreatment with α1-adrenoreceptor antagonist, prasozin (0.25 mmol/L), in median preoptic nucleus (MnPO; (b) on renal vasodilation induced by hypertonic saline (3 mol/L NaCl, 1.8 mL/kg), injected at time 0. Arrows indicate time of antagonist injection. Data represent the mean ± SEM. $^*p < 0.05$ compared with baseline; $^\dagger p < 0.05$ compared with control. RVC, renal vascular conductance; ca, anterior commissure; qo, optic chiasm. (Reprinted from *Auton Neurosci*, 117, Pedrino, G. R., C. T. Nakagawa Sera, S. L. Cravo, and Colombari, D. S. A., Anteroventral third ventricle lesions impair cardiovascular responses to intravenous hypertonic saline infusion, 9–16, Copyright 2005, with permission from Elsevier. Pedrino, G. R., L. R. Monaco, and S. L. Cravo. Renal vasodilation induced by hypernatremia: Role of alpha-adrenoceptors in the median preoptic nucleus. *Clin Exp Pharmacol Physiol.* 2009. 36:e83–9. Copyright Wiley-VCH Verlag GmbH & Co. KGaA. Reproduced with permission.)

adrenoreceptors in the MnPO (Figure 8.2c) abolishes the renal vasodilation induced by an intravenous infusion of hypertonic saline solution (Pedrino et al. 2009; Figure 8.2d). Together, these results suggest that noradrenergic afferents to the MnPO may represent a major element in the central pathways regulating water and sodium balance.

Among the major catecholaminergic medullary projections to the hypothalamic regions involved in regulating hydroelectrolytic balance and cardiovascular control, we highlight the tyrosine hydroxylase immunopositive neurons of the ventrolateral region of the medulla (VLM) and NTS (catecholaminergic groups A1/C1 and A2/C2, respectively). Numerous investigations have demonstrated the importance of these catecholaminergic bulbar neurons in the integration of mechanisms involved in autonomic, cardiovascular, and respiratory function (reviewed by Cravo et al. 2011).

The noradrenergic A1 and A2 neurons receive information from baroreceptors and cardiopulmonary receptors (Day and Sibbald 1990; Day et al. 1992; Spyer 1981) and project to the hypothalamic structures involved in circulating volume regulation. Thus, information about the circulating volume from these peripheral sensors are processed and transmitted to the hypothalamic centers via these noradrenergic neurons. Among the hypothalamic centers that receive dense noradrenergic projections, we highlight the MnPO and PVN (Blessing et al. 1982; Tanaka et al. 1997; Tucker et al. 1987). They receive massive noradrenergic projections from the A1 CVLM region (nearly 80% of the total projections they receive from CVLM) and from the A2-NTS (70% of the total projections they receive from NTS) (Tucker et al. 1987).

Consistent with the neuroanatomical data, *in vivo* electrophysiology studies have demonstrated that projections from A1 and A2 noradrenergic neurons to hypothalamic regions, particularly PVN and MnPO, are mostly excitatory and mediated by the activation of alpha-adrenergic receptors (Day et al. 1984; Kannan et al. 1984; Saphier 1993; Tanaka et al. 1992, 1997). Electrical stimulation of A1 and A2 regions excited, through activation of alpha-1 and alpha-2 adrenergic receptors, most of the magnocellular oxytocin- and vasopressin-secreting neurons, as shown by extracellular recordings in the PVN (Saphier 1993). Similar results were obtained with neurons located in the MnPO that project to the PVN in response to stimulation of A1 region (Tanaka et al. 1992). Lesions of the A1 neurons decreases the number of c-Fos positive vasopressin-secreting neurons in the PVN and SON in response to a reduced circulating volume (Buller et al. 1999). Similarly, other studies demonstrated that the lesion or inhibition of these noradrenergic neurons reduced the secretion of vasopressin induced by hypovolemic hemorrhaging (Head et al. 1987; Smith et al. 1995).

In addition to the control of hypothalamic secretory neurons, the A1 and A2 noradrenergic groups also act in the integration of mechanisms involved in behavior and autonomic cardiovascular responses. Specific lesion of A1 group increases the ingestion of 0.3 M sodium induced by subcutaneous administration of furosemide and captopril or 36 h of water deprivation (Colombari et al. 2008), a result consistent with the reduction of sodium appetite by injection of noradrenaline into the forebrain (Sugawara et al. 1999). Viral inactivation of the A2 group decreases water intake (Duale et al. 2007), and these noradrenergic neurons are activated by increases in osmolarity or plasma volume (Godino et al. 2005; Hochstenbach and Ciriello 1995; Howe et al. 2004). In accordance with these results, Hochstenbach and Ciriello (1994) recorded extracellular neuronal activity and identified increased firing by neurons located in NTS after sodium overload. Recently, it was demonstrated that specific lesions of the A2 neurons abolish the renal sympathoinhibition induced by a hypertonic saline infusion, which suggests the participation of these neurons in the autonomic response to hypernatremia (Pedrino et al. 2012).

Increase in plasma sodium concentration induces c-Fos expression in A1 noradrenergic neurons (Hochstenbach and Ciriello 1995). In agreement with these results, studies have demonstrated that specific lesioning of the noradrenergic group A1 (Figure 8.3a) by nanoinjection of anti-DβH-saporin into the CVLM region abolishes the renal vasodilation induced by sodium overload (Figure 8.3b) and reduces this

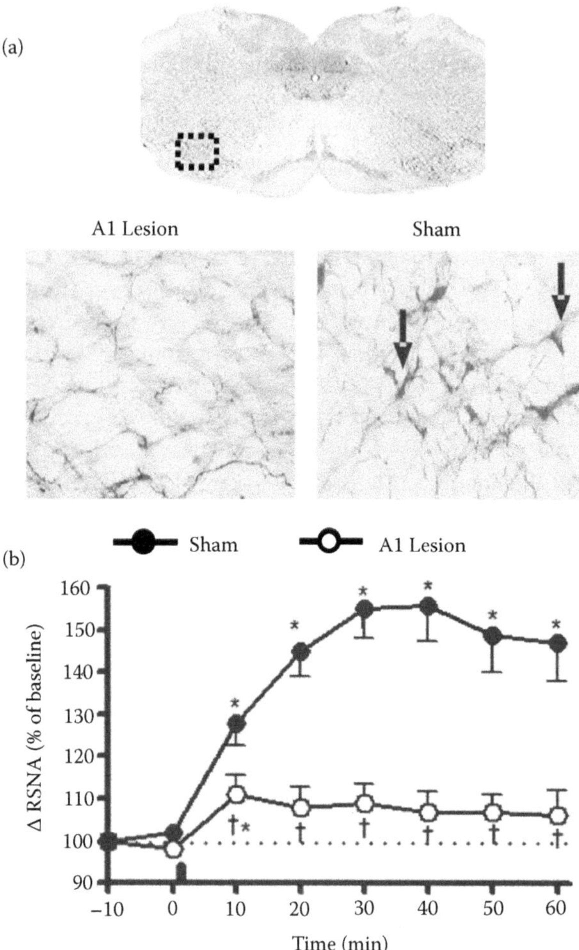

FIGURE 8.3 Effects of lesion of A1 noradrenergic neurons (a) on renal vasodilation (b) and sympathoinhibition (c) induced by hypertonic saline (HS; 3 mol/L NaCl, 1.8 mL/kg), injected at time 0. Arrows indicate TH-positive cells. Data represent the mean ± SEM. $^*p < 0.05$ compared with baseline; $^\dagger p < 0.05$ compared with control. RVC, renal vascular conductance; RSNA, renal sympathetic nerves activity. (Pedrino, G. R., I. Maurino, D. S. A. Colombari, and S. L. Cravo. Role of catecholaminergic neurones of the caudal ventrolateral medulla in cardiovascular responses induced by acute changes in circulating volume in rats. *Exp Physiol.* 2006. 91:995–1005. Copyright Wiley-VCH Verlag GmbH & Co. KGaA. Reproduced with permission. Reprinted from *Auton Neurosci*, 142, Pedrino, G. R., D. A. Rosa, W. S. Korim, and S. L. Cravo. Renal sympathoinhibition induced by hypernatremia: Involvement of A1 noradrenergic neurons, 55–63, Copyright 2008, with permission from Elsevier.)

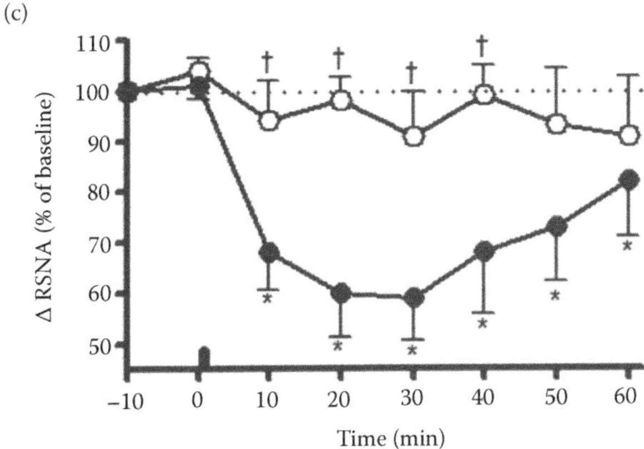

FIGURE 8.3 (*Continued*) Effects of lesion of A1 noradrenergic neurons (a) on renal vasodilation (b) and sympathoinhibition (c) induced by hypertonic saline (HS; 3 mol/L NaCl, 1.8 mL/kg), injected at time 0. Arrows indicate TH-positive cells. Data represent the mean ± SEM. *$p < 0.05$ compared with baseline; †$p < 0.05$ compared with control. RVC, renal vascular conductance; RSNA, renal sympathetic nerves activity. (Pedrino, G. R., I. Maurino, D. S. A. Colombari, and S. L. Cravo: Role of catecholaminergic neurones of the caudal ventrolateral medulla in cardiovascular responses induced by acute changes in circulating volume in rats. *Exp Physiol.* 2006. 91:995–1005. Copyright Wiley-VCH Verlag GmbH & Co. KGaA. Reproduced with permission. Reprinted from *Auton Neurosci*, 142, Pedrino, G. R., D. A. Rosa, W. S. Korim, and S. L. Cravo, Renal sympathoinhibition induced by hypernatremia: Involvement of A1 noradrenergic neurons, 55–63, Copyright 2008, with permission from Elsevier.)

response induced by volume expansion (Pedrino et al. 2006). Additionally, lesioning these neurons abolishes renal sympathoinhibition in response to hypernatremia (Figure 8.3c; Pedrino et al. 2008). Together, these results demonstrate that A1 neurons are involved in both humoral and autonomic mechanisms in response to hypernatremia because specifically inducing lesions in these neurons abolishes the vasodilation and renal sympathoinhibition induced by sodium overload. Thus, the fact that both lesion of the A1 group and blocking of the alpha adrenoreceptors of the MnPO nucleus abolish renal vasodilation in response to hypernatremia supports the hypothesis that the connections between the A1 neurons and MnPO nucleus mediate the adjustments induced by hypernatremia.

The various studies presented here corroborate the hypothesis that the noradrenergic neurons of the A1 and A2 groups activated by afferents from carotid afferents (baroreceptors and/or chemoreceptors) regulate endocrine, autonomic, and behavioral adjustments induced by changes in the circulating volume through efferent projections to the MnPO and PVN nuclei (Figure 8.4). Thus, these noradrenergic neurons are an integral part of the pathways involved in response to alterations of the plasma sodium concentration, and dysfunctions in these cells result in the inefficient functioning of these adjustments. Such dysfunctions could contribute to the genesis and aggravation of various pathological states, among which we highlight hypertension and diseases related to renal sodium retention, such as cirrhosis and cardiac failure.

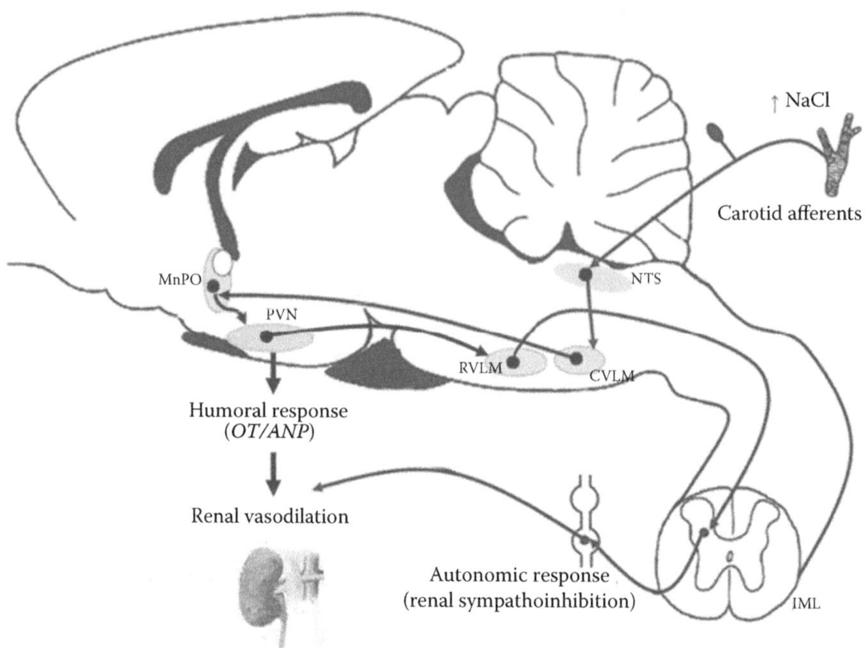

FIGURE 8.4 Schematic representation of neural pathways involved in control of autonomic, humoral, and cardiovascular adjustments induced by peripheral hypenatremia. MnPO, median preoptic nucleus; PVN, paraventricular nucleus; RVLM, rostral ventrolateral medulla; CVLM, caudal ventrolateral medulla; NTS, nucleus of the solitary tract; IML, intermediolateral cell column; OT, oxytocin; ANP, atrial natriuretic peptide.

ACKNOWLEDGMENTS

Studies described herein were funded in part by Coordenadoria de Aperfeiçoamento em Pesquisa (CAPES; #BEX1380/07-9), Conselho Nacional de Desenvolvimento Científico e Tecnológico (CNPq; #477832/2010-5) and Fundação de Amparo a Pesquisa do Estado de Goiás (FAPEG; #200910267000352). The authors gratefully acknowledge the following individuals for contributions toward generating the data contained within: Dr. Débora S. Colombari, Dr. Gerhardus H. M. Schoorlemmer, Dr. Willian S. Korim, Dr. Celisa T. N. Sera, Dr. André H. Freiria-Oliveira, and Marcio V. Rossi.

REFERENCES

Andresen, M. C., and D. L. Kunze. 1994. Nucleus tractus solitarius—gateway to neural circulatory control. *Annu Rev Physiol* 56:93–116.

Antunes-Rodrigues, J., M. de Castro, L. L. Elias, M. M. Valença, and S. M. McCann. 2004. Neuroendocrine control of body fluid metabolism. *Physiol Rev* 84:169–208.

Antunes-Rodrigues, J., B. H. Machado, H. A. Andrade et al. 1992. Carotid–aortic and renal baroreceptors mediate the atrial natriuretic peptide release induced by blood volume expansion. *Proc Natl Acad Sci USA* 89:6828–31.

Antunes-Rodrigues, J., U. Marubayashi, A. L. Favaretto, J. Gutkowska, and S. M. McCann. 1993. Essential role of hypothalamic muscarinic and alpha-adrenergic receptors in atrial natriuretic peptide release induced by blood volume expansion. *Proc Natl Acad Sci USA* 90:10240–4.

Antunes-Rodrigues, J., M. J. Ramalho, L. C. Reis et al. 1991. Lesions of the hypothalamus and pituitary inhibit volume-expansion-induced release of atrial-natriuretic-peptide. *Proc Natl Acad Sci USA* 88:2956–60.

Badoer, E., C. W. Ng, and R. De Matteo. 2003. Glutamatergic input in the PVN is important in renal nerve response to elevations in osmolality. *Am J Physiol Renal Physiol* 285:F640–50.

Barbosa, S. P., L. A. A. Camargo, W. A. Saad, A. Renzi, L. A. De Luca Jr., and J. V. Menani. 1992. Lesion of the anteroventral third ventricle region impairs the recovery of arterial pressure induced by hypertonic saline in rats submitted to hemorrhagic shock. *Brain Res* 587:109–14.

Bealer, S. L. 1983. Sodium excretion following lesions of preoptic recess periventricular tissue in the rat. *Am J Physiol Regul Integr Comp Physiol* 244:R815–22.

Bealer, S. L. 2000. Systemic angiotensin II and volume expansion release norepinephrine in the preoptic recess. *Brain Res* 864:291–7.

Blessing, W. W., C. B. Jaeger, D. A. Ruggiero, and D. J. Reis. 1982. Hypothalamic projections of medullary catecholamine neurons in the rabbit: A combined catecholamine fluorescence and HRP transport study. *Brain Res Bull* 9:279–86.

Bourque, C. W. 2008. Central mechanisms of osmosensation and systemic osmoregulation. *Nat Rev Neurosci* 9:519–31.

Buller, K. M., D. W. Smith, and T. A. Day. 1999. Differential recruitment of hypothalamic neuroendocrine and ventrolateral medulla catecholamine cells by non-hypotensive and hypotensive hemorrhages. *Brain Res* 834:42–54.

Ciriello, J., and M. M. Caverson. 1984. Ventrolateral medullary neurons relay cardiovascular inputs to the paraventricular nucleus. *Am J Physiol Regul Integr Comp Physiol* 246: R968–78.

Colombari, D. S. A., E. Colombari, O. U. Lopes, and S. L. Cravo. 1996. Effect of glutamatergic blockade in the nucleus tractus solitarii (NTS) on cardiovascular responses to volume load in anesthetized rats. *Soc Neurosci Abstr* 22:631.

Colombari, D. S. A., E. Colombari, O. U. Lopes, and S. L. Cravo. 2000. Afferent pathways in cardiovascular adjustments induced by volume expansion in anesthetized rats. *Am J Physiol Regul Integr Comp Physiol* 279:R884–90.

Colombari, D. S. A., and S. L. Cravo. 1999. Effects of acute AV3V lesions on renal and hindlimb vasodilation induced by volume expansion. *Hypertension* 34:762–7.

Colombari, D. S. A., G. R. Pedrino, A. H. Freiria-Oliveira, W. S. Korim, I. C. Maurino, and S. L. Cravo. 2008. Lesions of medullary catecholaminergic neurons increase salt intake in rats. *Brain Res Bull* 76:572–8.

Coote, J. H., Z. Yang, S. Pyner, and J. Deering. 1998. Control of sympathetic outflows by the hypothalamic paraventricular nucleus. *Clin Exp Pharmacol Physiol* 25:461–3.

Cravo, S. L., O. U. Lopes, and G. R. Pedrino. 2011. Involvement of catecholaminergic medullary pathways in cardiovascular responses to acute changes in circulating volume. *Braz J Med Biol Res* 44:877–82.

Crystal, G. J., J. Gurevicius, S. J. Kim, P. K. Eckel, E. F. Ismail, and M. R. Salem. 1994. Effects of hypertonic saline solutions in the coronary circulation. *Circ Shock* 42:27–38.

Cunningham, E. T., and P. E. Sawchenko. 1988. Anatomical specificity of noradrenergic inputs to the paraventricular and supraoptic nuclei of the rat hypothalamus. *J Comp Neurol* 274:60–76.

Day, T. A., A. V. Ferguson, and L. P. Renaud. 1984. Facilitatory influence of noradrenergic afferents on the excitability of rat paraventricular nucleus neurosecretory cells. *J Physiol* 355:237–49.

Day, T. A., and J. R. Sibbald. 1990. Involvement of the A1 cell group in baroreceptor inhibition of neurosecretory vasopressin cells. *Neurosci Lett* 113:156–62.

Day, T. A., J. R. Sibbald, and D. W. Smith. 1992. A1 neurons and excitatory amino acid receptors in rat caudal medulla mediate vagal excitation of supraoptic vasopressin cells. *Brain Res* 594:244–52.

de Almeida Costa, E. F., G. R. Pedrino, O. U. Lopes, and S. L. Cravo. 2009. Afferent pathways involved in cardiovascular adjustments induced by hypertonic saline resuscitation in rats submitted to hemorrhagic shock. *Shock* 32:190–3.

de Felippe Jr., J., J. Timoner, I. T. Velasco, O. U. Lopes, and M. Rocha e Silva Jr. 1980. Treatment of refractory hypovolaemic shock by 7.5% sodium chloride injections. *Lancet* 2:1002–4.

DiBona, G. F., and U. C. Kopp. 1997. Neural control of renal function. *Physiol Rev* 77:75–197.

Duale, H., H. Waki, P. Howorth, S. Kasparov, A. G. Teschemacher, and J. F. Paton. 2007. Restraining influence of A2 neurons in chronic control of arterial pressure in spontaneously hypertensive rats. *Cardiovasc Res* 76:184–93.

Fitzsimons, J. T. 1998. Angiotensin, thirst, and sodium appetite. *Physiol Rev* 78:583–686.

Gallego, R., C. Eyzaguirre, and L. Monti-Bloch. 1979. Thermal and osmotic responses of arterial receptors. *J Neurophysiol* 42:665–80.

Gazitua, S., J. B. Scott, B. Swindall, and F. J. Haddy. 1971. Resistance responses to local changes in plasma osmolality in three vascular beds. *Am J Physiol* 220:384–91.

Giraldelo, C. M., E. Colombari, J. V. Menani, and J. F. Fracasso. 1989. Central lesion abolishes the beneficial effect of hypertonic saline on circulatory shock produced by compound 48/80 in rats. *Braz J Med Biol Res* 22:1029–32.

Godino, A., A. Giusti-Paiva, J. Antunes-Rodrigues, and L. M. Vivas. 2005. Neurochemical brain groups activated after an isotonic blood volume expansion in rats. *Neuroscience* 133:493–505.

Goetz, K. L., A. S. Hermreck, G. L. Slick, and H. S. Starke. 1970. Atrial receptors and renal function in conscious dogs. *Am J Physiol* 219:1417–23.

Gonzalez, C., L. Almaraz, A. Obeso, and R. Rigual. 1994. Carotid body chemoreceptors: From natural stimuli to sensory discharges. *Physiol Rev* 74:829–98.

Haanwinckel, M. A., L. L. K. Elias, A. L. V. Favaretto, J. Gutkowska, S. M. McCann, and J. Antunes-Rodrigues. 1995. Oxytocin mediates atrial-natriuretic-peptide release and natriuresis after volume expansion in the rat. *Proc Natl Acad Sci USA* 92:7902–6.

Head, G. A., A. W. Quail, and R. L. Woods. 1987. Lesions of A1 noradrenergic cells affect AVP release and heart rate during hemorrhage. *Am J Physiol Heart Circ Physiol* 253:H1012–7.

Hines, T., G. M. Toney, and S. W. Mifflin. 1994. Responses of neurons in the nucleus tractus solitarius to stimulation of heart and lung receptors in the rat. *Circ Res* 74:1188–96.

Hochstenbach, S. L., and J. Ciriello. 1994. Effects of plasma hypernatremia on nucleus tractus solitarius neurons. *Am J Physiol Regul Integr Comp Physiol* 266:R1916–21.

Hochstenbach, S. L., and J. Ciriello. 1995. Plasma hypernatremia induces c-Fos activity in medullary catecholaminergic neurons. *Brain Res* 674:46–54.

Hosoya, Y., Y. Sugiura, N. Okado, A. D. Loewy, and K. Kohno. 1991. Descending input from the hypothalamic paraventricular nucleus to sympathetic preganglionic neurons in the rat. *Exp Brain Res* 85:10–20.

Howe, B. M., S. B. Bruno, K. A. Higgs, R. L. Stigers, and J. T. Cunningham. 2004. FosB expression in the central nervous system following isotonic volume expansion in unanesthetized rats. *Exp Neurol* 187:190–8.

Huang, W., S. L. Lee, and M. Sjoquist. 1995. Natriuretic role of endogenous oxytocin in male rats infused with hypertonic NaCl. *Am J Physiol Regul Integr Comp Physiol* 268:R634–40.

Kannan, H., H. Yamashita, and T. Osaka. 1984. Paraventricular neurosecretory neurons: Synaptic inputs from the ventrolateral medulla in rats. *Neurosci Lett* 51:183–8.

Kreimeier, U., U. B. Bruckner, S. Niemczyk, and K. Messmer. 1990. Hyperosmotic saline dextran for resuscitation from traumatic–hemorrhagic hypotension: Effect on regional blood flow. *Circ Shock* 32:83–99.

Kuramochi, G., and I. Kobayashi. 2000. Regulation of the urine concentration mechanism by the oropharyngeal afferent pathway in man. *Am J Nephrol* 20:42–7.

Lang, R. E., H. Thölken, D. Ganten, F. C. Luft, H. Ruskoaho, and T. Unger. 1985. Atrial natriuretic factor—a circulating hormone stimulated by volume loading. *Nature* 314:264–6.

Lopes, O. U., V. Pontieri, M. Rocha e Silva, and I. T. Velasco. 1981. Hyperosmotic NaCl and severe hemorrhagic shock: Role of the innervated lung. *Am J Physiol* 241:H883–90.

Lovick, T. A., S. Malpas, and M. T. Mahony. 1993. Renal vasodilatation in response to acute volume load is attenuated following lesions of parvocellular neurons in the paraventricular nucleus in rats. *J Auton Nerv Syst* 43:247–55.

May, C. N., R. M. McAllen, and M. J. McKinley. 2000. Renal nerve inhibition by central NaCl and ANG II is abolished by lesions of the lamina terminalis. *Am J Physiol Regul Integr Comp Physiol* 279:R1827–33.

Mazzoni, M. C., P. Borgstrom, K. E. Arfors, and M. Intaglietta. 1988. Dynamic fluid redistribution in hyperosmotic resuscitation of hypovolemic hemorrhage. *Am J Physiol Heart Circ Physiol* 255:H629–37.

Mazzoni, M. C., P. Borgstrom, M. Intaglietta, and K. E. Arfors. 1990. Capillary narrowing in hemorrhagic shock is rectified by hyperosmotic saline–dextran reinfusion. *Circ Shock* 31:407–18.

McKinley, M. J., B. Lichardus, J. G. McDougall, and R. S. Weisinger. 1992. Periventricular lesions block natriuresis to hypertonic but not isotonic NaCl loads. *Am J Physiol Renal Physiol* 262:F98–107.

Milsom, W. K., and M. L. Burleson. 2007. Peripheral arterial chemoreceptors and the evolution of the carotid body. *Respir Physiol Neurobiol* 157:4–11.

Moon, P. F., M. A. Hollyfield-Gilbert, T. L. Myers, T. Uchida, and G. C. Kramer. 1996. Fluid compartments in hemorrhaged rats after hyperosmotic crystalloid and hyperoncotic colloid resuscitation. *Am J Physiol Renal Physiol* 270:F1–8.

Morris, M., and N. Alexander. 1988. Baroreceptor influences on plasma atrial natriuretic peptide (ANP): Sinoaortic denervation reduces basal levels and the response to an osmotic challenge. *Endocrinology* 122:373–5.

Morris, M., S. M. McCann, and R. Orias. 1977. Role of transmitters in mediating hypothalamic control of electrolyte excretion. *Can J Physiol Pharmacol* 55:1143–54.

Nishida, Y., I. Sugimoto, H. Morita, H. Murakami, H. Hosomi, and V. S. Bishop. 1998. Suppression of renal sympathetic nerve activity during portal vein infusion of hypertonic saline. *Am J Physiol Regul Integr Comp Physiol* 274:R97–103.

Pedrino, G. R., I. Maurino, D. S. A. Colombari, and S. L. Cravo. 2006. Role of catecholaminergic neurones of the caudal ventrolateral medulla in cardiovascular responses induced by acute changes in circulating volume in rats. *Exp Physiol* 91:995–1005.

Pedrino, G. R., L. R. Monaco, and S. L. Cravo. 2009. Renal vasodilation induced by hypernatremia: Role of alpha-adrenoceptors in the median preoptic nucleus. *Clin Exp Pharmacol Physiol* 36:e83–9.

Pedrino, G. R., C. T. Nakagawa Sera, S. L. Cravo, and D. S. A. Colombari. 2005. Anteroventral third ventricle lesions impair cardiovascular responses to intravenous hypertonic saline infusion. *Auton Neurosci* 117:9–16.

Pedrino, G. R., D. A. Rosa, W. S. Korim, and S. L. Cravo. 2008. Renal sympathoinhibition induced by hypernatremia: Involvement of A1 noradrenergic neurons. *Auton Neurosci* 142:55–63.

Pedrino, G. R., M. V. Rossi., G. H. Schoorlemmer, O. U. Lopes, and S. L. Cravo. 2011. Cardiovascular adjustments induced by hypertonic saline in hemorrhagic rats: Involvement of carotid body chemoreceptors. *Auton Neurosci* 160:37–41.

Pedrino, G. R., A. H. Freiria-Oliveira, D. A. Rosa, D. S. A. Colombari, and S. L. Cravo. 2012. A2 noradrenergic lesions prevent renal sympathoinhibition induced by hypernatremia in rats. *PlosOne* 7:e37587.

Penfield, W. G. 1919. The treatment of severe and progressive hemorrhage by intravenous injections. *Am J Physiol* 48:121–32.

Ramsay, D. J. 1991. Water: Distribution between compartments and in relationship to thirst. In *Thirst: Physiological and Pschological Aspects*, ed. Ramsay, D. J., and D. A. Booth, 23–24. London: Springer-Verlag.

Rauch, A. L., M. F. Callahan, V. M. Buckalew, and M. Morris. 1990. Regulation of plasma atrial natriuretic peptide by the central nervous system. *Am J Physiol Regul Integr Comp Physiol* 258:R531–5.

Rocha e Silva, M., G. A. Negraes, A. M. Soares, V. Pontieri, and L. Loppnow. 1986. Hypertonic resuscitation from severe hemorrhagic shock: Patterns of regional circulation. *Circ Shock* 19:165–75.

Saper, C. B., and D. Levisohn. 1983. Afferent connections of the median preoptic nucleus in the rat: Anatomical evidence for a cardiovascular integrative mechanism in the anteroventral third ventricular (AV3V) region. *Brain Res* 288:21–31.

Saphier, D. 1993. Electrophysiology and neuropharmacology of noradrenergic projections to rat PVN magnocellular neurons. *Am J Physiol Regul Integr Comp Physiol* 264:R891–902.

Sawchenko, P. E., and L. W. Swanson. 1981. Central noradrenergic pathways for the integration of hypothalamic neuroendocrine and autonomic responses. *Science* 214:685–7.

Sawchenko, P. E., and L. W. Swanson. 1983. The organization of forebrain afferents to the paraventricular and supraoptic nuclei of the rat. *J Comp Neurol* 218:121–44.

Sera, C. T. N., D. S. A. Colombari, and S. L. Cravo. 1999. Afferent pathways involved in the renal vasodilation induced by hypertonic saline. *Hypertension* 33:1315.

Sjoquist, M., W. Huang, E. Jacobsson, O. Skott, E. M. Stricker, and A. F. Sved. 1999. Sodium excretion and renin secretion after continuous versus pulsatile infusion of oxytocin in rats. *Endocrinology* 140:2814–8.

Smith, D. W., J. R. Sibbald, S. Khanna, and T. A. Day. 1995. Rat vasopressin cell responses to simulated hemorrhage: Stimulus-dependent role for A1 noradrenergic neurons. *Am J Physiol Regul Integr Comp Physiol* 268:R1336–42.

Soares, T. J., T. M. Coimbra, A. R. Martins et al. 1999. Atrial natriuretic peptide and oxytocin induce natriuresis by release of cGMP. *Proc Natl Acad Sci USA* 96:278–3.

Spyer, K. M. 1981. Neural organization and control of the baroreceptor reflex. *Rev Physiol Biochem Pharmacol* 88:24–124.

Strange, K. 1993. *Cellular and Molecular Physiology of Cell Volume Regulation*, 1st ed. Boca Raton, FL: CRC Press.

Sugawara, A. M., T. T. Miguel, L. B. De Oliveira, J. V. Menani, and L. A. De Luca Jr. 1999. Noradrenaline and mixed α_2-adrenoreceptor/imidazoline-receptor ligands: Effects on sodium intake. *Brain Res* 839:227–34.

Tanaka, J., Y. Hayashi, T. Watai, Y. Fukami, Y., R. Johkoh, and S. Shimamune. 1997. A1 noradrenergic modulation of AV3V inputs to PVN neurosecretory cells. *Neuroreport* 8:3147–50.

Tanaka, J., J. Nishimura, F. Kimura, and M. Nomura. 1992. Noradrenergic excitatory inputs to median preoptic neurones in rats. *Neuroreport* 3:946–8.

Thrasher, T. N., and L. C. Keil. 1998. Arterial baroreceptors control blood pressure and vasopressin responses to hemorrhage in conscious dogs. *Am J Physiol Regul Integr Comp Physiol* 275:R1843–57.

Toney, G. M., and S. D. Stocker. 2010. Hyperosmotic activation of CNS sympathetic drive: Implications for cardiovascular disease. *J Physiol* 588:3375–84.

Toney, G. M., G. R. Pedrino, G. D. Fink, and J. W. Osborn. 2010. Does enhanced respiratory-sympathetic coupling contribute to peripheral neural mechanisms of angiotensin II-salt hypertension? *Exp Physiol* 95:587–94.

Tucker, D. C., C. B. Saper, D. A. Ruggiero, and D. J. Reis. 1987. Organization of central adrenergic pathways: I. Relationships of ventrolateral medullary projections to the hypothalamus and spinal cord. *J Comp Neurol* 259:591–603.

Velasco, I. T., and R. C. Baena. 2004. The role of the vagus nerve in hypertonic resuscitation of hemorrhagic shocked dogs. *Braz J Med Biol Res* 37:419–25.

Velasco, I. T., R. C. Baena, M. Rocha e Silva, M. I. and Loureiro. 1990. Central angiotensinergic system and hypertonic resuscitation from severe hemorrhage. *Am J Physiol Heart Circ Physiol* 259:H1752–8.

Velasco, I. T., V. Pontieri, M. Rocha e Silva, and O. U. Lopes. 1980. Hyperosmotic NaCl and severe hemorrhagic shock. *Am J Physiol Heart Circ Physiol* 239:H664–73.

Verbalis, J. G., M. P. Mangione, and E. M. Stricker. 1991. Oxytocin produces natriuresis in rats at physiological plasma concentrations. *Endocrinology* 128:1317–22.

Weiss, M. L., D. E. Claassen, T. Hirai, and M. J. Kenney. 1996. Nonuniform sympathetic nerve responses to intravenous hypertonic saline infusion. *J Auton Nerv Syst* 57:109–15.

Younes, R. N., F. Aun, R. M. Tomida, and D. Birolini. 1985. The role of lung innervation in the hemodynamic response to hypertonic sodium chloride solutions in hemorrhagic shock. *Surgery* 98:900–6.

Zardetto-Smith, A. M., and A. K. Johnson. 1995. Chemical topography of efferent projections from the median preoptic nucleus to pontine monoaminergic cell groups in the rat. *Neurosci Lett* 199:215–9.

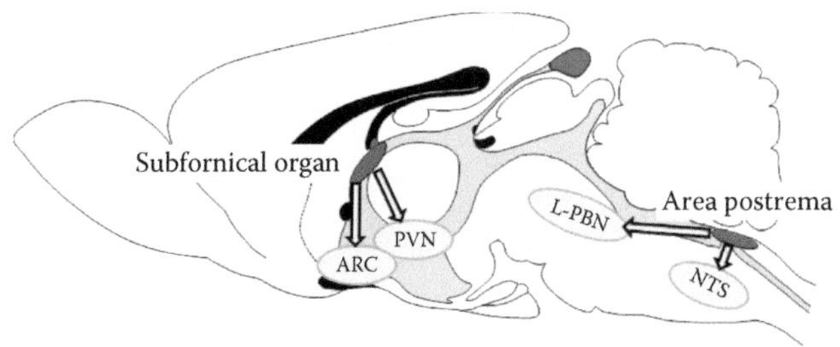

FIGURE 2.1 This schematic midsagittal section through the rat brain illustrates location of two of circumventricular organs, subfornical organ and area postrema. It also highlights some of their primary efferent connections to arcuate (ARC) and paraventricular (PVN) nuclei for the subfornical organ, or nucleus tractus solitarius (NTS) and lateral parabrachial nucleus (L-PBN) for the area postrema.

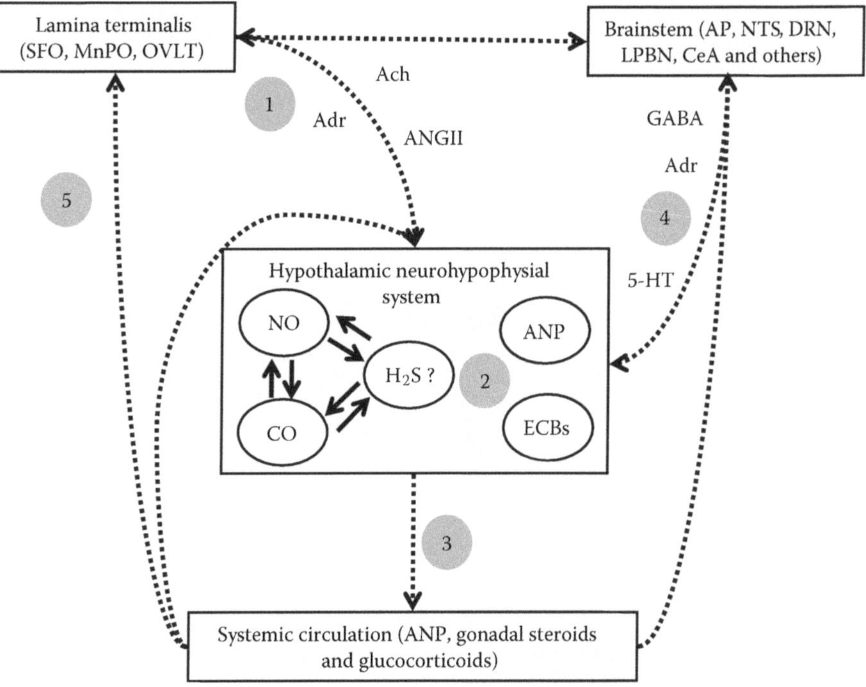

FIGURE 3.1 Schematic illustration of hypothalamic–neurohypophysial system, its main connections and central/peripheral control. Each number indicates a mechanism discussed in the literature: Mechanism 1 (Antunes-Rodrigues et al. 2004), Mechanism 2 (Baldissera et al. 1989; Gomes et al. 2004, 2010; Reis et al. 2007; Ruginsk et al. 2010; Ventura et al. 2002, 2005), Mechanism 3 (Antunes-Rodrigues et al. 1991; Haanwinkel et al. 1995), Mechanism 4 (Godino et al. 2005, 2010; Margatho et al. 2007, 2008, 2009), and Mechanism 5 (Durlo et al. 2004; Lauand et al. 2007; Mecawi et al. 2007, 2008, 2011; Ruginsk et al. 2007; Vilhena-Franco et al. 2011).

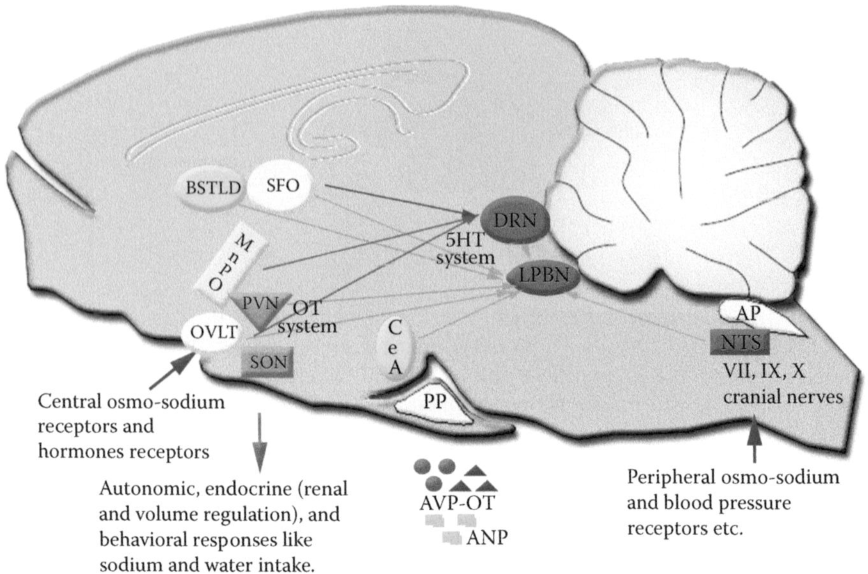

FIGURE 9.1 Neurochemical circuits involved in fluid balance regulation.

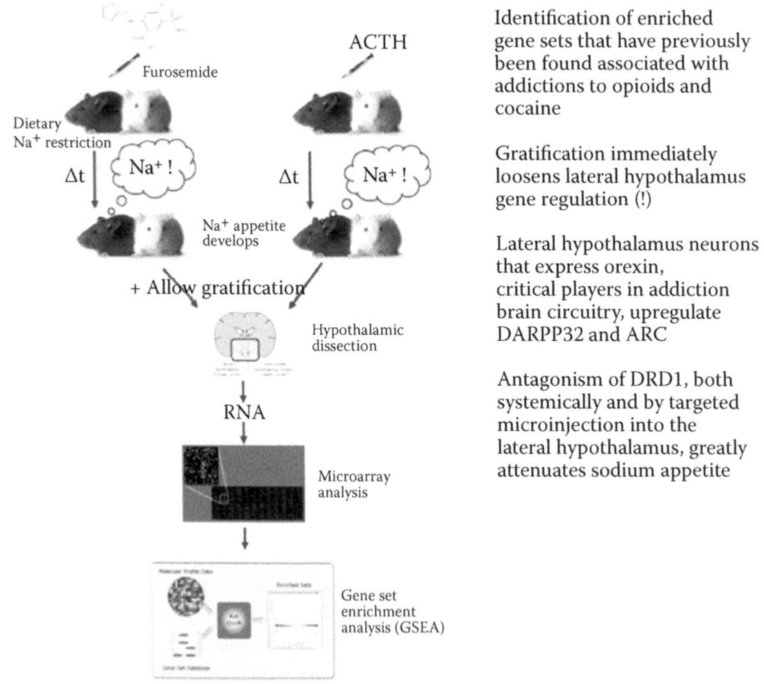

FIGURE 12.1 Left-hand illustration depicts experimental workflow; right-hand side statements reiterate critical results obtained.

9 Neurochemical Circuits Subserving Fluid Balance and Baroreflex

A Role for Serotonin, Oxytocin, and Gonadal Steroids

Laura Vivas, Andrea Godino, Carolina Dalmasso, Ximena E. Caeiro, Ana F. Macchione, and Maria J. Cambiasso

CONTENTS

9.1 INTRODUCTION

Changes in body water/sodium balance are tightly controlled by the central nervous system (CNS) to avoid abnormal cardiovascular function and the development of pathological states. Every time there is a disturbance in extracellular sodium concentration or body sodium content, there is also a change in extracellular fluid volume and, depending on its magnitude, this can be associated with an adjustment in arterial blood pressure (BP). The process of sensory integration takes place in different nuclei, with diverse phenotypes and at different levels of the CNS. To control those several changes, the CNS receives continuous input about the status of extracellular fluid osmolarity, sodium concentration, sense of taste, fluid volume, and BP (Figure 9.1). Signals detected by taste receptors, peripheral osmo-sodium, volume receptors, and arterial/cardiopulmonary baroreceptors reach the nucleus of the solitary tract (NTS) by the VIIth, IXth, and Xth cranial nerves. The other main brain entry of the information related to fluid and cardiovascular balance are the lamina terminalis (LT) and one of the sensory circumventricular organs (CVOs), the area postrema (AP). The LT, consisting of the median preoptic nucleus (MnPO) and the other two sensory CVOs—i.e., subfornical organ (SFO) and organum vasculosum of the lamina terminalis (OVLT) —is recognized as a site in the brain that is crucial for the physiological regulation of hydroelectrolyte balance. The SFO and OVLT lack a blood–brain barrier and contain cells that are sensitive to humoral signals, such as changes in plasma and cerebrospinal fluid sodium concentration (Vivas et al. 1990), osmolality (Sladek and Johnson 1983), and angiotensin II (ANG II) levels (Ferguson and Bains 1997; Simpson et al. 1978). Such unique features make the SFO and OVLT key brain regions for sensing the status of the body fluids and electrolytes. Humoral and neural signals that arrive to the two main brain entries—that is, the CVOs of the LT and within the hindbrain the AP-NTS—activate a central circuit that includes integrative areas such as the MnPO, the paraventricular (PVN), the supraoptic (SON), lateral parabrachial nucleus (LPBN), dorsal raphe nucleus (DRN), and neurochemical systems such as the angiotensinergic, vasopressinergic, oxytocinergic (OT), and serotonergic (5-HT) systems (Figures 9.1 and 9.2). Once these signals act on the

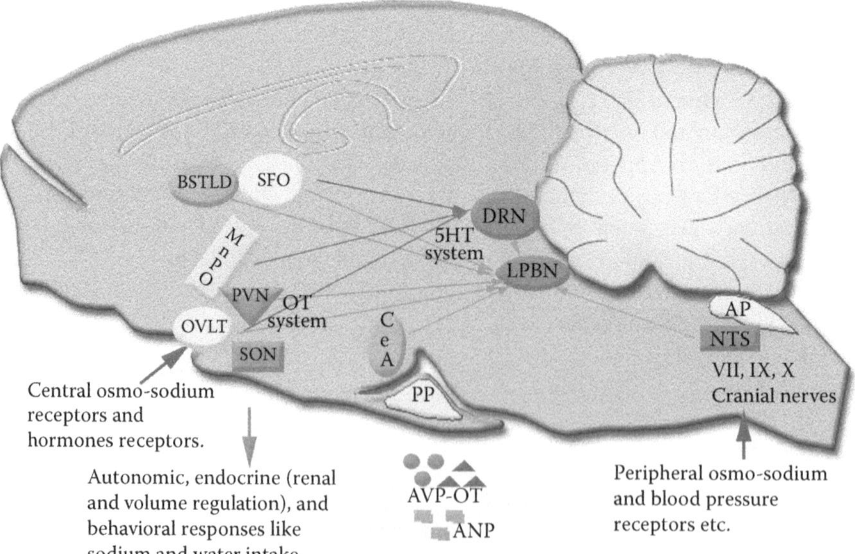

FIGURE 9.1 **(See color insert.)** Neurochemical circuits involved in fluid balance regulation.

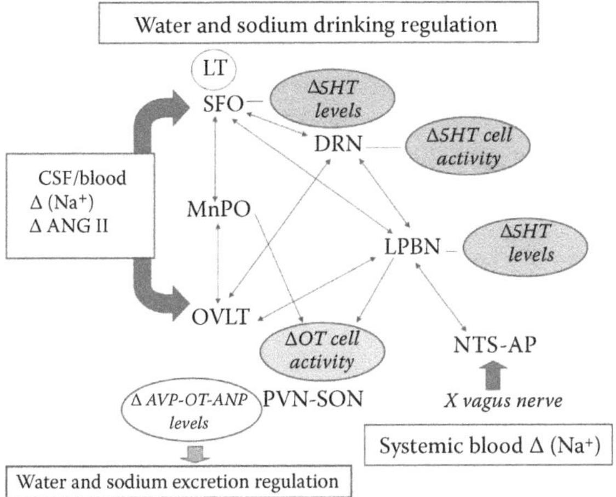

FIGURE 9.2 Schematic representation of brain angiotensinergic and serotonergic circuit interactions that may regulate thirst and salt appetite.

above-mentioned neurochemical networks, they trigger appropriate sympathetic, endocrine, and behavioral responses. Therefore, after a body fluid deficit, water and sodium intake and excretion need to be controlled to minimize disturbances of hydromineral homeostasis. In this context, hypovolemia and hyponatremia induced by body fluid depletion stimulate central and peripheral osmo–sodium receptors, taste receptors, volume and arterial/cardiopulmonary baroreceptors, and the renin–angiotensin system (RAS). This latter system, for example, acts mainly through the sensory CVOs and/or the AP to activate brain neural pathways that elevate BP, release vasopressin and aldosterone (ALDO), increase renal sympathetic nerve activity, and increase the ingestion of water and sodium. Among these responses, sodium appetite constitutes an important homeostatic behavior involved in seeking out and acquiring sodium from the environment. Under normal circumstances, the average daily intake of sodium in animals exceeds what is actually needed; however, when they are challenged by environmental (e.g., increased ambient temperature), physiological (e.g., exercise, pregnancy and lactation), or pathophysiological (e.g., emesis, diarrhea, adrenal, or kidney insufficiency) conditions, endocrine and autonomic mechanisms primarily target the kidney, to influence the rate of water and sodium loss, and the vasculature, to maintain arterial BP. Afterward, a behavioral mechanism such as sodium appetite is the means by which sodium loss to the environment is ultimately restored (Geerling and Loewy 2008). It is important to note that in humans, salt appetite is permanently enhanced after perinatal sodium loss (Crystal and Berstein 1995, 1998; Leshem 2009), but putative sodium loss in adults due to, for example, hemorrhage, dehydration, or breastfeeding, does not increase salt appetite significantly; thus, the existence of sodium appetite as a result of sodium loss in adult humans remains controversial (Bertino et al. 1982; Beauchamp et al. 1983, 1987; Leshem 2009).

This review will focus on evidence from our laboratory for neurophysiological mechanisms that regulate sodium balance. Specifically, it tries to answer how the brain elicits sodium appetite in response to hyponatremia/hypovolemia associated with sodium depletion, which areas are activated after sodium depletion, how the brain controls the inhibition of this behavior once the deficit is compensated (satiety phase), and what role brain neurochemical groups have for endocrine responses. We close the chapter by analyzing the effects of gonadal hormones and sex chromosome complement (SCC) on sodium appetite and cardiovascular function, respectively.

9.2 MAPPING BRAIN NUCLEI INVOLVED IN APPETITIVE AND SATIETY PHASES OF SODIUM APPETITE

Previous studies from our laboratory have shown that acute sodium depletion by peritoneal dialysis (PD) produces a rapid and significant drop in sodium concentration in serum and CSF within 1–4 h after PD, rising gradually until 20–24 h later when the animals not only recover normal extracellular sodium values (possibly by body sodium reservoirs) but sodium appetite also becomes evident (Ferreyra and Chiaraviglio 1977). In addition, the dialyzed animals have shown a significant decrease in blood volume immediately after PD, returning to control values 12 h later (Ferreyra and Chiaraviglio 1977).

As with any motivated behavior, sodium appetite has two phases: the appetitive phase is the flexible or adaptable behavior that an animal or person adopts, before the motivational goal is found, whereas the satiety or consummatory phase that follows is elicited only by the goal stimulus, and thus consummates the appetitive phase (Berridge 2004). We have investigated the brain areas and neurochemical systems involved in both the appetitive (24 h after PD and before the intake test) and satiety (after induced sodium intake) phases of sodium appetite (Franchini and Vivas 1999; Franchini et al. 2002; Godino et al. 2007; Johnson et al. 1999; Vivas et al. 1995). In the above-mentioned studies, we have distinguished the spatial brain pattern of c-*fos* expression during the arousal or the satiation of sodium appetite stimulated by PD. This approach was selected because Fos, the nuclear protein product of the immediately early gene c-*fos*, has been used as a marker of neural activation in response to a wide variety of stimuli. The analysis of the pattern of c-*fos* expression in the CNS, preceding and after depletion-induced sodium ingestion, potentially identifies the individual cellular components of functional neural networks accompanying the presence of sodium appetite and then its satiety. During sodium depletion or the appetitive phase of sodium appetite, the number of Fos-ir neurons increased in the SFO, OVLT, and central and medial extended amygdala (ExA), whereas it decreased within the 5-HT neurons of the DRN, suggesting their participation in the genesis of sodium appetite. Moreover, after hypertonic sodium consumption induced by PD, Fos activity increased in the different cell groups of the NTS, LPBN, AP, and MnPO; OT cells of the SON and PVN; and 5-HT neurons of the DRN, indicating their involvement in the inhibition of sodium appetite. However, during this satiety phase, some areas activated after sodium depletion, such as the LT and ExA, also showed increased activity. This evidence can be interpreted as remaining depletion-induced elevation of Fos, or the Fos activity is the result of stimulation caused by the entry to the body of a hypertonic sodium solution during sodium access, that activates thirst since the animals did not ingest water during the intake test.

9.2.1 Lamina Terminalis

Sodium-sensitive channels are expressed in glial cells (ependimary cells and astrocytes) of the SFO and OVLT as a response to physiological increase in extracellular sodium concentration (Noda 2007; Noda and Hiyama 2005; Watanabe 2000). Moreover, the presence of atrial natriuretic peptide (ANP) and ANG II receptors has been described in neurons of both nuclei (Allen et al. 2000; Brown and Czarnecki 1990; Lenkei et al. 1998). In our first studies, we observed that Fos immunoreactive neurons (Fos-ir), detected by immunohistochemistry, first appeared in these nuclei 60 min after PD, increased gradually in the next 4 h, and remained high for 27 h after PD. Fos-ir cells were distributed throughout the SFO body, with the core of the posterior sections being preferentially activated, whereas Fos-ir neurons occurred around the periphery of OVLT (annular disposition) (Franchini and Vivas 1999; Vivas et al. 1995). The present evidence supports previous results showing increased production of Fos within the OVLT and SFO after intravenous infusion of ANG II (McKinley et al. 1995), and shows that the spatial distribution of Fos-ir cells in this condition is similar to that observed in sodium-depleted animals by PD (Franchini and Vivas 1999). It is also important to note that this spatial pattern of Fos-ir cells is different from those observed in the SFO and OVLT after acute intravenous infusion of hypertonic NaCl or sucrose (Oldfield et al. 1991). In these latter cases, Fos is detected, for example, in many neurons in the dorsal cap of the OVLT with only a few neurons in the perimeter of the SFO.

In a more recent study, we observed that c-*fos* expression at OVLT level is dependent on the tonicity of the solution consumed after PD, because depleted animals with isotonic solution access had very low activation, the same neuronal activation as sham-dialyzed animals, whereas sodium-depleted rats that consumed hypertonic NaCl had the highest level of activation, probably because of the stimulation of sodium-sensitive cells (Godino et al. 2007). Consistent with these data, the OVLT lesion attenuated the oxytocin plasma increase observed after intra-atrial infusion of hypertonic solution (Negoro et al. 1988), and the electrical activity of the SON increased after hyperosmotic OVLT stimulation, but not after isosmotic stimulation (Richard and Bourque 1992), suggesting that the OVLT plays a functional role in the osmoregulation of neurohypophyseal hormone release after hypertonic drinking. Furthermore, angiotensin intracerebroventricular (icv) infusion experiments have also suggested that the SFO and OVLT do not play the same roles. Infusions of ANG into the SFO produced water drinking without saline intake, but infusions in the OVLT and the ventral part of the MnPO produced both water and saline drinking. It is concluded that ANG acting in the OVLT, and the most ventral part of the median preoptic nucleus, is important for ANG-induced salt appetite (Fitts and Masson 1990).

As previously shown (Franchini and Vivas 1999; Godino et al. 2007; Vivas et al. 1995), the sodium-depletion enhancement of Fos expression observed in SFO cells remains until sodium access and does not change after hypertonic or isotonic sodium consumption. A possible explanation of this delayed deactivation process after sodium repletion may be associated with the gradual decrease in the circulating ANG II levels until they reach the baseline values (Houpt et al. 1998; Johnson

1985; Tordoff et al. 1991; Vivas et al. 1995). A well-characterized function of the SFO is its role as a CNS target site for circulating ANG II; this octapeptide activates 70% of all SFO neurons with very few inhibitory neural responses (Ferguson and Bains 1996). The LT is also formed by another structure, the MnPO; however, the activation of Fos in the MnPO has a different pattern because in our previous studies MnPO only increased its neuronal activity during the satiety phase of sodium appetite (Franchini and Vivas 1999). Previous lesion studies demonstrated that animals with ventral MnPO damage exhibit a chronic and robust hyperdipsia and spontaneous sodium appetite, which occurs only at night (Gardiner and Stricker 1985; Gardiner et al. 1986). The brain-damaged rats had normal sodium concentrations, renin activities, and ALDO levels in plasma during basal maintenance conditions, and they conserved sodium in urine when maintained on a sodium-deficient diet. However, in another study lesion (Fitts et al. 1990) including the most ventral part of the median preoptic nucleus and the OVLT, it was observed to reduce saline intake induced by treatment with chronic oral captopril or sodium depletion without affecting water intake stimulated by different treatments. Taken together, these results suggest that the MnPO or local fibers of passage may play an important role in the tonic and phasic control of water and sodium intake.

9.2.2 Extended Amygdala

A critical region integrating and modulating the hormone and neural information related to the control of sodium appetite is the ExA complex. The ExA is formed by what appears to be a continuum of structures, including mainly the central and medial amygdaloid nuclei (Ce and Me, respectively) and their extensions in the lateral and medial divisions of the bed nucleus of the stria terminalis (BSTL and BSTM, respectively) (Alheid et al. 1995; De Olmos et al. 1985; Johnson et al. 1999). The concept of the ExA was initially proposed by Johnston (1923), who postulated that the amygdala and bed nucleus of the stria terminalis were once a single structure that becomes separated in mammals by development of the internal capsule. Cytochemoarchitectural, tract tracing, immunocytochemical, and functional studies indicate that the BSTL-Ce and the region of the BSTM-Me have structural, hodological neurochemical, and functional similarities (Aldheid et al. 1995; De Olmos et al. 1985; Johnston 1923; Price et al. 1987). The lateral portions of the BNST and the CeA are reciprocally connected with several nuclei involved in cardiovascular control and fluid balance (Holstege et al. 1985; Moga et al. 1989; Sofroniew 1983). In addition, the amygdala and bed nucleus of the stria terminalis also receive afferent input from the structures of the LT, as well as visceral and somatic inputs via the LPBN (Alden et al. 1994; Veinante and Freund-Mercier 1998). Most studies on the role of this region in sodium ingestion have focused on the CeA and MeA. Previous work from this laboratory showed that amygdala damage or transection of ventral amygdalofugal pathway decreased sodium intake stimulated by PD-sodium depletion (Chiaraviglio 1971). Later on, different studies found that lesions of the CeA greatly impair the salt appetite responses to deoxycorticosterone acetate (DOCA), sodium depletion, subcutaneous (sc) administration of yohimbine, and to icv ANG II (Galaverna et al. 1992, 1993; Reilly et al. 1994; Zardetto-Smith et al. 1994). Similar to CeA lesions, ablation of the BST also inhibited sodium appetite stimulated by different treatments (Reilly et al. 1993, 1994; Zardetto-Smith et al. 1994). On the other hand, damage to the MeA by electrolytic lesions impairs or abolishes mineralocorticoid-induced sodium appetite while not affecting intake induced by adrenalectomy or by acute sodium depletion. These latter types of salt appetite are dependent on a central action of ANG II (Nitabach et al. 1989; Schulkin et al. 1989). There is also evidence indicating that cells within the MeA are essential for this steroid-induced salt appetite, because cell body damage produced by ibotenic acid in this nucleus impairs ALDO-induced salt intake without interfering with sodium depletion-induced or angiotensin-dependent salt appetites (Zhang et al. 1993). Moreover adrenal steroid implants (ALDO or DOCA) in the MeA produced a rapid arousal of specific sodium intake, and central administration of mineralocorticoid antagonist (RU28318) or mineralocorticoid receptor antisense inhibited salt appetite stimulated by systemic ALDO and DOCA, but not by adrenalectomy (Reilly et al. 1993; Sakai et al. 1996).

We have investigated the role of the ExA in sodium appetite control by using Fos immunohistochemistry, giving thus the first evidence at cellular resolution and without implicating invasive maneuvers about the involvement of specific confined groups along the ExA components (central and medial division) in PD-induced sodium appetite (Johnson et al. 1999). Body sodium depletion induced by PD produced a pattern of highly localized and intense Fos immunoreactive (Fos-ir) cells within specific sectors of the central and medial division of the ExA. Compared with the control groups, the largest increase in c-*fos* activation was found in the central ExA division, specifically in the central subdivision of the lateral part of the central amygdaloid nucleus and its continuation in the dorsal part of the lateral bed nucleus of the stria terminalis. Along the medial division of the ExA complex, we also found the activated correspondent sectors of the medial amygdala and the medial bed nucleus, precisely the anterodorsal and caudal–ventral parts of the anterior medial amygdaloid nucleus and the intermediate subdivision of the posterior part of the medial bed nucleus. We have also analyzed the participation of the ExA complex during the satiety phase of sodium appetite induced by PD, and the results showed that there was a major increase in c-*fos* expression in depleted animals with access to hypertonic NaCl solutions compared with the isotonic NaCl access group (Godino et al. 2007).

In summary, the medial ExA is one of the regions of the brain with the highest uptake of aldosterone (De Nicola et al. 1992; McEwen et al. 1986), thus a likely site where ALDO exerts its action on sodium intake regulation (Nitabach et al. 1989; Sakai et al. 1996; Schulkin et al. 1989; Reilly et al. 1993; Zhang et al. 1993). Otherwise, ANG II immunoreactivity is found in the central ExA (Lind and Ganten 1990), and the CeA contains angiotensinergic as well as mineralocorticoid receptors, being postulated as a possible site of interaction between ANG II and ALDO to synergistically induce sodium appetite in particular conditions of body sodium deficit (Fluharty and Epstein 1983; Thornton and Nicolaidis 1994).

In conclusion, these results demonstrate that an acute body sodium loss induces a specific stimulation of ExA complex, verifying the major role of these particular components in a functional neural network that receive, and integrate inputs derived from electrolyte–fluid sensory systems.

9.2.3 Hindbrain

It was previously described by Houpt et al. (1998) and then confirmed by our studies (Franchini and Vivas 1999) that different levels of the NTS show increased activity after hypertonic sodium access induced by sodium depletion (satiety phase of salt appetite), but this nucleus shows no activation during the appetitive phase. Together, these studies suggest this nucleus can integrate peripheral taste and visceral sensory signals during sodium depletion. This sodium-related input originates primarily from the anterior portion of the tongue and is carried to the brain through the chorda tympani branch of VII cranial nerve. These fibers probably play a strong role in the induction of sodium appetite through the suppression of the aversive aspects of sodium taste. Specifically, dietary sodium deprivation leads to decreases in the responses of chorda tympani fibers to sodium, but not to sucrose, hydrochloride acid, or quinine hydrochloride, thus allowing sodium consumption (Contreras 1978; Contreras and Frank 1979). Our study also indicated that a restricted population of neurons in the AP, the medial level of the NTS (mNTS and NTS adjacent to AP), and the LPBN (external subdivision) express c-*fos* after induced hypertonic sodium consumption (Franchini and Vivas 1999; Godino et al. 2007). These data are consistent with previous lesion studies showing that ablation of the AP and immediately adjacent medial NTS markedly enhanced the *ad libitum* intake of saline solutions (need-free sodium appetite) and increase stimulated sodium appetite (Contreras and Stetson 1981; Edwards et al. 1993). In addition, pretreatment with bilateral injections of serotonergic receptor antagonist into the LPBN significantly increases salt intake induced by either icv ANG II or sodium depletion (Menani et al. 1996). Collectively, these results have led to the hypothesis that the NTS, the AP, and the LPBN are components of a hindbrain inhibitory circuit modulating sodium and fluid ingestion. Of the many hindbrain areas that potentially mediate the inhibition of sodium appetite behavior, it is likely that the NTS plays a central role. As the major relay site for taste, the NTS may exert control over both food and fluid ingestion through its neural projections to hindbrain somatic and autonomic motor nuclei, including the ventrolateral medulla, the vagal complex, and the LPBN (Van Giersbergen et al. 1992). Thus, induced expression of c-*fos* in the medial NTS after stimulated sodium ingestion could be correlated with the inhibition of sodium intake. The NTS receives viscerosensory inputs from volume receptors in the atriovenous junction, the stomach, and other abdominal viscera, from baroreceptors via the carotid sinus, and from hepatic osmoreceptors via the hepatic branch of the vagus (Carlson et al. 1997; Morita et al. 1997; Van Giersbergen et al. 1992). Hence, it is plausible that, in our model, gastric mechanoreceptor activation, hepatic osmosodium receptor stimulation, and/or cholecystokinin release contribute to the satiety signal and mirror c-*fos* expression seen in these hindbrain structures.

The data are consistent also with previous studies in which gastric distension increased the Fos-positive neurons and the electrical activity of the NTS/AP and LPBN cells (Baird et al. 2001; Sabbatini et al. 2004; Suemori et al. 1994). For example, hypertonic saline ingestion by DOCA-treated rats may be inhibited by two presystemic signals, that is, gastrointestinal distension and osmolality increase, whereas during DOCA-induced isotonic NaCl ingestion in the gastrointestinal distension may provide the only signal that inhibits further intake (Stricker et al. 2007). Our results could be explained as an additive effect of gastric stretch and hyperosmotic stimulation in the PD-hypertonic group, above all taking into account that vagal and splanchnic afferent nerves carry information sensed by hepatic osmo–sodium receptors, which is projected to the NTS/AP and the LPBN in the brainstem (Kharilas and Rogers 1984; Kobashi et al. 1993; Tordoff et al. 1986). Accordingly, vagotomized and NTS/AP-lesioned rats drank larger volumes of concentrated saline solutions compared with control animals. NTS/AP-lesioned rats also have an attenuated increase of vasopressin and oxytocin release in response to intravenous hypertonic saline infusion (Curtis et al. 1999; Huang et al. 2000; Stricker et al. 2001; Tordoff et al. 1986). Moreover, the c-*fos* expression in the NTS/AP and LPBN increased after intragastric hypertonic sodium infusion in nondeprived animals (Carlson et al. 1997). Together, these results suggest that this brainstem neural pathway mediates peripheral satiety and osmoregulatory signals that modulate fluid intake and neurohypophyseal hormone secretion.

9.3 SEROTONIN AND OXYTOCIN: NEUROCHEMICAL PROCESSING OF HYPERTONICITY AND SODIUM APPETITE

Early studies from Dr. Chiaraviglio's laboratory (Munaro and Chiaraviglio 1981) showed that a body sodium overload increases the hypothalamic levels of serotonin (5-HT). Later, systemic injections of 5-HT antagonists (e.g., dexfenfluramine) were shown to increase need-free and need-induced sodium intake (Neill and Cooper 1989; Rouah-Rosilio et al. 1994). Finally, studies combining focused lesions and 5-HT agonists and antagonists conclusively demonstrated that brain 5-HT circuits involving the DRN (a brain source of serotonin) and LPBN (which receives neuronal projections from DRN and other sources) exert an inhibitory control on sodium appetite (Menani and Johnson 1995; Menani et al. 1996; Reis 2007; Olivares et al. 2003). Immunohistochemical work from our laboratory supported the notion of a DRN–LPBN inhibitory control of sodium intake by showing that neuronal activity in these nuclei is associated with sodium balance and sodium appetite. Fos-ir in many of the DRN subdivision cells was decreased by PD and increased when the animals were either at near sodium balance or in the process of restoring it by ingesting a 2% solution of NaCl (Franchini et al. 2002).

As previously noted, inhibitory mechanisms of sodium appetite also involve the oxytocinergic system. Thus, saline ingestion occurs when circulating levels of OT are suppressed, whereas it is inhibited when OT release is stimulated (Stricker and Verbalis 1986). Moreover, systemic OT administration produces renal sodium excretion; however, it does not inhibit salt appetite in sodium-deficient rats. In contrast, central OT administration inhibits angiotensinergic and PEG-induced sodium appetite (Stricker and Verbalis 1987, 1996), and centrally injected OT receptor antagonists increase sodium appetite stimulated by ANG II (Blackburn et al. 1992). More recently, it has been demonstrated that OT knockout mice show a significant increase in spontaneous and induced sodium consumption compared with wild type controls (Amico et al. 2001; Puryear et al. 2001).

Our immunohistochemical studies are also consistent with an inhibitory role for oxytocin in response to hypertonicity because hypertonic NaCl ingestion induced by PD increased the activity of oxytocinergic SON and PVN cells, similar to what it does for DRN cells (Franchini and Vivas 1999). However, in contrast to DRN (Franchini et al. 2002), PD *per se* induced no change in Fos-ir of oxytocinergic SON and PVN cells.

Because of the differences observed in the neural activity of 5-HT and OT cells in the PD model, and to find out whether this activity is a consequence of the sodium satiation process or the stimulation caused by the entry of a hypertonic sodium solution into the body during sodium access, we analyzed the number of Fos-5-HT- and Fos-OT-immunoreactive neurons in the DRN and the PVN and SON, respectively, after isotonic vs. hypertonic NaCl intake induced by PD (Godino et al. 2007). As expected, body sodium status was equally restored by ingesting iso- or hypertonic sodium chloride solution during the satiety phase of sodium appetite, and the 5-HT neurons of DRN were activated after induced sodium ingestion, independently of their tonicity (iso- or hyper NaCl), whereas the hypothalamic OT neural activity and the oxytocin plasma concentration were only stimulated after hypertonic sodium ingestion (Godino et al. 2007).

Therefore, we may postulate the 5-HT system as a sodium satiety marker whose neurons were activated after body sodium status was reestablished, suggesting that this system is activated under conditions of satiety. Otherwise, the OT system maybe a marker of hypertonic stimulation, because the activity of the OT PVN and SON nuclei neurons and plasma OT release were directly correlated with the ingestion of hypertonic sodium solution during induced consumption, independently of the satiety condition, suggesting that this system is involved in the processing of hyperosmotic signals.

In summary, in light of the anatomic and functional evidence, it is reasonable to postulate the presence of 5-HT pathways with cell bodies in the DRN that project to the LPBN as well as PVN (e.g., OT neurons) and other forebrain structures (ExA, CVOs), and that they act to exert both tonic and phasic inhibitory tone in the control of sodium intake; to be precise, under conditions of satiety, the raphe 5-HT cells act tonically to inhibit sodium intake (Reis 2007), and they are also activated in the process of sodium ingestion to phasically increase inhibitory control and thereby limit excess sodium intake (Figure 9.2).

9.4 NEURAL PATHWAYS INTERACTIONS BETWEEN APPETITIVE AND SATIETY SYSTEMS

The cerebral structures involved in controlling the excitatory appetitive and inhibitory or satiety phases of sodium intake are likely to be interconnected with one another, constituting a neural network that integrates associated information (Fitzsimons 1998; Johnson and Thunhorst 2007). Our previous evidence indicates that modulation of salt appetite involves interactions between the CVO receptive areas and inhibitory hindbrain serotonergic circuits (Figures 9.1 and 9.2) (Badauê-Passos et al. 2007; Godino et al. 2007, 2010). That is, for normal sodium appetite sensation, and consequently for appropriate salt drinking after sodium depletion, the hyponatremia and the released ANG II should act centrally both to activate brain osmo–sodium and angiotensinergic receptors that stimulate salt appetite, and also to inhibit inhibitory brain 5-HT mechanisms, thus removing a "braking" mechanism. The central 5-HT circuits underlying this interaction mainly include bidirectional connections between the CVOs, 5-HT neurons of the DRN, and 5-HT terminals within the LPBN (Figure 9.2; Andrade-Franzé et al. 2010; Castro et al. 2003; Cavalcante-Lima et al. 2005a,b; Colombari et al. 1996; Menani and Johnson 1995; Menani et al. 1996, 1998 a,b, 2000; Olivares et al. 2003; Lima et al. 2004; Tanaka et al. 1998, 2001, 2004, 2003b).

Lind (1986) has anatomically demonstrated a neural angiotensin connection originating in the SFO and projecting to the DRN. ANG II injected via the carotid artery, or into the SFO, enhances the electrical activity of SFO neurons that project to the DRN (Tanaka et al. 1998, 2003b). A microdialysis study (Tanaka et al. 2003) indicates that ANG II activation of SFO neurons projecting to the DRN results in inhibition of DRN neurons and reduced local 5-HT release in the SFO. This suggests that neurons in the SFO monitor the circulating levels of ANG II and send this information to the DRN. A comparable projection from the MnPO to the DRN may play a similar role (Zardetto-Smith et al. 1995).

Our recent connectional studies using retrograde tracers in sodium-depleted rats ingesting salt suggest that structures of the LT inform the DRN and LPBN of sodium status or sodium consumption, and/or of volume expansion by a descending neural pathway. In this way, cells within the LT may contribute to inhibitory mechanisms involving 5-HT neurons in the DRN and the release of 5-HT within the LPBN, which limit the intake of sodium and prevent excess expansion of extracellular volume (Badauê-Passos et al. 2007; Godino et al. 2010; Margatho et al. 2008). In these morphofunctional studies using the retrograde tracer, fluorogold (FG), with Fos we found significantly increased numbers of Fos–FG double-immunolabeled neurons in the LT and several other brain areas previously involved in the control of water and saline drinking and excretion after fluid depletion (Badaue-Passos et al. 2007; Godino et al. 2010). In these studies, the retrograde tracer was injected into the DRN or the LPBN approximately 10 days before sodium depletion experiments. Subsequently, the rats were sodium depleted and were allowed to rehydrate by drinking water and 2% NaCl. Increased numbers of double-labeled neurons were found in the OVLT, SFO, and the MnPO of the LT after the rats drank water and saline in the case of FG injection into the DRN (Badauê-Passos et al. 2007). These results suggest that during the reestablishment of water and sodium balance, neurons of the LT that are monosynaptically connected with the DRN become significantly stimulated by fluid ingestion. These neurons then send information to the DRN, resulting in modulation of the behavioral response and inhibiting further sodium intake. The number of double-labeled Fos–FG neurons in the LT increased after sodium consumption following sodium depletion. In other experiments using a similar approach, FG was injected into the LPBN (Godino et al. 2010). We observed that specific groups of neurons along the LT, PVN, ExA, insular cortex, NTS and 5-HT cells of the DRN are both directly connected to the LPBN, and are significantly activated in response to water and sodium ingestion after PD. During the appetitive phase of sodium appetite, the organism needs to acquire sodium salt from the environment to recover lost sodium and ultimately restore natremia, plasma osmolality, and plasma volume. Additionally, with the onset of sodium intake, an inhibitory signal is gradually required to avoid overingestion of sodium. This inhibitory signal represents the drive to achieve sodium satiety and is characterized by interruption of the previously motivated salt intake. Control of sodium appetite is attributed in part to serotonergic pathways of the DRN and the LPBN.

In these latter studies, we have also analyzed the associated endocrine response, specifically oxytocin and ANP plasma release, because both hormones have been implicated in the regulatory response to fluid reestablishment. In this regard, our data clearly demonstrated that induced sodium ingestion in PD rats produced a significant increase in plasma OT and ANP concentrations compared with control (sham) dialyzed animals with access to 2% NaCl. We have previously mentioned the involvement of the oxytocinergic system in sodium appetite control, and these data once more confirm our hypothesis about the role of the oxytocinergic system in body fluid regulation, signaling the entry to the body of a hypertonic sodium solution during sodium intake (Godino et al. 2007). The oxytocin released would then act at cardiac level to stimulate ANP release (Haanwinckel et al. 1995) with both hormones acting at the kidney level, inducing renal diuresis/natriuresis and also antagonizing the central and peripheral ANG II system, thus preventing sodium overload. With regard to involvement of the ANPergic system, peripheral and central ANPs are also seen to be modulated by the RAS antagonism (Zavala et al. 2004) and, although plasma ANP concentràtion tended to decrease after sodium depletion, this reduction did not reach a significant level.

9.5 ESTROGEN-DEPENDENT INHIBITION OF SODIUM APPETITE: A ROLE FOR SEROTONIN

As discussed in Chapter 3, estradiol inhibits sodium appetite by opposing facilitatory ANG II mechanisms (Mecawi et al. 2008). Moreover, estradiol may also reduce sodium intake by altering the gustatory processing of sodium taste because the electrophysiological responses of the chorda tympani nerve to oral NaCl were blunted by estrogen treatment in ovariectomized female rats. This suggests that females are less sensitive to concentrated NaCl solutions during high estrogen conditions (Curtis and Contreras 2006). We checked if estrogen could also interact with the serotonergic inhibitory system in normally cycling and ovariectomized female rats.

Estrogenic modulation of serotonergic system activity may be involved in both tonic and phasic inhibition of sodium appetite, as several studies have shown that estrogen modulates tryptophan–hydroxylase enzyme activity and expression (Donner and Handa 2009; Hiroi et al. 2006; Sanchez et al. 2005). Our recently published results suggest an interaction between facilitatory neurons of the OVLT and inhibitory serotonergic cells of the DRN (Dalmasso et al. 2011), because estradiol induced Fos activation in serotonergic cells of the DRN and reduced it in OVLT neurons (Figure 9.3). In other words, hormonal status during the estrous cycle and estradiol replacement after ovariectomy changes the neural activity induced by sodium depletion in cells of the OVLT and serotonergic cells of the DRN. In particular, there is an interesting correlation between a ~50% reduction in sodium appetite and

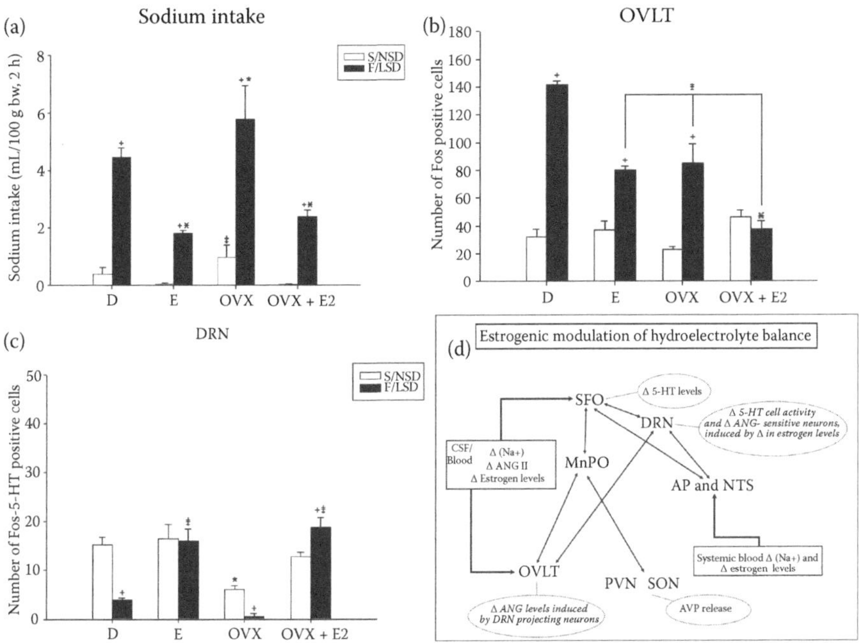

FIGURE 9.3 Estrogen, sodium appetite, and neural markers: (a) Sodium intake (mL/100 g bw; 2 h) of diestrus (D), estrus (E), ovariectomized (OVX) and ovariectomized + estradiol replacement (OVX + E2) groups, "treated with saline/normal sodium diet (S/NSD; white bars) or furosemide/ low sodium diet (F/LSD; black bars). Vertical lines represent SEM. +: $p < 0.001$ compared with S/NSD. ‡: $p < 0.05$ compared with E-S/NSD y OVX + E_2-S/NSD. ӿ: $p < 0.05$ compared with D-F/LSD, OVX-F/LSD. *: $p < 0.005$ compared with D-F/LSD. (b) Average number of Fos immunoreactive (Fos-ir) neurons in the OVLT of diestrus (D), estrus (E), ovariectomized (OVX) and ovariectomized + estradiol replacement (OVX + E_2) groups, treated with saline/normal sodium diet (S/NSD) or furosemide/low sodium diet (F/LSD). Vertical lines represent SEM. +: $p < 0.05$ compared with S/NSD. ‡: $p < 0.05$ compared with D-F/LSD. ӿ: $p < 0.01$ compared with E-F/LSD and OVX-F/LSD. (c) Average number of double-immunolabeled Fos-serotonin (Fos-5-HT) neurons in the DRN of diestrus (D), estrus (E), ovariectomized (OVX) and ovariectomized + estradiol replacement (OVX + E_2) groups, treated with saline/normal sodium diet (S/NSD) or furosemide/low sodium diet (F/LSD). (d) Schematic representation showing possible mechanisms that may contribute to inhibitory action of estrogen on induced sodium intake in female rats. Vertical lines represent SEM. +: $p < 0.01$ compared with S/NSD. *: $p < 0.01$ compared with D-S/NSD, E-S/NSD and OVX + E_2-S/NSD. ‡: $p < 0.01$ compared with D-F/LSD and OVX-F/LSD. (Reprinted with permission from Dalmasso, C. et al., *Physiol Behav*, 104, 398–407, 2011).

the neural activity found in the (possibly facilitatory) neurons of the OVLT and the inhibitory serotonergic neurons of the DRN. Thus, taking into account our previous observations in males, our results in females show that the expected sodium depletion–induced activity of the OVLT is absent in ovariectomized rats treated with estradiol, whereas the usual inhibitory tonic activity of serotonergic neurons of the DRN increases or remains unchanged, rather than decreases, after sodium depletion.

Taking into account these results and that: (1) serotonin and the DRN have been implicated in the inhibitory control of salt intake in males (Badauê-Passos et al. 2007; Cavalcante-Lima et al. 2005a,b; Franchini et al. 2002; Godino et al. 2007, 2010), (2) estrogen has an inhibitory effect on sodium consumption in females (Dalmasso et al. 2011), (3) signals arriving from the LT evoked by fluid depletion-induced sodium ingestion interact with this inhibitory serotonergic system (Badauê-Passos et al. 2007), and (4) estrogen modulates 5-HT synthesis and release (Robichaud and Debonnel 2005; Rubinow et al. 1998; Sanchez et al. 2005), it is possible to postulate serotonergic system involvement in the inhibitory action of estrogen on sodium appetite in female rats. Another possibility, also considered in the schematic representation shown in Figure 9.3d, is that changes in estrogen levels may modulate ANG-sensitive DRN-projecting neurons in the OVLT, to suppress their response to circulating ANG II and consequently inhibit sodium intake. Our results also demonstrate that estradiol influences vasopressinergic neural activity and the associated diuresis after fluid depletion, before drinking. The increased vasopressinergic activity observed in the hypothalamic cells of animals in E and OVX+E2 rats is consistent with previous studies showing an increase in plasma AVP concentration as well as AVP-mRNA in female rats in estrus and in OVX rats after estradiol replacement (Crofton et al. 1985; Crowley et al. 1978; Peysner and Forsling 1990). As shown in the schematic representation, estrogen may be acting directly on vasopressinergic neurons, modulating estrogen receptor beta density expressed in these neurons (Sladek and Somponpun 2008; Somponpun and Sladek 2003; Tanaka et al. 2002). Likewise, estrogen may be modulating its alpha receptor density localized in the SFO and OVLT and—by means of these projections—be acting on vasopressinergic neurons (Grassi et al. 2010; Kensicki et al. 2002; Menani et al. 1998a; Somponpun and Sladek 2004; Tanaka et al. 2001, 2003a). On the other hand, taking into account our data regarding vasopressin neural activity and the inhibitory action of estrogen on induced sodium intake, it is possible to speculate that central vasopressin, or vasopressin-mediated fluid retention during high estrogen states may be another physiological mechanism limiting sodium ingestion (Sato et al. 1997).

In summary, the main finding of these results is a serotonergic system involvement, as a possible mechanism in the inhibitory action of estrogen on induced sodium appetite, which may involve an interaction between excitatory neurons of the OVLT and inhibitory serotonin cells of the DRN key brain cells underlying the responses to hyponatremia and hypovolemia.

9.6 SEX CHROMOSOME COMPLEMENT: ANG II AND GENDER-RELATED DIFFERENCES IN BAROREFLEX

The arterial baroreceptor reflex is a major negative feedback mechanism involved in stabilization of perfusion pressure, and changes in baroreflex control of heart rate (HR) have been described in physiological and pathophysiological states. Clinical and basic findings indicate a sexually dimorphic baroreflex control of HR. The acute administration of ANG II in normotensive male and female patients induces increases in BP of similar magnitude; however, in men, bradycardic baroreflex response is blunted relative to that observed in women (Gandhi et al. 1998). Likewise, in intact male mice, the slope of ANG II–induced baroreflex bradycardia is significantly less than that evoked by phenylephrine (PE), whereas no differences are observed in the response to both pressor agents in gonadectomized female mice (Pamidimukkala et al. 2003). Although previous studies have shown a facilitatory role of estrogen and testosterone on the baroreflex control of HR (El-Mas et al. 2001; Pamidimukkala et al. 2003), classical hormonal manipulations have failed to cause sex reversal of the ANG II–bradycardic baroreflex response.

A growing body of evidence indicates that some sexually dimorphic traits cannot be entirely explained solely as a result of gonadal steroid action, but may also be ascribed to differences in SCC. Males and females carry a different complement of sex chromosome genes and are influenced throughout life by different genomes. Genetic and/or hormone pathways may thus act independently or interact (synergistically/antagonistically) in modulating sexual dimorphic development (Arnold and Chen 2009; Arnold et al. 2004; Cambiasso et al. 1995; De Vries et al. 2002).

In this context, we have investigated whether SCC modulates bradycardic baroreflex response and contributes to the ANG II–bradycardic baroreflex sex differences (Caeiro et al. 2011). To this end, we used the four core genotype (FCG) mouse model, in which the effect of gonadal sex and SCC is dissociated (Figure 9.4), allowing comparisons of sexually dimorphic traits among XX and XY females as well as in XX and XY males.

In conscious gonadectomized (GDX) free moving mice, we evaluated baroreflex regulation of HR in response to changes in BP evoked by PE (1.0 mg/mL), and ANG II (100 μg/mL). Our findings revealed that the ANG II–bradycardic baroreflex sexual dimorphism response may be ascribed to differences in sex chromosomes, indicating an XX-SCC facilitatory bradycardic baroreflex control of HR. Moreover, the results showed that the PE-baroreflex bradycardic response depends on the complex interaction between SCC and gonadal steroids during critical periods of development in fetal and neonatal life (Caeiro et al. 2011). ANG II infusion in GDX-XY male mice induced a blunted bradycardic response when compared to PE administration, whereas GDX-XX female, GDX-XX male, and GDX-XY female mice showed the same bradycardic baroreflex response to both PE and ANG II. Mice with XX-SCC but with different gonadal sex (GDX-XX male and GDX-XX female mice) showed the same bradycardic baroreflex response. Moreover, the comparison of female mice with different SCC (GDX-XX female vs. GDX-XY female) showed an attenuated

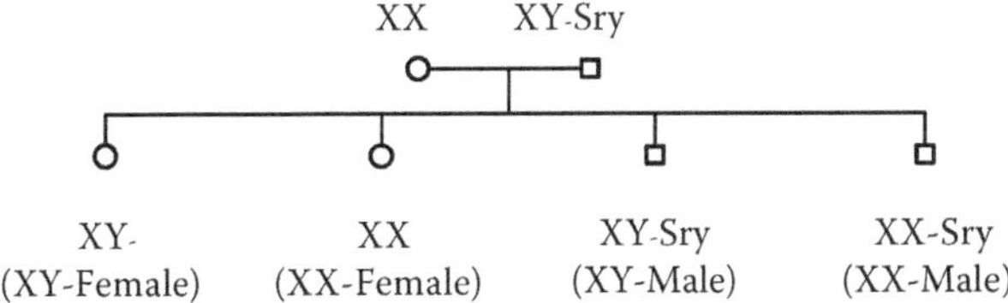

FIGURE 9.4 Schematic representation of four core genotype mouse models, in which effect of gonadal sex and sex chromosome complement is dissociated. (Reprinted with permission from Caeiro, X.E. et al., *Hypertension*, 58, 505–511, 2011.)

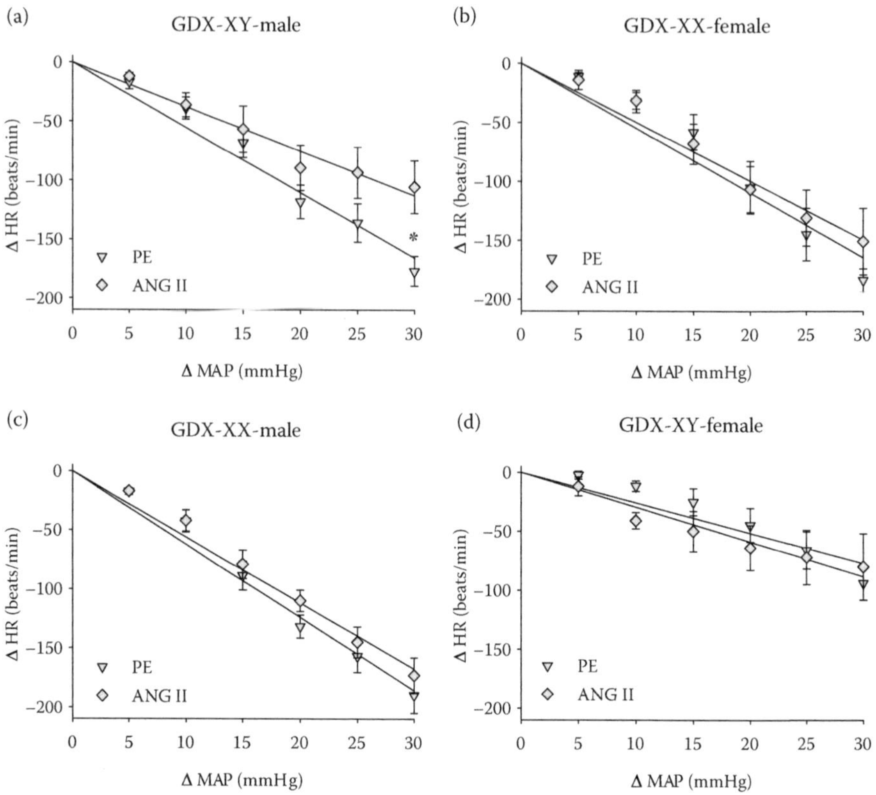

FIGURE 9.5 Comparative reflex bradycardic baroreflex responses to phenylephrine (PE) and angiotensin II (ANG II) infusion in MF1 gonadectomized (GDX) mice of the four core genotype (FCG). Graphs show mean relationship lines relating peak changes in heart rate (HR; delta HR) to increases in blood pressure (BP) induced by both ANG II and PE in mice of the FCG mouse model. $*p \leq 0.05$, significant differences between bradycardic response to ANG II and PE infusion; GDX-XY male ($n = 6$), GDX-XX male ($n = 6$), GDX-XY female ($n = 6$), and GDX-XX female ($n = 5$) mice. (Reprinted with permission from Caeiro, X.E. et al., *Hypertension*, 58, 505–511, 2011.)

baroreflex response to both pressor agents in GDX-XY female mice, indicating an XX-SCC facilitatory bradycardic baroreflex effect (Figure 9.5).

Previous studies conducted in patients and spontaneously hypertensive rats have demonstrated an association of the Y chromosome with high BP (Ellis et al. 2000; Ely and Turner 1990). More recently, Ji et al. (2010) using the FCG mouse model, have shown that, after 2 weeks of ANG II infusion, mean arterial pressure is greater in GDX-XX than in GDX-XY mice. In the current study, using the same mouse model, we found that acute ANG II infusion induces a blunted bradycardic response in GDX-XY compared with GDX-XX mice, indicating that the acute administration of ANG II produced, regardless of the gonadal phenotype, a different baroreflex response depending on the genetic sex.

Although these results described in the previous paragraph appear to be contradictory, it is important to note that chronic infusion of ANG II triggers regulatory responses associated with neuroendocrine compensatory mechanisms. In particular, changes in ANG II receptor expression attributed to increases in ANG II levels have been reported. *In vitro* and *in vivo* studies have shown an upregulation of central AT1R expression in response to physiological increases in plasma ANG II levels induced by water deprivation and sodium depletion (Barth and Gerstberger 1999; Chen et al. 2003; Sanvitto et al. 1997). Studies carried out by Wei et al. (2009) have also demonstrated that the sc infusion of a low dose of ANG II for 4 weeks induces an increase in AT1R mRNA expression in the SFO, and PVN, although no significant effect on BP is observed. Moreover, increases in AT1R expression have been reported in pathophysiological states, such as hypertension (Saavedra et al. 1986). Thus, changes in AT1R expressions in central brain areas attributed to chronic ANG II infusion may differentially influence the activity and responsiveness of the RAS system.

It is important to point out that the AT2R gene (*Agtr2*) is located in the X chromosome (De Gasparo et al. 2000; Koike et al. 1994), whereas the AT1R gene (*Agtr1*) is localized in an autosome chromosome (Arnold et al. 2004; Szpirer et al. 1993). Thus, it is tempting to speculate that genes residing on the sex chromosomes (which are asymmetrically inherited between males and females) could well be influential in eliciting and maintaining sex-bias phenotypes (Davies and Wilkinson 2006). If this is the case, differential transcription or expression of the AT1R/AT2R might be responsible for ANG II sex-biased differences.

9.7 CONCLUDING COMMENTS

In summary, we can postulate that mechanisms controlling fluid and cardiovascular balance rely on many different neurochemical systems acting on similar neural pathways in parallel and simultaneously or in different moments, in relation to the changing physiological body conditions. Besides these mechanisms, there is the influence of gonadal hormones and a direct modulation by sex chromosomes, the latter an important subject for future research. Elucidating the foundational sources of sexually dimorphic traits may offer important insights into designing improved oriented sex-tailored therapeutic treatments for cardiovascular and renal diseases.

ACKNOWLEDGMENTS

This work was, in part, supported by grants from the CONICET, ANPCyT, MinCyT Roemmers, FUCIBICO, and SECyT Universidad Nacional de Córdoba. Carolina Dalmasso and Ana Fabiola Macchione were recipients of CONICET fellowships.

REFERENCES

Alden, M., J. M. Besson, and J. F. Bernard. 1994. Organization of the efferent projections from the pontine parabrachial area to the bed nucleus of the stria terminalis and neighboring regions: A PHA-L study in the rat. *J Comp Neurol* 341:289–14.

Alheid, G. F., J. de Olmos, and C. A. Beltramino. 1995. Amygdala and extended amygdala. In *The Rat Nervous System*, 2nd edn, ed. G. Paxinos, 495–578. San Diego: Academic Press.

Allen, A. M., J. Zhuo, and F. A. Mendelsohn. 2000. Localization and function of angiotensin AT1 receptors. *Am J Hypertens* 13:31S–8S.

Amico, J. A., M. Morris, and R. R. Vollmer. 2001. Mice deficient in oxytocin manifest increased saline consumption following overnight fluid deprivation. *Am J Physiol Regul Integr Comp Physiol* 281:R1368–73.

Andrade-Franzé, G. M., C. A. Andrade, L. A. De Luca Jr., P. M. De Paula, and J. V. Menani. 2010. Lateral parabrachial nucleus and central amygdala in the control of sodium intake. *Neuroscience* 165:633–41.

Arnold, A. P., and X. Chen. 2009. What does the "four core genotypes" mouse model tell us about sex differences in the brain and other tissues? *Front Neuroendocrinol* 30:1–9.

Arnold, A. P., J. Xu, W. Grisham, X. Chen, Y. H. Kim, and Y. Itoh. 2004. Minireview: Sex chromosomes and brain sexual differentiation. *Endocrinology* 145:1057–62.

Badauê-Passos Jr., D., A. Godino, A. K. Johnson, L. Vivas, and J. Antunes-Rodrigues. 2007. Dorsal raphé nuclei integrate allostatic information evoked by depletion-induced sodium ingestion. *Exp Neurol* 206:86–94.

Baird, J. P., J. B. Travers, and S. P. Travers. 2001. Parametric analysis of gastric distension responses in the parabrachial nucleus. *Am J Physiol Regul Integr Comp Physiol* 281:R1568–80.

Barth, S. W., and R. Gerstberger. 1999. Differential regulation of angiotensinogen and AT1A receptor mRNA within the rat subfornical organ during dehydration. *Brain Res* 64:151–64.

Berridge, K. C. 2004. Motivation concepts in behavioral neuroscience. *Physiol Behav* 81:179–209.

Beauchamp, G. K., M. Bertino, K. Engelman. 1983. Modification of salt taste. *Ann Intern Med* 98:763–9.

Beauchamp, G. K., M. Bertino, and K. Engelman. 1987. Failure to compensate decreased dietary sodium with increased table salt usage. *JAMA* 258:3275–8.

Bertino, M., G. K. Beauchamp, and K. Engelman. 1982. Long-term reduction in dietary sodium alters the taste of salt. *Am J Clin Nutr* 36:1134–44.

Blackburn, R. E., A. D. Demko, G. E. Hoffman, E. M. Stricker, and J. G. Verbalis. 1992. Central oxytocin inhibition of angiotensin-induced salt appetite in rats. *Am J Physiol Regul Integr Comp Physiol* 263:R1347–53.

Brown, J., and A. Czarnecki. 1990. Autoradiographic localization of atrial and brain natriuretic peptide receptors in rat brain. *Am J Physiol Regul Integr Comp Physiol* 258:R57–63.

Caeiro, X. E., F. R. Mir, L. M. Vivas, H. F. Carrer, and M. J. Cambiasso. 2011. Sex chromosome complement contributes to sex differences in bradycardic baroreflex response. *Hypertension* 58:505–11.

Cambiasso, M. J., H. Díaz, A. Cáceres, and H. F. Carrer. 1995. Neuritogenic effect of estradiol on rat ventromedial hypothalamic neurons co-cultured with homotopic or heterotopic glia. *J Neurosci Res* 42:700–9.

Carlson, S. H., A. Beitz, and J. W. Osborn. 1997. Intragastric hypertonic saline increases vasopressin and central Fos immunoreactivity in conscious rats. *Am J Physiol Regul Integr Comp Physiol* 272:R750–8.

Castro, L., R. Athanazio, M. Barbetta et al. 2003. Central 5- HT2B/2C and 5-HT3 receptor stimulation decreases salt intake in sodium depleted rats. *Brain Res* 981:151–9.

Cavalcante-Lima, H. R., D. Badaue-Passos Jr., W. de-Lucca Jr. et al. 2005a. Chronic excitotoxic lesion of the dorsal raphe nucleus induces sodium appetite. *Braz. J. Med. Biol. Res* 38:1669–75.

Cavalcante-Lima, H. R., H. R. Lima, R. H. Costa-e-Sousa et al. 2005b. Dipsogenic stimulation in ibotenic DRN-lesioned rats induces concomitant sodium appetite. *Neurosci Lett* 374:5–10.

Chiaraviglio, E. 1971. Amygdaloid modulation of sodium chloride and water intake in the rat. *J Comp Physiol Psychol* 76:401–7.

Chen, Y., M. J. da Rocha, and M. Morris. 2003. Osmotic regulation of angiotensin AT1 receptor subtypes in mouse brain. *Brain Res* 965:35–44.

Colombari, D. S. A., J. V. Menani, and A. K. Johnson. 1996. Forebrain angiotensin type 1 receptors and parabrachial serotonin in the control of NaCl and water intake. *Am J Physiol Regulatory Integrative Comp Physiol* 271:R1470–6.

Contreras, R. J. 1978. Salt taste and disease. *Am J Clin Nutr* 31:1088–97.

Contreras, R. J., and M. Frank. 1979. Sodium deprivation alters neural responses to gustatory stimuli. *J Gen Physiol* 73:569–94.

Contreras, R., and P. Stetson. 1981. Changes in salt intake after lesions of the area postrema and the nucleus of the solitary tract in rats. *Brain Res* 211:355–66.

Crystal, S. R., and I. L. Bernstein. 1995. Morning sickness: Impact on offspring salt preference. *Appetite* 25:231–40.

Crystal, S. R., and I. L. Bernstein. 1998. Infant salt preference and mother's morning sickness. *Appetite* 30:297–307.

Crofton, J., P. Baer, L. Share, and D. Brooks. 1985. Vasopressin release in male and female rats: Effects of gonadectomy and treatment with gonadal steroid hormones. *Endocrinology* 117:1195–200.

Crowley, W. R., T. L. O'Donohue, J. M. George, and D. M. Jacobowitz. 1978. Changes in pituitary oxytocin and vasopressin during the estrous cycle and after ovarian hormones, evidence for mediation by norepinephrine. *Life Sciences* 23:2579–86.

Curtis, K. S., and R. J. Contreras. 2006. Sex differences in electrophysiological and behavioral responses to NaCl taste. *Behav Neurosci* 120:917–24.

Curtis, K. S., W. Huang, A. F. Sved, J. G. Verbalis, and E. M. Stricker. 1999. Impaired osmoregulatory responses in rats with area postrema lesion. *Am J Physiol Regul Integr Comp Physiol* 277: R209–19.

Dalmasso, C., J. L. Amigone, and L. Vivas. 2011. Serotonergic system involvement in the inhibitory action of estrogen on induced sodium appetite in female rats. *Physiol and Behav* 104:398–407.

Davies, W., and L. S. Wilkinson. 2006. It is not all hormones: Alternative explanations for sexual differentiation of the brain. *Brain Res* 1126:36–45.

De Gasparo, M., K. J. Catt, T. Inagami, J. W. Wright, and T. Unger. 2000. International union of pharmacology: XXIII. The angiotensin II receptors. *Pharmacol Rev* 52:415–72.

De Nicola, A. F., C. Grillo, and S. Gonzalez. 1992. Physiological, Biochemical and Molecular mechanisms of salt appetite control by mineralocorticoid action in brain. *Braz J Med Biol Res* 25:1153–62.

De Olmos, J., G. F. Aldheid, and C. Beltramino. 1985. Amygdala. In *The Rat Nervous System*, ed. G. Paxinos, 223–234, Vol. 1. Forebrain and Midbrain. Orlando, FL: Academic Press.

De Vries, G. J., E. F. Rissman, R. B. Simerly et al. 2002. A model system for study of sex chromosome effects on sexually dimorphic neural and behavioral traits. *J Neurosci* 22:9005–14.

Donner, N., and R. J. Handa. 2009. Estrogen receptor beta regulates the expression of tryptophan-hydroxylase 2 mRNA within serotonergic neurons of the rat dorsal raphe nuclei. *Neuroscience* 163:705–18.

Edwards, G. L., T. G. Beltz, J. D. Power, and A. K. Johnson. 1993. Rapid-onset "need-free" sodium appetite after lesions of the dorsomedial medulla. *Am J Physiol Regul Integr Comp Physiol* 264:R1242–7.

Ellis, J. A., M. Stebbing, and S. B. Harrap. 2000. Association of the human Y chromosome with high blood pressure in the general population. *Hypertension* 36:731–3.

El-Mas, M. M., E. A. Afify, M. M. M. El-Din, A. G. Omar, and F. M. Sharabi. 2001. Testosterone facilitates the baroreceptor control of reflex bradycardia: Role of cardiac sympathetic and parasympathetic components. *J Cardiovasc Pharmacol* 38:754–63.

Ely, D. L., and M. E. Turner. 1990. Hypertension in the spontaneously hypertensive rat is linked to the Y chromosome. *Hypertension* 16:277–81.

Ferguson, A. V., and J. S. Bains. 1996. Electrophysiology of the circumventricular organs. *Front Neuroendocrinol* 17:440–75.

Ferreyra, M. D., and E. Chiaraviglio. 1977. Changes in volemia and natremia and onset of sodium appetite in sodium depleted rats. *Physiol Behav* 19:197–201.

Fitts, D. A., and D. B. Masson. 1990. Preoptic angiotensin and salt appetite. *Behav Neurosci* 104:643–50.

Fitts, D. A., D. S. Tjepkes, and R. O. Bright. 1990. Salt appetite and lesions of the ventral part of the ventral median preoptic nucleus. *Behav Neurosci* 104:818–27.

Fitzsimons, J. T. 1998. Angiotensin, thirst, and sodium appetite. *Physiol Rev* 78:583–686.

Fluharty, S. J., and A. N. Epstein. 1993. Sodium appetite elicited by intracerebroventricular infusion of angiotensin II in the rat: II. Synergistic interaction with systemic mineralocorticoids. *Behav Neurosci* 97:746–58.

Franchini, L., and L. Vivas. 1999. Fos induction in rat brain neurons after sodium consumption induced by acute body sodium depletion. *Am J Physiol Regul Integr Comp Physiol* 276:R1180–7.

Franchini, L., A. K. Johnson, J. de Olmos, and L. Vivas. 2002. Sodium appetite and Fos activation in serotonergic neurons. *Am J Physiol Regul Integr Comp Physiol* 282:R235–43.

Galaverna, O. G., L. A. De Luca Jr., J. Shulkin, S. Yao, and A. N. Epstein. 1992. Deficits in NaCl ingestion after damage to the central nucleus of the amygdala in the rat. *Brain Res Bull* 28:89–98.

Galaverna, O. G., R. J. Seeley, K. C. Berridge, H. J. Grill, A. N. Epstein, and J. Schulkin. 1993. Lesions of the central nucleus of the amygdale: Effects on taste reactivity, taste aversion learning and sodium appetite. *Behav Brain Res* 59:11–7.

Gandhi, S. K., J. Gainer, D. King, and N. J. Brown. 1998. Gender affects renal vasoconstrictor response to ANG I and ANG II. *Hypertension* 31:90–6.

Gardiner, T. W., and E. M. Stricker. 1985. Hyperdipsia in rats after electrolytic lesions of nucleus medianus. *Am J Physiol Regul Integr Comp Physiol* 248:R214–23.

Gardiner, T. W., J. R. Jolley, A. H. Vagnucci, and E. M. Stricker. 1986. Enhanced sodium appetite in rats with lesions centered on nucleus medianus. *Behav Neurosci* 100:531–5.

Geerling, J. C., and A. D. Loewy. 2008. Central regulation of sodium appetite. *Exp Physiol* 93:177–209.

Godino, A., L. O. Margatho, X. E. Caeiro, J. Antunes-Rodrigues, and L. Vivas. 2010. Activation of lateral parabrachial afferent pathways and endocrine responses during sodium appetite regulation. *Exp Neurol* 221:275–84.

Godino, A., L. A. De Luca Jr., J. Antunes-Rodrigues, and L. Vivas. 2007. Oxytocinergic and serotonergic systems involvement in sodium intake regulation: Satiety or hypertonicity markers? *Am J Physiol Regul Integr Comp Physiol* 293:R1027–36.

Grassi, D., M. A. Amorim, L. M. Garcia-Segura, and G. Panzica. 2010. Estrogen receptor alpha is involved in the estrogenic regulation of arginine vasopressin immunoreactivity in the supraoptic and paraventricular nuclei of ovariectomized rats. *Neurosci Lett* 474:135–9.

Haanwinckel, M. A., L. K. Elias, A. L. Favaretto, J. Gutkowska, S. M. McCann, and J. Antunes-Rodrigues. 1995. Oxytocin mediates atrial natriuretic peptide release and natriuresis after volume expansion in the rat. *Proc Natl Acad Sci USA* 92:7902–6.

Hiroi, R., R. A. McDevitt, and J. F. Neumaier. 2006. Estrogen selectively increases tryptophan hydroxylase-2 mRNA expression in distinct subregions of rat midbrain raphe nucleus: Association between gene expression and anxiety behavior in the open field. *Biol Psychiatry* 60:288–95.

Holstege, G., L. Meiners, and K. Tan. 1985. Projections of the bed nucleus of the stria terminalis to the mesencephalon, pons, and medulla oblongata in the cat. *Exp Brain Res* 58:379–91.

Houpt, T. A., G. P. Smith, T. H. Joh, and S. P. Frankmann. 1998. C-fos like immunoreactivity in the subfornical organ and nucleus of the solitary tract following salt intake by sodium depleted rats. *Physiol Behav* 63:505–10.

Huang, W., A. F. Sved, and E. M. Stricker. 2000. Vasopressin and oxytocin release evoked by NaCl loads are selectively blunted by area postrema lesions. *Am J Physiol Regul Integr Comp Physiol* 278:R732–40.

Ji, H., W. Zheng, X. Wu et al. 2010. Sex chromosome effects unmasked in angiotensin II-induced hypertension. *Hypertension* 55:1275–82.

Johnson, A. K., J. De Olmos, C. V. Pastuskovas, A. M. Zardetto-Smith, and L. Vivas. 1999. The extended amygdala and salt appetite. *Ann N Y Acad Sci* 877:258–80.

Johnson, A. K. 1985. The periventricular anteroventral third ventricle (AV3V): Its relationship with the subfornical organ and neural system involved in the maintaining body fluid homeostasis. *Brain Res Bull* 15:595–601.

Johnston, A. K., and R. L. Thunhorst. 2007. The neuroendocrinology, neurochemistry and molecular biology of thirst and salt appetite. In *Handbook of Neurochemistry and Molecular Neurobiology,* 3rd edn, ed. A. Lajtha, and J. D. Blaustein, 641–687. Berlin: Springer-Verlag.

Johnston, J. B. 1923. Further contributions to the study of the evolution of the forebrain. *J Comp Neurol* 35:337–481.

Kensicki, E., G. Dunphy, and D. Ely. 2002. Estradiol increases salt intake in female normotensive and hypertensive rats. *J Appl Physiol* 93:479–83.

Kharilas, P. J., and R. C. Rogers. 1984. Rat brainstem neurons responsive to changes in portal blood sodium concentration. *Am J Physiol Regul Integr Comp Physiol* 247:R792–9.

Kobashi, M., H. Ichikawa, T. Sugimoto, and A. Adachi. 1993. Response of neurons in the solitary tract nucleus, area postrema and lateral parabrachial nucleus to gastric load of hypertonic saline. *Neurosci Lett* 158:47–50.

Koike, G., M. Horiuchi, T. Yamada, C. Szpirer, H. J. Jacob, and V. J. Dzau. 1994. Human type 2 angiotensin II receptor gene: Cloned, mapped to the X chromosome, and its mRNA is expressed in the human lung. *Biochem Biophys Res Commun* 30:1842–50.

Lenkei, Z., M. Palkovits, P. Corvol, and C. Llorens-Cortes. 1998. Distribution of angiotensin type-1 receptor messenger RNA expression in the adult rat brain. *Neuroscience* 82:827–41.

Leshem, M. 2009. Biobehavior of the human love of salt. *Neurosci Biobehav Rev* 33:1–17.

Lima, H. R., H. R. Cavalcante-Lima, P. L. Cedraz-Mercez et al. 2004. Brain serotonin depletion enhances the sodium appetite induced by sodium depletion or beta-adrenergic stimulation. *An Acad Bras Cienc* 76:85–92.

Lind, R. W. 1986. Bi-directional, chemically specified neural connections between the subfornical organ and the midbrain raphe system. *Brain Res* 384:250–61.

Lind, W., and D. Ganten. 1990. Angiotensin. In *Handbook of Chemical Neuroanatomy, Vol.9: Neuropeptides in the CNS. Part II.*, ed. A. Bjorklund, T. Hokfelt, and M. J. Kuhar. Elsevier: Amsterdam.

Margatho, L. O., A. Godino, F. R. Oliveira, L. Vivas, and J. Antunes-Rodrigues. 2008. Lateral parabrachial afferent areas and serotonin mechanisms activated by volume expansion. *J Neurosci Res* 86:3613–21.

McEwen, B. S., L. Lambdin, T. C. Rainbow, and A. F. De Nicola. 1986. Aldosterone effects on salt appetite in adrenalectomized rats. *Neuroendocrinol*ogy 43:38–43.

McKinley, M. J., E. Badoer, L. Vivas, and B. J. Oldfield. 1995. Comparison of c-fos expression in the forebrain of conscious rats after intravenous or intracerebroventricular angiotensin. *Brain Res Bull* 37:131–7.

Mecawi, A. S., A. Lepletier, I. G. Araujo, F. V. Fonseca, and L. C. Reis. 2008. Oestrogenic influence on brain AT1 receptor signalling on the thirst and sodium appetite in osmotically stimulated and sodium-depleted female rats. *Exp Physiol* 93:1002–10.

Menani, J. V., D. S. A. Colombari, T. G. Beltz, R. L. Thunhorst, and A. K. Johnson. 1998a. Salt appetite: Interaction of forebrain angiotensinergic and hindbrain serotonergic mechanisms. *Brain Res* 801:29–35.

Menani, J. V., L. A. De Luca Jr., and A. K. Johnson. 1998b. Lateral parabrachial nucleus serotonergic mechanisms and salt appetite induced by sodium depletion. *Am J Physiol Regulatory Integrative Comp Physiol* 274:R555–60.

Menani, J. V., L. A. De Luca Jr., R. L. Thunhorst, and A. K. Johnson. 2000. Hindbrain serotonin and the rapid induction of sodium appetite. *Am J Physiol Regulatory Integrative Comp Physiol* 279:R126–131.

Menani, J. V., and A. K. Johnson. 1995. Lateral parabrachial serotonergic mechanisms: Angiotensin-induced pressor and drinking responses. *Am J Physiol Regulatory Integrative Comp Physiol* 269:R1044–9.

Menani, J. V., R. L. Thunhorst, and A. K. Johnson. 1996. Lateral parabrachial nucleus and serotonergic mechanisms in the control of salt appetite in rats. *Am J Physiol Regulatory Integrative Comp Physiol* 270:R162–8.

Moga, M. M., C. B. Saper, and T. S. Gray. 1989. Bed nucleus of the stria terminalis: Cytoarchitecture, immunohistochemistry, and projection to the parabrachial nucleus in the rat. *J Comp Neurol* 283:315–32

Morita, H., Y. Yamashita, Y. Nishimida, M. Tokuda, O. Hatase, and H. Hosomi. 1997. Fos induction in rat brain neurons after stimulation of the hepatoportal Na-sensitive mechanism. *Am J Physiol Regulatory Integrative Comp Physiol* 272:R913–23.

Munaro, N., and E. Chiaraviglio. 1981. Hypothalamic levels and utilization of noradrenaline and 5-hydroxytryptamine in the sodium depleted rat. *Pharmacol Biochem Behav* 15:1–5.

Negoro, H., T. Higuchi, Y. Tadokoro, and K. Honda. 1988. Osmoreceptor mechanisms for oxytocin release in the rat. *Jpn J Physiol* 38:19–31.

Neill, J. C., and S. J. Cooper. 1989. Selective reduction by serotonergic agents of hypertonic saline consumption in rats: Evidence for possible 5-HT1C receptor mediation. *Psychopharmacology* 99:196–201.

Nitabach, M., J. Schulkin, and A. N. Epstein. 1989. The medial amygdala is a part of a mineralocorticoid-sensitive circuit controlling NaCl intake in the rat. *Behav Brain Res* 35:127–34.

Noda, M. 2007. Hydromineral neuroendocrinology: Mechanism of sensing sodium levels in the mammalian brain. *Exp Physiol* 92:513–522.

Noda, M., and T. Y. Hiyama. 2005. Sodium-level-sensitive sodium channel and salt-intake behavior. *Chem Senses* 30(Suppl 1):i44–5.

Oldfield, B. J., R. J. Bicknell, R. M. McAllen, R. S. Weisinger, and M. J. McKinley. 1991. Intravenous hypertonic saline induces Fos immunoreactivity in neurons throughout the lamina terminalis. *Brain Res* 561:151–6.

Olivares, E. L., R. H. Costa-E-Sousa, H. R. Cavalcante-Lima, H. R. Lima, P. L. Cedraz-Mercez, and L. C. Reis. 2003. Effect of electrolytic lesion of the dorsal raphe nucleus on water intake and sodium appetite. *Braz J Med Biol* Res 36:1709–16.

Pamidimukkala, J., J. A. Taylor, W. V. Welshons, D. B. Lubahn, and M. Hay. 2003. Estrogen modulation of baroreflex function in conscious mice. *Am J Physiol Regul Integr Comp Physiol* 284:R983–9.

Peysner, K., and M. L. Forsling. 1990. Effect of ovariectomy and treatmentwith ovarian steroids on vasopressin release and fluid balance in the rat. *J Endocrinol* 124:277–84.

Puryear, R., K. V. Rigatto, J. A. Amico, and M. Morris. 2001. Enhanced salt intake in oxytocin deficient mice. *Exp Neurol* 171:323–8.

Reilly, J. J., D. B. Mamani, J. Schulkin, B. S. McEwen, and R. R. Sakai. 1993. Effect of amygdala adrenal steroid implants in rat brain on sodium intake (Abstract). *Appetite* 21:201.

Reilly, J. J., R. Maki, J. Nardozzi, and J. Schulkin. 1994. The effects of lesions of the bed nucleus of the stria terminalis on sodium appetite. *Acta Neurobiol Exper* 54:253–7.

Reis, L. C. 2007. Role of the serotoninergic system in the sodium appetite control. *An Acad Bras Cienc* 79:261–83.

Richard, D., and C. W. Bourque. 1992. Synaptic activation of rat supraoptic neurons by osmotic stimulation of the organum vasculosum of the lamina terminalis. *Neuroendocrinology* 55:609–11.

Robichaud, M., and G. Debonnel. 2005. Oestrogen and testosterone modulate the firing activity of dorsal raphe nucleus serotonergic neurones in both male and female rats. *J Neuroendocrinol* 17:179–85.

Rouah-Rosilio, M., M. Orosco, and S. Nicolaidis. 1994. Serotoninergic modulation of sodium appetite in the rat. *Physiol Behav* 55:811–6.

Rubinow, D. R., P. J. Schmidt, and C. A. Roca. 1998. Estrogen–serotonin interactions: Implications for affective regulation. *Biol Psychiatry* 44:839–50.

Saavedra, J. M., F. M. Correa, M. Kurihara, and K. Shigematsu. 1986. Increased number of angiotensin II receptors in the subfornical organ of spontaneously hypertensive rats. *J Hypertension* 4:S27–S30.

Sabbatini, M., E. Molinari, C. Grossini, D. A. Mary, G. Vacca, and M. Cannas. 2004. The pattern of c-Fos immunoreactivity in the hindbrain of the rat following stomach distension. *Exp Brain Res* 157:315–23.

Sakai, R. R., L. Y. Ma, D. M. Zhang, B. S. McEwen, and S. J. Fluharty. 1996. Intracerebral administration of mineralocorticoid receptor antisense oligonucleotides attenuate adrenal steroid-induced salt appetite in rats. *Neuroendocrinology* 64:426–9.

Sanchez, R. L., A. P. Reddy, M. L. Centeno, J. A. Henderson, and C. L. Bethea. 2005. A second tryptophan hydroxylase isoform, TPH-2 mRNA, is increased by ovarian steroids in the raphe region of macaques. *Brain Res Mol Brain Res* 135:194–203.

Sanvitto, G. L., O. Johren, W. Hauser, and J. M. Saavedra. 1997. Water regulation upregulates ANG II AT1 binding and mRNA in rat subfornical organ and anterior pituitary. *Am J Physiol Endocrinol Metab* 273:E156–63.

Sato, M. A., A. M. Sugawara, J. V. Menani, and L. A. De Luca Jr. 1997. Idazoxan and the effect of intracerebroventricular oxytocin or vasopressin on sodium intake of sodium-depleted rats. *Regul Pept* 69:137–42.

Schulkin, J., J. Marini, and A. N. Epstein. 1989. A role for the medial region of the amygdala in mineralocorticoid-induced salt hunger. *Behav Neurosci* 103:178–85.

Simpson, J. B., A. N. Epstein, and J. S. Camargo Jr. 1978. Localization of receptors for the dipsogenic action of angiotensin II in the subfornical organ of rat. *J Comp Physiol Psychol* 92:581–601.

Sladek, C. D., and A. K. Johnson. 1983. Effect of anteroventral third ventricle lesions on vasopressin release by organ-cultured hypothalamo-neurohypophyseal explants. *Neuroendocrinology* 37:78–84.

Sladek, C. D., and S. J. Somponpun. 2008. Estrogen receptors: Their roles in regulation of vasopressin release for maintenance of fluid and electrolyte homeostasis. *Front Neuroendocrinol* 29:114–27.

Sofroniew, M. V. 1983. Direct reciprocal connections between the bed nucleus of the stria terminalis and dorsomedial medulla oblongata: Evidence from immunohistochemical detection of tracer proteins. *J Comp Neurol* 213:399–405.

Somponpun, S. J., and C. D. Sladek. 2003. Osmotic regulation of estrogen receptor-β in rat vasopressin and oxytocin neurons. *J Neuroscience* 23:4261–9.

Somponpun, S. J., and C. D. Sladek. 2004. Depletion of oestrogen receptor-β expression in magnocellular arginine vasopressin neurones by hypovolaemia and dehydration. *J Neuroendocrinol* 16:544–9.

Stricker, E. M., and J. G. Verbalis. 1986. Interaction of osmotic and volume stimuli in regulation of neurohypophyseal secretion in rats. *Am J Physiol Regul Integr Comp Physiol* 250:R267–75.

Stricker, E. M., and J. G. Verbalis. 1987. Central inhibitory control of sodium appetite in rats: Correlation with pituitary oxytocin secretion. *Behav Neurosci* 101:560–7.

Stricker, E. M., and J. G. Verbalis. 1996. Central inhibition of salt appetite by Oxytocin in rats. *Regul Pept* 66:83–85.

Stricker, E. M., M. A. Bushey, M. L. Hoffmann, M. McGhee, A. M. Cason, and J. C. Smith. 2007. Inhibition of NaCl appetite when DOCA-treated rats drink saline. *Am J Physiol Regul Integr Comp Physiol* 292:R652–62.

Stricker, E. M., C. F. Craver, K. S. Curtis, K. A. Peacock-Kinzing, A. F. Sved, and J. C. Smith. 2001. Osmoregulation in water-deprived rats drinking hypertonic saline: Effects of area postrema lesions. *Am J Physiol Regul Integr Comp Physiol* 280:R831–42.

Suemori, K., M. Kobashi, and A. Adachi. 1994. Effects of gastric distension and electrical stimulation of dorsomedial medulla on neurons in parabrachial nucleus of rats. *J Auton Nerv Syst* 48:221–9.

Szpirer, C., M. Riviere, J. Szpirer et al. 1993. Chromosomal assignment of human and rat hypertension candidate genes: Type 1 angiotensin II receptor genes and the SA gene. *J Hypertens* 11:919–25.

Tanaka, J., K. Kariya, H. Miyakubo, K. Sakamaki, and M. Nomura. 2002. Attenuated drinking response induced by angiotensinergic activation of subfornical organ projections to the paraventricular nucleus in estrogen-treated rats. *Neurosci Lett* 324:242–6.

Tanaka, J., T. Okumura, K. Sakamaki, and H. Miyakubo. 2001. Activation of serotonergic pathways from the midbrain raphe system to the subfornical organ by hemorrhage in the rat. *Exp Neurol* 169:156–62.

Tanaka, J., Y. Hayashi, K. Yamato, H. Miyakubo, and M. Nomura. 2004. Involvement of serotonergic systems in the lateral parabrachial nucleus in sodium and water intake: A microdialysis study in the rat. *Neurosci Lett* 357:41–44.

Tanaka, J., H. Miyakubo, S. Fujisawa, and M. Nomura. 2003a. Reduced dipsogenic response induced by angiotensin II activation of subfornical organ projections to the median preoptic nucleus in estrogen-treated rats. *Exp Neurology* 197:83–9.

Tanaka, J., K. Kariya, and M. Nomura. 2003b. Angiotensin II reduces serotonin release in the rat subfornical organ area. *Peptides* 24:881–7.

Tanaka, J., A. Ushigome, K. Hori, and M. Nomura. 1998. Responses of raphe nucleus projecting subfornical organ neurons to angiotensin II in rats. *Brain Res Bull* 45:315–8.

Thornton, S. N., and S. Nicolaidis. 1994. Long term mineralocorticoid-induced changes in rat neuron properties plus interaction of aldosterone and AII. *Am J Physiol Regul Integr Comp Physiol* 266:R564–71.

Tordoff, M. G., S. J. Fluharty, and J. Schulkin. 1991. Physiological consequences of NaCl ingestion by Na^+ depleted rats. *Am J Physiol Regul Integr Comp Physiol* 261:R289–95.

Tordoff, M. G., J. Schulkin, and M. I. Friedman. 1986. Hepatic contribution to satiation of salt appetite in rats. *Am J Physiol Regul Integr Comp Physiol* 251:R1095–1102.

Van Giersbergen, P. L. M., M. Palkovits, and W. De Jong. 1992. Involvement of neurotransmitters in the nucleus tractus solitarii in cardiovascular regulation. *Physiol Rev* 72: 789–824.

Veinante, P., and M. J. Freund-Mercier. 1998. Intrinsic and extrinsic connections of the rat central extended amygdala: An in vivo electrophysiological study of the central amygdaloid nucleus. *Brain Res* 794:188–98.

Vivas, L., E. Chiaraviglio, and H. F. Carrer. 1990. Rat organum vasculosum laminae terminalis in vitro: Responses to changes in sodium concentration. *Brain Res* 519:294–300.

Vivas, L., C. V. Pastuskovas, and L. Tonelli. 1995. Sodium depletion induces Fos immunoreactivity in circumventricular organs of the lamina terminalis. *Brain Res* 679:34–41.

Watanabe, E., A. Fujikawa, H. Matsunaga et al. 2000. Nax2/NaG channel is involved in control of salt intake behavior in the central nervous system. *J Neurosci* 20:7743–51.

Wei, S. G., Y. Yu, Z. H. Zhang, and R. B. Felder. 2009. Angiotensin II upregulates hypothalamic AT1 receptor expression in rats via the mitogen-activated protein kinase pathway. *Am J Physiol Heart Circ Physiol* 296: H1425–33.

Zardetto-Smith, A. M., T. G. Beltz, and A. K. Johnson. 1994. Role of the central nucleus of the amygdala and bed nucleus of the stria terminalis in the experimentally-induced salt appetite. *Brain Res* 645:123–34.

Zavala, L., Y. Barbella, Y., and A. J. Israel. 2004. Neurohumoral mechanism in the natriuretic action of intracerebroventricular administration of renin. *J Renin Angiotensin Aldosterone Syst* 5:39–44.

Zhang, D. M., A. N. Epstein, and J. Shulkin. 1993. Medial region of the amygdala: Involvement in adrenal-steroid-induced salt appetite. *Brain Res* 600:20–6.

10 Brain Serotonergic Receptors and Control of Fluid Intake and Cardiovascular Function in Rats

Josmara B. Fregoneze, Hilda S. Ferreira, and Carla Patricia N. Luz

CONTENTS

10.1 INTRODUCTION

Body fluid homeostasis is maintained by a steady-state interchange of water between extracellular fluid (ECF) and intracellular fluid compartments. The main source of body water in terrestrial vertebrates is from drinking and, once in the body, it is distributed between body fluid compartments. The mechanisms that keep the body fluid osmolarity within its narrow range (280–300 mOsm/L) depend on matching the water volume excreted by the kidneys and the fluid intake volume. Sodium is the main electrolyte in ECF and sodium loss or gain is usually accompanied by an increase or decrease in water in this compartment in order to maintain sodium concentration. Body sodium and fluid balance is achieved through mechanisms that control sodium intake and sodium urinary excretion. The relationship between water and sodium in ECF may change both the osmolarity and volume of this compartment. Whereas the osmolarity of ECF is regulated by water intake and renal water excretion, the volume is controlled by the sodium content in the ECF, which is determined by the amount of sodium intake and the amount of sodium excreted in the urine (Verbalis 2003).

Changes in the water and sodium content of ECF may result in severe consequences to the cardiovascular system. Furthermore, some pathological cardiovascular conditions may lead to changes in body fluid homeostasis. Increased sodium levels in ECF increase the effective circulating volume, leading to an enhancement in cardiac output and blood pressure (BP). On the other hand, heart failure may decrease water and sodium urinary excretion, promoting a fluid disorder in the body. Therefore, it is unsurprising to find overlapping mechanisms controlling cardiovascular function and body fluid homeostasis.

Multiple sensory signals that trigger thirst and sodium appetite in response to dehydration are basically produced by hyperosmolarity and/or hypovolemia of the ECF (Fitzsimons 1998; Johnson 2007; McKinley and Johnson 2004; Stricker and Sved 2000; Chapters 2, 3, 4, and 9 of this book). Such sensory information reaches the brain to facilitate or inhibit responses that correct changes in body fluid–mineral balance and control cardiovascular function, thus activating a neural network of noradrenergic, cholinergic, angiotensinergic, GABAergic, vasopressinergic, oxytocinergic, and serotonergic pathways (Johnson 2007).

10.2 OVERVIEW OF BRAIN SEROTONERGIC SYSTEM

Serotonin [5-hydroxytryptamine (5-HT)] is among the most ancient signaling molecules in the phylogenetic scale and has been identified in the central nervous system (CNS) of invertebrates since Coelenterata (Hydra), Platyhelminthes (flatworm, *Bipalium* sp.), Annelida (Pheretima Communissima), Arthropoda (sea louse, crayfish, crab and cricket), Mollusca and Echinodermata, and Protochordata (Fujii and Takeda 1988). The first identification of 5-HT in the CNS of mammals was made by Twarog and Page (1953), findings considered an important landmark in neuroscience. In vertebrates, especially in humans and rats, different effects of serotonin on the CNS have been observed, controlling, for example, sleep–wake cycle, appetite, nociception, stress, sexual behavior, blood coagulation, blood pressure, and hydroelectrolytic balance (Peroutka 1994). Changes in serotonin metabolism are associated with several disorders, including obsessive–compulsive disorder, panic disorder, depression, anxiety, eating disorders, social phobia, drug abuse and addiction, migraine, hypertension, pulmonary hypertension, emesis, and irritable bowel syndrome (Hoyer et al. 2002; Saxena 1995).

Serotonin-containing neurons in the dorsal raphe nucleus (DRN), median raphe nucleus (MRN), and raphe centralis superior (B7–B9 groups) provide extensive serotonergic innervations to telencephalon and diencephalon, whereas the intermediate and posterior groups (B1–B6 groups) send local projections at the pons, and descendent projections to the mesencephalon, medulla and spinal cord (Jacobs and Azmitia 1992; Molliver 1987).

Serotonin receptors are found in platelets, lungs, heart, gastrointestinal tract, and vascular endothelium, and are widespread throughout the peripheral and central nervous systems. They are grouped into seven distinct families (5-HT_1 to 5-HT_7), including 15 structurally and pharmacologically distinct subtypes (Barnes and

Sharp 1999; Green 2006; Hoyer et al. 1994, 2002). Most of the serotonin receptors act through G-protein coupling, with the exception of the 5-HT_3 receptor, which is a ligand-gated ion channel (Barnes and Sharp 1999). The G-protein–coupled receptors induce changes in the intracellular levels of cAMP, with the exception of the 5-HT_2 receptor family, which modulates C-phospholipase function (Hoyer et al. 2002). It is noteworthy that the high expression of serotonin receptors in brain areas is related to hydroelectrolytic balance and cardiovascular control (Saxena 1995).

The 5-HT_1 serotonin receptor family is composed of five different subtypes negatively coupled to adenylate cyclase via G-protein: 5-HT_{1A}, 5-HT_{1B}, 5-HT_{1D}, 5-HT_{1E}, and 5-HT_{1F} (Hoyer et al. 2002; Barnes and Sharp 1999; Saxena 1995). These receptors are found in various regions of the CNS, including the septum, amygdala (AMY), hippocampus, basal ganglia, thalamus, frontal cortex, raphe nuclei, and hypothalamus (Barnes and Sharp 1999; Bruinvels et al. 1994; Hoyer et al. 2002; Lanfumey and Hamon 2004; Pazos and Palacios 1985). The receptor subtypes 5-HT_{1A}, 5-HT_{1B} and 5-HT_{1D} may act either as autoreceptors, reducing the synthesis and release of serotonin, or as heteroreceptors, controlling the release of other neurotransmitters such as glutamate, gamma-aminobutyric acid (GABA), acetylcholine (ACh), dopamine (DA), and norepinephrine (NOR) (Barnes and Sharp 1999; Bockaert and Pin 1998; Hoyer et al. 2002; Pauwels 1997).

The family of 5-HT_2 receptors is composed of three different subtypes—5-HT_{2A}, 5-HT_{2B}, and 5-HT_{2C}—positively coupled to C-phospholipase and A2-phospholipase that change intracellular calcium levels (Barnes and Sharp 1999; Bockaert and Pin 1998; Hoyer et al. 2002; Saxena 1995; Van Oekelen et al. 2003). The 5-HT_{2A} and 5-HT_{2C} receptors subtypes are found in several brain areas such as the cortex, hippocampus, hypothalamus, nucleus accumbens (NAc), cerebellum, AMY, basal ganglia, and thalamus (Barnes and Sharp 1999; Hoyer et al. 2002; Leysen 2004; Van de Kar et al. 2001; Van Oekelen et al. 2003), whereas the 5-HT_{2B} receptors are found in the medial amygdala (MeA), the cerebellum, the septum, and the lateral hypothalamus (Hoyer et al. 2002; Duxon et al. 1997). In addition, there is a strong expression of 5-HT_{2C} receptor subtypes in the choroid plexus (Barnes and Sharp 1999; Hoyer et al. 2002; Leysen 2004; Van de Kar et al. 2001; Van Oekelen et al. 2003). Some studies have shown that activation of these receptors in the brain can interfere with the control of food intake (Foster-Schubert and Cummings 2006; Nonogaki et al. 2006), locomotion (Landry and Guertin 2004), anxiety (De Mello Cruz et al. 2005; Hackler et al. 2006; Wood 2003) and hormone secretion of ACTH, corticosterone, oxytocin (OT), renin, and prolactin (Bagdy et al. 1992; Van de Kar et al. 2001).

Unlike other serotonin receptors, 5-HT_3 receptors belong to the superfamily of ligand-gated ion channels (Barnes and Sharp 1999; Boess and Martin 1994; Faerber et al. 2007; Gaster and King 1997; Hoyer et al. 2002; Joshi et al. 2006; Melis et al. 2006; Saxena 1995). The 5-HT_3 receptor is a pentameric protein, and the five subunits have been cloned and classified from A to E (Hayrapetyan et al. 2005; Joshi et al. 2006; Maricq et al. 1991; Melis et al. 2006; Reeves and Lummis 2006), although only the subunits A and B appear to be present in functional 5-HT_3 receptors, constituting 5HT_3 receptors homo-pentamers 5-HT_{3A} subunit [i.e., (5HT_{3A})] and heteromers 5-HT_{3A} and 5-HT_{3B} subunit [i.e., 5-HT_{3AB}] (Maricq et al. 1991; Niesler et al. 2003). Activation of 5-HT_3 receptors leads to rapid depolarization by opening nonspecific cation channels allowing the entry of Na^+ and Ca^{2+} and the output of K^+ (Hoyer et al. 2002). These receptors are found in various regions of the CNS such as hippocampus, nucleus of the solitary tract (NTS), area postrema (AP), AMY, nucleus of the vagus dorsomotor, and cerebral cortex; all the same their physiological role remains elusive (Barnes and Sharp 1999; Hoyer et al. 2002; Morales et al. 1998).

The 5-HT_4 receptors are coupled to G-protein and positively coupled to adenylate cyclase increasing the intracellular levels of cAMP (Saxena 1995; Uphouse 1997; Barnes and Sharp 1999; Hoyer et al. 2002). Currently, 10 isoforms (A to J) of the 5-HT_4 receptor have been identified in three different species: rats, guinea pigs, and monkeys (Alex and Pehek 2007; Vilaró et al. 2005). In the CNS, these receptors are distributed heterogeneously with the highest expression being in the limbic and nigrostriatal areas (Alex and Pehek 2007; Patel et al. 1995; Vilaró et al. 2005; Waeber et al. 1996). The information about the physiological responses induced through activation of the central 5-HT_4 receptor are still scarce; however, it seems that these receptors may modulate the release of various neurotransmitters such as ACh, DA, GABA, and serotonin (Barnes and Sharp 1999; Hoyer et al. 2002).

The family of 5-HT_5 receptors is composed of two putative subtypes, referred to as 5-HT_{5a} and 5-HT_{5b} (Hoyer et al. 2002; Nelson 2004; Wesolowska 2002), which are G-protein–coupled (Carson et al. 1996; Hoyer et al. 2002; Nelson 2004; Wesolowska 2002). These receptors are found in the hippocampus, hypothalamus, AMY, striatum, thalamus, cerebellum and cortex, habenula, and DRN (Hoyer et al. 2002; Matthes et al. 1993; Oliver et al. 2000; Pasqualetti et al. 1998; Wesolowska 2002). The physiological function and the pharmacological characteristics of 5-HT_5 receptors remain to be clarified.

The 5-HT_6 receptor is coupled to the G-protein and positively coupled to adenylate cyclase, increasing the intracellular cAMP levels (Barnes and Sharp 1999; Hoyer et al. 2002; Mitchell and Neumaier 2005; Wesolowska 2002). These receptors are highly expressed in areas such as the striatum, NAc, olfactory tubercle, and cortex, and moderately expressed in the AMY, hypothalamus, thalamus, cerebellum, and hippocampus (Hoyer et al. 2002; Mitchell and Neumaier 2005; Wesolowska 2002). Some studies have shown that the blockade of 5-HT_6 receptors may modulate the release of other neurotransmitters such as DA, ACh, NOR, GABA, and glutamate (Mitchell and Neumaier 2005).

The 5-HT_7 receptor family is the most recently identified. These receptors are also positively coupled to G-protein and increase intracellular cAMP levels. The density of these receptors is high in the hypothalamus, thalamus, AMY, hippocampus, cortex, and DRN (Hoyer et al. 2002; Thomas and Hagan 2004; Wesolowska 2002). Four different isoforms of the 5-HT_7 receptor are expressed in rodents and humans. However, up to the present, these isoforms do not seem to differ in their pharmacological profile, signal transduction, or distribution. The 5-HT_7 receptors appear to play a physiological role in thermoregulation and circadian rhythm (Glass et al. 2003; Guscott et al. 2003). In addition, these receptors are present in glutamatergic neurons at the raphe nuclei, suggesting that they modulate the activity of serotonergic neurons (Harsing et al. 2004).

Although there are a large number of drugs for different subtypes of serotonin receptors, there is still much controversy regarding the selectivity of the agonists and antagonists so far described. The lack of strict selectivity of serotonergic agents has hampered a clear definition of the physiological role of many serotonin receptor subtypes. However, these pharmacological tools have contributed to identify the

TABLE 10.1
Pharmacological Data about the Agonists and Antagonists Serotonergic Presented in this Review Summarizing the Effects of Central Injections of Each Serotonergic Agent on Water Intake, Salt Intake and Blood Pressure

Drugs	Receptors Selectivity[a]	Name[b]	Brain Site of Injection	Effects		
				Water Intake	Salt Intake	Blood Pressure
		Agonists				
8-OH-DPAT	5-HT_{1A} (9.4)	7-(Dipropylamino)-5,6,7,8-tetrahydronaphthalen-1-ol	ICV	?	?	Decrease
	5-HT_{1D} (7.3)		LPBN	?	Increase	?
	5-HT_{7} (7.6)		PVN	?	Decrease	?
	5-HT_{1F} (5.8)		Septal area	?	Decrease	?
	5-ht_{5a} (5.7)		DRN	?	Increase	Decrease
	5-HT_{2A} (5.6)		MRN	?	?	Decrease
	5-HT_{2C} (5.6)		Raphe pallidus	?	?	Decrease
	5-ht_{1E} (5.5)		Raphe obscurus	?	?	Increase
	5-HT_{2B} (5.4)		Preoptic area	?	?	Increase
	5-HT_{1B} (6.2)		RVLM	?	?	Decrease
L-694,247	5-HT_{1A} (9.3)	*N*-[4-[[5-[3-(2-Aminoethyl)-1*H*-indol-5-yl]-1,2,4-oxadiazol-3-yl] methyl] phenyl]methanesulfonamide	ICV	Decrease	?	?
	5-HT_{1B} (9.2)					
	5-HT_{1D} (9.0)					

(*continued*)

TABLE 10.1 (Continued)
Pharmacological Data about the Agonists and Antagonists Serotonergic Presented in this Review Summarizing the Effects of Central Injections of Each Serotonergic Agent on Water Intake, Salt Intake and Blood Pressure

Drugs	Receptors Selectivity[a]	Name[b]	Brain Site of Injection	Effects		
				Water Intake	Salt Intake	Blood Pressure
		Agonists				
mCPP	$5\text{-}HT_{2B}$ (8.5)	1-(3-Chlorophenyl)piperazine	ICV	Decrease	Decrease	Increase
	$5\text{-}HT_{2C}$ (8.5)		MeA	?	NC	?
	$5\text{-}HT_{2A}$ (7.5)		CeA	?	NC	?
	$5\text{-}HT_{1D}$ (6.7)		LPBN	?	Decrease	?
	$5\text{-}ht_{1E}$ (5.4)					
MK-212	$5\text{-}HT_{2B}$ (6.8)	2-Chloro-6-piperazin-1-ylpyrazine	ICV	Decrease	?	?
	$5\text{-}HT_{2C}$ (7.0)					
	$5\text{-}HT_{2A}$ (6.0)					
DOI	$5\text{-}HT_{2A}$ (9.2)	1-(4-Iodo-2,5-dimethoxyphenyl) propan-2-amine	ICV	?	?	Increase
	$5\text{-}HT_{2B}$ (7.7)		LPBN	Decrease	?	?
	$5\text{-}HT_{2C}$ (8.6)					
	$5\text{-}HT_{1F}$ (5.8)					
	$5\text{-}ht_{1e}$ (5.8)					
DOB	$5\text{-}HT_{2A}$ (9.2)	1-(4-Bromo-2,5-dimethoxyphenyl) propan-2-amine	NTS	?	?	Decrease
	$5\text{-}HT_{2C}$ (8.9)					
	$5\text{-}HT_{2B}$ (7.6)					
Quipazine	$5\text{-}HT_{2B}$ (7.1)	2-Piperazin-1-ylquinoline	CVLM	?	?	Decrease
	$5\text{-}HT_{2A}$ (6.9)					
	$5\text{-}HT_{2C}$ (7.3)					

mCPBG	5-HT$_{3AB}$ (7.0) 5-HT$_{3A}$ (7.2)	2-(3-Chlorophenyl)-1-(diaminomethylidene)guanidine	ICV	Decrease	Decrease	Decrease
			MeA	?	Decrease	?
			CeA	?	Decrease	?
			LPBN	?	Increase	?
			NTS	?	?	Increase
			MS/vDB	?	?	NC
		Antagonists				
Methysergide	5-HT$_{2C}$ (9.1) 5-HT$_{2B}$ (9.4) 5-HT$_{1D}$ (8.9) 5-HT$_{2A}$ (8.4) 5-HT$_{1F}$ (8.2) 5-HT$_{7}$ (7.8) 5-HT$_{1B}$ (7.6) 5-ht5a (7.0) 5-ht$_{1e}$ (6.8) 5-HT$_{6}$ (6.8)	(4*R*,7*R*)-*N*-[(2*S*)-1-Hydroxybutan-2-yl]-6,11-dimethyl-6,11-diazatetracyclo[7.6.1.0^{2,7}.0{12,16}]hexadeca-1(16),2,9,12,14-pentaene-4-carboxamide	LPBN	?	Increase	?
SDZ-SER082	5-HT$_{2C}$ (8.1) 5-HT$_{2B}$ (6.7) 5-HT$_{2A}$ (6.3)	4-Methyl-4,9-diazatetracyclo[7.6.1.0^{2,7}.0^{12,16}]hexadeca-1(16),12,14-triene	ICV	NC	NC	NC
			MeA	?	Decrease	?
			CeA	?	NC	?
			LPBN	?	NC	?
LY-278,584	5-HT$_{3}$	1-Methyl-*N*-(8-methyl-8-azabicyclo[3.2.1]-oct-3-yl)-1*H*-indazole-3-carboxamide	ICV	NC	NC	?
			LPBN	?	NC	?
GR 113808	5-HT$_{4}$ (10.3)	[1-(2-Methanesulfonamidoethyl)piperidin-4-yl]methyl 1-methylindole-3-carboxylate	ICV	Decrease[c] Increase[d]	?	?

(continued)

TABLE 10.1 (Continued)
Pharmacological Data about the Agonists and Antagonists Serotonergic Presented in this Review Summarizing the Effects of Central Injections of Each Serotonergic Agent on Water Intake, Salt Intake and Blood Pressure

Drugs	Receptors Selectivity[a]	Name[b]	Brain Site of Injection	Effects		
				Water Intake	Salt intake	Blood Pressure
		Antagonists				
SB 204070	5-HT_4 (10.4)	(1-Butylpiperidin-4-yl)methyl 5-amino-6-chloro-2,3-dihydro-1,4-benzodioxine-8-carboxylate	ICV	Decrease[c] Increase[d]	?	?
Ondansetron	5-HT_{3A} (8.3) 5-HT_{3AB} (7.8)	9-Methyl-3-[(2-methylimidazol-1-yl) methyl]-2,3-dihydro-1*H*-carbazol-4-one	ICV	NC	NC	Increase
			MeA	?	NC	?
			CeA	?	NC	?
			MS/vDB	?	?	Increase
Ketanserin	5-HT_{2A} (9.7) α_{1B}-ADR (8.2) α_{1A}-ADR (8.2) α_{1D}-ADR (7.8) 5-HT_{1D} (7.5) 5-HT_{2C} (7.5) D_1 (6.7) 5-HT_{2B} (6.7) 5-HT_7 (6.5) D_5 (5.6) 5-HT_{1A} (5.0) 5-ht_{5a} (4.7) 5-HT_{1B} (5.4)	3-[2-[4-(4-Fluorobenzoyl)piperidin-1-yl] ethyl]-1*H*-quinazoline-2,4-dione	ICV	?	Decrease	?

SB269970	5-HT_7 (?) 5-ht_{5a} (?) α-ADR (?) D (?)	(2*R*)-1-[(3-Hydroxyphenyl)sulfonyl]-2-(2-(4-methyl-1-piperidinyl)ethyl) pyrrolidine	NTS	?	?	Increase

Note: CeA, central amygdala; CVLM, caudal ventrolateral medulla; DRN, dorsal raphe nucleus; ICV, intracerebroventricular; LPBN, lateral parabrachial nucleus; MeA, medial amygdala; MRN, median raphe nucleus; MS/vDB medial septum/vertical limb of the diagonal band complex; NC, no change; NTS, nucleus of the solitary tract; PVN, paraventricular nucleus; RVLM, rostral–ventrolateral medulla.

For the effect of serotonergic agents on water intake, we considered only the data obtained in experimental protocols specific to induce water intake; we did not include the data when water intake was associated with salt intake. For the effect of serotonergic agents on blood pressure, we considered only the data obtained in baseline condition.

[a] Receptor selectivity according to IUPHAR (International Union of Pure and Applied Chemistry) Database. The numbers between parentheses represent the pKi unit for the drug affinity for each serotonin receptor accordingto IUPHAR Database (Andrade et al. 2011; Peters et al. 2011).

[b] Chemical names of the drugs according to IUPAC rules.

[c] During hyperosmolarity protocol to induce water intake.

[d] During hypovolemia protocol to induce water intake.

functional role of central serotonin in the CNS. Table 10.1 summarizes the data on serotonergic agonists and antagonists that have been used to investigate the participation of serotonin receptors in the control of thirst, salt appetite, and blood pressure.

10.3 SEROTONIN AND FLUID INTAKE

10.3.1 Serotonin and Thirst

The role of serotonin in controlling hydroelectrolytic balance was first described in the 1970s when Tangaprégasson's group observed an increase in water intake and a natriuretic effect in rats after an electrolytic lesion of the DRN and MRN (Tangaprégasson et al. 1974). In addition, specific brain serotonin depletion with *p*-chlorophenylalanine methyl ester induced a dipsogenic effect (Reis et al. 1994). Furthermore, excitotoxic lesion of the DRN with ibotenic acid produces an increase in water intake in water-deprived rats (Cavalcante-Lima et al. 2005). These data favor the idea of an inhibitory role of the central serotonergic system in water intake. Interestingly, when administered by peripheral routes, serotonin, as well as its agonists, seems to induce an increase in water intake probably as a consequence of peripheral angiotensin II (AII) release (Rowland et al. 1994; Simansky 1995).

The use of specific antagonists and agonists for the different serotonin receptors has allowed the information to be refined on the role of the brain serotonergic pathways in the control of water intake. Here, we describe the participation of 5-HT_{1A}, 5-HT_{1D}, 5-HT_2, and 5-HT_3 receptors in the control of water intake.

Activation of central postsynaptic 5-HT_{1D} receptors located in regions around the third ventricle leads to a significant decrease in water intake both in dehydrated rats and in those receiving central injections of AII, carbachol (cholinergic agonist), and isoproterenol (β-adrenoceptor agonist) (De Castro-e-Silva et al. 1997). 5-HT_{1D} receptors are widely distributed throughout the rat brain, and they seem to inhibit serotonin release acting as autoreceptors (Hoyer et al. 1994) or mediate neurotransmitter release by nonserotonergic neurons when acting as postsynaptic heteroreceptors (Harel-Dupas et al. 1991). In the first case, the inhibitory effect of the 5-HT_{1D} agonist (L694,247) may be attributable to diminished serotonin release, indicating a stimulatory role of brain serotonin in drinking behavior. As most evidence suggests that central serotonergic pathways inhibit water intake (as mentioned above), it is possible that the effects observed after 5-HT_{1D} receptor stimulation may be due mainly to preferential activation of postsynaptic heteroreceptors. Activation of 5-HT_{1A} receptors in specific brain areas appears to exert the same inhibitory pattern on water intake, as observed after central stimulation of the 5-HT_{1D} receptors. Indeed, administration of 8-OH-DPAT (8-hydroxy-2-(di-*n*-propylamino)tetralin), a 5-HT_{1A} agonist, into the paraventricular nucleus (PVN) and lateral septal area (LSA) decreases water intake in water-deprived (24 h) animals (De Arruda Camargo et al. 2010b; De Souza Villa et al. 2008).

Intracerebroventricular administration of MK-212 (2-chloro-6-(1-piperazinyl) pyrazine), a serotonergic 5-HT_2 agonist, significantly reduces water intake induced by water deprivation and by beta-adrenergic, angiotensinergic, and cholinergic stimulation in rats (Reis et al. 1990a,b, 1992). Similarly, central 5-HT_2 receptor activation by mCPP (1-(3-chlorophenyl)piperazine) inhibits water intake in water-deprived, hyperosmotic, and hypovolemic rats (Castro et al. 2002a). It is interesting to note that mCPP may reduce ACh release in hippocampal synaptosomes (Bolaños-Jimenez et al. 1994; Harel-Dupas et al. 1991) and that 5-HT_2 receptors are present in GABAergic interneurons that normally inhibit ACh release in many areas of the brain (Morilak et al. 1993). Thus, it is valid to suggest that ACh inhibition may constitute a mechanism that could explain the reduction in water intake seen in hyperosmotic animals treated with mCPP. 5-HT_2 receptor activation at the lateral parabrachial nucleus (LPBN) by DOI (1-(4-iodo-2,5-dimethoxyphenyl)propan-2-amine(2*S*)-1-(4-iodo-2,5-dimethoxyphenyl)propan-2-aminev), a 5-$HT_{2A/2C}$ agonist, inhibits water intake induced by central AII stimulation (Menani and Johnson 1995). Ritanserin, a serotonergic antagonist that blocks 5-HT_2 receptors as well as other serotonergic receptors, has been used to treat sleep and mood disturbances and alcohol dependence (Bakish et al. 1993; Johnson et al. 1996; Reyntjens et al. 1986; Strauss and Klieser 1991). Intraperitoneal administration of ritanserin increases water intake in rats. This response seems to be dependent on brain adrenergic activity, because a locus coeruleus lesion abolishes the dipsogenic effect of ritanserin (Lu et al. 1992).

Differently from the predominant inhibitory role the other receptors have for water intake, activation of 5-HT_4 receptors facilitate AII effects and inhibit cholinergic pathways action on water intake (Castro et al. 2000). Additionally, it has been found that central blockade of 5-HT_4 receptors by third ventricle injections of two different selective 5-HT_4 antagonists, GR113808 and SB204070, enhances water intake of hypovolemic animals, but inhibits water intake of hyperosmotic animals (Castro et al. 2001). The effects of the 5-HT_4 antagonists suggests that these receptors may exert a positive drive on water intake due to hyperosmolarity and a negative one on fluid intake induced by hypovolemia, suggesting a dual role for 5-HT_4 in the control of thirst. Cholinergic pathways seem to trigger water intake after hyperosmolarity, whereas angiotensinergic pathways activate drinking after hypovolemia (Johnson and Thunhorst 1997). As the blockade of central 5-HT_4 receptors reduces water intake due to hyperosmolarity, it seems reasonable to suggest that 5-HT_4 activation somehow potentiates cholinergic circuitries related to thirst-generating mechanisms. There are anatomical and neurochemical data to support such an interaction, since connections between serotonergic and cholinergic pathways have been reported (Khateb et al. 1993). Central 5-HT_4 receptors increase cholinergic activity (Siniscalchi et al. 1999), and the role of 5-HT_4 receptors in memory processes seems to be mediated by some cholinergic step (Galeotti et al. 1998). Thus, it is not a surprise that the antagonist of 5-HT_4 receptors increased water intake induced by cholinergic activation; however, in contrast, it would be expected that the same antagonist would increase the water intake induced by hyperosmolarity, which was not the case. Similar reasoning applies to the effects of antagonists on thirst induced by hypovolemia versus that induced by AII. A possible explanation for these discrepancies between the effects of the two types of dehydration and exogenous activation of brain receptors may reside in the multiple mechanisms activated by dehydration.

The 5-HT_3 receptors are the only serotonergic receptors coupled to a voltage-gated channel. Acting on these receptors, serotonin induces depolarization by opening membrane ion channels, leading to fast modulatory actions on the release of several neurotransmitters (Derkach et al. 1989; Yakel and Jackson 1988) including serotonin itself (Blier et al. 1993). Third ventricle injections of *m*-CPBG (*meta*-chlorophenylbiguanide), a selective 5-HT_3 receptor agonist, significantly reduced water intake elicited by three different stimuli: hyperosmolarity, hypovolemia, and double dehydration (in water-deprived animals) (Castro et al. 2002b). Therefore, activation of central 5-HT_3 receptors inhibits drinking behavior to the different thirst inducers indistinctively, repeating the same pattern of inhibition of thirst shown by the activation of most brain serotonin receptors. The antidipsogenic effect of 5-HT_3 receptor stimulation during hyperosmolarity could involve cholinergic pathways because under these particular circumstances, intracellular loss of water supervenes and plasma hyperosmolarity is perceived by brain osmoreceptors triggering the activation of the central cholinergic pathways that leads to two essential corrective responses: water intake and natriuresis (Fitzsimons 1972).

Central 5-HT_3 receptor activation induces a decrease in ACh release in several preparations of cortical synaptosomes and enthorinal cortex in rats and humans (Barnes et al. 1989; Crespi et al. 1997; Maura et al. 1992). Thus, it is reasonable to suggest that water intake inhibition after central 5-HT_3 receptor stimulation may result from a decrease in the activity of the cholinergic pathways directly involved with the expression of drinking motivation and behavior.

The inhibition of water intake by injections of *m*-CPBG into the third ventricle of hypovolemic animals may be dependent on changes in the activity of the angiotensinergic pathways. Hypovolemia evokes a significant increase in circulating AII (Abdelaal et al. 1976) that acts on circumventricular structures, inducing drinking behavior (Johnson and Thunhorst 1997). To the best of our knowledge, no studies have been published concerning a direct neuroanatomical and/or neurochemical interaction between 5-HT_3 receptors and angiotensin in the nervous system. However, functional interaction between these systems seems to occur because central administration of *m*-CPBG reduces water intake in animals whose drinking is stimulated by the pharmacological activation of central angiotensinergic pathways (Castro et al. 2002b). Thus, it is reasonable to suggest that the stimulation of central 5-HT_3 receptors somehow disrupts the functional integrity of the mechanism(s) normally triggered by brain AII, which are essential for the expression of drinking behavior during hypovolemia. Furthermore, pharmacological stimulation of central 5-HT_3 receptors by *m*-CPBG was able to reduce water intake in fluid-deprived animals (Castro et al. 2002b). Under these circumstances, thirst is mainly generated by the action of blood-borne AII at the circumventricular structures and by activation of central cholinergic pathways. Indeed, the blockade of both angiotensinergic and cholinergic systems is necessary to inhibit water intake in dehydrated animals (Saavedra 1992).

10.3.1.1 Conclusion

Activation of most brain serotonin receptors results in an antidipsogenic action, albeit under different conditions. At least three types of 5-HT receptors—central 5-HT_{1D}, 5-HT_2, and 5-HT_3 receptors—inhibit water intake through a mechanism that involves the disruption of functional activity of brain angiotensinergic and cholinergic pathways. Conversely, the activation of central 5-HT_4 receptors promotes a dual effect, potentiating the dipsogenic effect of AII and inhibiting drinking induced by central cholinergic activation.

10.3.2 Serotonin and Salt Appetite

As discussed in Chapters 9 and 11, important sources of serotonin for the control of sodium appetite or "salt hunger" reside in the brainstem, particularly in the DRN. Such control seems more complex than just simple inhibition as suggested by the study of selective pharmacological intervention. Essentially the same three types of brain serotonergic receptors (5-HT_1, 5-HT_2, 5-HT_3) shown to control water intake also control sodium appetite, but their autoreceptor and heteroreceptor functions, controlling the release of serotonin and other neurotransmitters (Barnes and Sharp 1999), may mediate inhibition and facilitation of sodium appetite depending on which area of the brain they are activated.

Initial studies showed that peripheral administration of serotonergic agents such as dexfenfluramine (which enhances serotonin transmission) and fluoxetine (a serotonin uptake inhibitor) suppresses hypertonic saline intake in water-deprived and sodium-depleted rats and also decreases need-free sodium intake that occurs in the absence of any sodium deficit (Neil and Cooper 1989; Rouah-Rosilio et al. 1994). Later, the inhibitory effect of brain serotonin on salt appetite has been confirmed by intracerebroventricular administration of serotonergic agents, which decrease salt intake in sodium-depleted rats (Castro et al. 2003).

Injection of a 5-HT_1 agonist, 8-OH-DPAT, into the lateral septal area and PVN reduces hypertonic saline solution intake in sodium-depleted animals (De Arruda Camargo et al. 2010a; De Souza Villa et al. 2007). The effect in the PVN is consistent with an increase in c-Fos expression in both serotonergic neurons in the DRN and in oxytocinergic cells in the PVN in response to sodium intake induced by peritoneal dialysis (Godino et al. 2007). Recall that oxytocin also inhibits sodium appetite (Antunes-Rodrigues et al. 2004; Chapters 3 and 9, but see Fitts et al. 2003). Additionally, 5-HT_1 receptors are expressed throughout the magnocellular regions of the PVN in the oxytocinergic neurons, and activation of these receptors in the PVN increases plasma OT (Jorgensen et al. 2003; Zhang et al. 2004). On the other hand, injections of 8-OH-DPAT into the LPBN or DRN increase salt intake, probably by inhibiting serotonin release (De Gobbi et al. 2005; Fonseca et al. 2009).

The brain activation of 5-HT_2 receptors by intracerebroventricular administration of the agonist mCPP reduces the salt intake induced by sodium depletion. This inhibitory effect is blocked by a specific antagonist, SDZ-SER 082 ((+)-*cis*-4,5,7*a*,8,9,10,11,11*a*-octahydro-7*H*-10-methylindolo[1,7-*bc*][2,6]-naphthyridine). However, when the antagonist is administered alone, it does not produce any change in salt intake in sodium-depleted rats (Castro et al. 2003). On the other hand, a nonspecific 5-HT_2 antagonist, ketanserin, when injected intraperitoneally or intracerebroventricularly, inhibited salt appetite induced by deoxycorticosterone acetate or by sodium depletion (Gentili et al. 1991). This contradiction may be explained by the difference in the pharmacological agents used and in the protocol applied to induce salt appetite.

A diverse scenario is observed when 5-HT_2 receptors are activated in specific brain areas. Activation of these receptors at the MeA by injection of the 5-HT_2 receptors agonist, mCPP, produces no change in salt intake in sodium-depleted rats. However, administration of the 5-HT_2 receptors antagonist, SDZ-SER 082, reduces sodium appetite in this model (Luz et al. 2006). These data suggest the existence of a facilitatory drive of serotonin acting on 5-HT$_2$ receptors in the MeA. On the other hand, administration of the same agonist at the LPBN decreases salt intake in sodium-depleted rats. Injection of the SDZ-SER 082 blocks the inhibitory action of mCPP in the LPBN, but when injected alone no effect is found (De Gobbi et al. 2007).

The reciprocal connection between LPBN and CeA may play a role in the regulation of salt intake (Jhamandas et al. 1996; Norgren 1995; Takeuchi et al. 1982). Bilateral lesions of the central amygdala (CeA) reduce salt intake induced by LPBN injections of methysergide (a nonspecific serotonergic antagonist) (Andrade-Franzé et al. 2010). Curiously, when mCPP and SDZ-SER 082 are injected into the CeA they are unable to alter salt intake in sodium-depleted rats (Luz et al. 2007). It may be feasible to conclude that 5-HT$_2$ receptors at the CeA do not participate in the control of salt appetite. However, the functional integrity of the CeA may be important for the facilitatory action of methysergide, a 5-HT$_{2A/2C}$ antagonist, at the LPBN on salt intake (Andrade-Franzé et al. 2010). In addition, administration of methysergide into the LPBN has been shown to increase the expression of c-*fos* in areas related to salt intake such as bed nucleus of the stria terminalis (BST), subfornical organ (SFO), supraoptic nucleus (SON), organ vasculosum of the lamina terminalis (OVLT), NTS, median preoptic nucleus (MnPO), PVN, and AP (Davern and McKinley 2010). The inhibitory action of 5-HT$_2$ receptors at the LPBN appears to depend on the atrial natriuretic peptide (ANP) and OT, but not on alpha$_2$-adrenergic/imidazoline receptors in the forebrain areas (Margatho et al. 2002, 2007).

The pharmacological stimulation of the central 5-HT$_3$ receptors through an intracerebroventricular injection of *m*-CPBG reduces salt intake in sodium-depleted rats. Pretreatment with the 5-HT$_3$ antagonist, LY-278,584 (1-methyl-*N*-(8-methyl-8-azabicyclo[3.2.1]-oct-3-yl)-1*H*-indazole-3-carboxamide), eliminates the antinatriorexigenic response of *m*-CPBG, although intracerebroventricular injections of LY-278,584 alone do not change salt intake in this model (Castro et al. 2003). A reduction in salt intake in sodium-depleted rats is also observed after the pharmacological activation of 5-HT$_3$ receptors located in the MeA and CeA, an effect impaired by pretreatment with the selective 5-HT$_3$ receptor antagonist, ondansetron (Luz et al. 2006, 2007). Activation of 5-HT$_3$ receptors in the LPBN by PBG [1-(diaminomethylidene)-2-phenylguanidine] injections increases salt intake, whereas pretreatment with the 5-HT$_3$ receptor antagonist (LY-278,584) eliminates the effects of *m*-CPBG, whereas LY-278,584 has no effect on sodium intake when injected by itself (De Gobbi et al. 2007). The binding of 5-HT$_3$ receptors has been demonstrated in several brainstem and forebrain nuclei, and these receptors may be located at both pre- and postsynaptic terminals, modulating the release of neurotransmitters such as DA, ACh, glutamate, and GABA (for review see Barnes and Sharp 1999; Chameau and Van Hooft 2006; Pratt et al. 1990). Therefore, it is reasonable to propose that serotonergic modulation of those neurotransmitters in the different brain areas may be one of the mechanisms modulating the control of salt intake in the CNS. However, further studies are required to confirm this hypothesis.

10.3.2.1 Conclusion

The brain serotonergic system participates through 5-HT$_1$, 5-HT$_2$, and 5-HT$_3$ receptors acting on multiple locations to control sodium appetite. Stimulation of 5-HT$_1$

and 5-HT_3 receptors in certain areas of the forebrain (septal area, PVN, MeA, CeA) inhibits salt intake but enhances salt intake at the LPBN and DRN. On the other hand, activation of 5-HT_2 receptors at the MeA presents a facilitatory drive, whereas at the LPBN this drive is inhibitory.

It is important to recall that neurophysiological and behavioral responses may be different in sodium-depleted and sodium-replete animals (Hill and Mistretta 1990; Jacobs et al. 1988; Schulkin 1982). Most of the studies used sodium depletion protocols to induce salt intake for investigation of the participation of brain serotonergic receptors in the control of salt intake. Therefore, brain serotonin seems to be important in the homeostatic regulation of body sodium, and further studies should be performed to investigate the role of serotonin in "need-free salt intake." Serotonin has been implicated in the reward mechanism, particularly the 5-HT_1, 5-HT_2, and 5-HT_3 receptors acting at the NAc, ventral tegmental area, and AMY. It would appear that 5-HT_2 receptors usually inhibit reward-related behaviors, whereas 5-HT_3 receptors reinforce properties of drug abuse (for review, see Hayes and Greenshaw 2011). Therefore, it is reasonable to suggest that the brain's serotonergic system controls behavior associated with reward-related salt appetite, but further studies are required to confirm this hypothesis.

10.4 SEROTONIN AND BLOOD PRESSURE

Blood volume, cardiac output, arterial resistance, and blood pressure (BP) adapt to different metabolic situations ensuring adequate blood flow for microcirculation. An important factor controlling effective circulating volume is the amount of sodium in the ECF. Changes in blood volume may result in alterations in BP and tissue perfusion. Thus, central integrated mechanisms to control water and sodium intake and BP are essential for life (Stachenfeld 2008; Stricker and Sved 2000). Indeed, an overlapping brain network controlling water and sodium intake and blood pressure has been shown (Dampney et al. 2002; De Gobbi et al. 2008). Among the brain neurotransmitters, serotonin has been implicated in the control of all these parameters. In this section, we review the role of the different types of serotonin receptors in the control of cardiovascular function and present a summary of the results produced in our laboratory on this subject.

In the periphery, serotonin affects cardiovascular system targeting the heart and blood vessels. The effects of the systemic administration of serotonin on the cardiovascular system are mediated mainly by five serotonin receptors: 5-HT_1, 5-HT_2, 5-HT_3, 5-HT_4, and 5-HT_7 (De Vries et al. 1996; Martin 1994; Saxena 1995; Saxena and Villalón 1991; Villalón et al. 1996, 1997a). Usually, an intravenous injection of serotonin induces a triphasic response characterized by initial hypotension followed by an increase in BP and then a stable hypotension (Côté et al. 2004; Villalón et al. 1997b). The initial hypotension is induced by a bradycardic reflex (von Bezold-Jarisch reflex) mediated by activation of 5-HT_3 receptors present at afferent cardiac nerve endings of the vagus (Yusuf et al. 2003). The pressor phase seems to be a result of the vasoconstriction induced by activation of 5-HT_2 receptors in blood vessels and of the positive ionotropic and chronotropic effects mediated by 5-HT_4 receptor activation in cardiac myocytes (Côté et al. 2004). The final hypotensive phase may involve the activation of different serotonin receptors at both central and peripheral level (Villalón et al. 1997b). The increase in BP obtained by intravenous administration of serotonin may be dependent on the integrity of circumventricular organs such as the SFO and the OVLT. In fact, a lesion in the anteroventral region of the third ventricle attenuates the hypertensive response induced by peripheral administration of serotonin (Muntzel et al. 1996).

Brain administration of serotonin and its analogs produces variegated responses: bradycardia or tachycardia, hypotension or hypertension, and vasodilatation or vasoconstriction (Kuhn et al. 1980). The effect of central serotonin stimulation on the cardiovascular system is mediated mainly by 5-HT_1, 5-HT_2, and 5-HT_3 receptors (Ramage and Villalón 2008).

Some contradictory findings have been published in the literature concerning the effect of 5-HT_{1A} receptors on cardiovascular control. Stimulation of brain 5-HT_{1A} receptor produces hypotension and bradycardia in normotensive and spontaneously hypertensive rats (Buisson-Defferier and Van de Buuse 1992; Dabiré et al. 1987). The hypotension and bradycardia induced by central activation of 5-HT_{1A} receptor may be the result of either the stimulation of heteroreceptors in the rostral–ventrolateral medulla (RVLM) (Clement and McCall 1990; Valenta and Singer 1990) or autoreceptors in the DRN (Connor and Higgins 1990). Additionally, the hypotension and bradycardia may be mediated by a decrease in sympathetic activity (Fozard et al. 1987; Gradin et al. 1985) and an increase in vagal tone to the heart (McCall and Clement 1994; McCall et al. 1987). Depending on the brain area, the stimulation 5-HT_{1A} receptors may trigger different sympathetic responses. Indeed, injections of the agonist 8-OH-DPAT into raphe magnus and pallidus, dorsal raphe, and rostral ventrolateral medulla inhibit sympathetic activity and decrease blood pressure (McCall and Clement 1994), whereas injections into the raphe obscurus (Dreteler et al. 1991) and preoptic area (Szabo et al. 1998) promote an increase in sympathetic activity and blood pressure.

The importance of 5-HT_{1A} receptors in the control of cardiac function has been reevaluated. Some of the cardiovascular effects attributed to the stimulation of 5-HT_{1A} receptors may be attributable to the activation of other serotonergic receptors such as the 5-HT_7 receptors (Barnes and Sharp 1999; McCall and Clement 1994). The main agonist of 5-HT_{1A} receptors available, 8-OH-DPAT, may also interact with other receptors such as the 5-$HT_{1B/1D}$, 5-$HT_{2A/B/C}$, and 5-HT_7 serotonergic receptors (Hoyer et al. 1994; Knight et al. 2004; Krobert et al. 2001). The use of more selective procedures targeting serotonin receptors, such as drugs and genomic interventions, will help to clarify the role of 5-HT_1 receptors in cardiovascular function. The peripheral activation of 5-HT_2 receptors promotes generalized sympathoexcitation and a rise in blood pressure. Some results in the literature suggest that this may be attributable to an increase in total peripheral resistance, because stimulation of the 5-HT_2 receptors, located in the vascular endothelial cells, promotes arterial vasoconstriction (Chandra and Chandra 1993; Dabiré et al. 1990; Meller et al. 1992). Furthermore, peripheral activation of the 5-HT_2 receptors has been shown to induce positive chrono- and inotropic effects and to inhibit ANP secretion leading to an increase in cardiac output and blood volume that may also contribute to the increase in BP (Cao et al. 2003; Laer et al. 1998; Mertens et al. 1993).

The definition of 5-HT_2 receptors actions at the brain on cardiovascular control is difficult, because the data obtained in experiments using cats and rats are contradictory. In rats, 5-HT_2 receptor activation causes sympathetic inhibition, whereas in cats it enhances sympathetic activity (Anderson et al. 1992, 1995). Also, central 5-HT_2 receptor activation in rats, but not in cats, increases circulating vasopressin (AVP) levels (Brownfield et al. 1988; Montes and Johnson 1990; Steardo and Iovino 1986). In view of the fact that AVP enhances baroreflex sensitivity (Koshimizu et al. 2006; Oikawa et al. 2007), it seems that the initial conclusions of sympathoinhibition based on Anderson's experiments in rats are precipitated. In fact, the 5-HT_2 receptors activation induces sympathoexcitation, which is masked by the concomitant AVP release and its consequent baroreflex-mediated sympathoinhibition.

The cardiovascular effects induced by central injection of 5-HT_2 receptor agonists have been attributed to the 5-HT_{2A} receptor subtype (Ramage 2001). However, it is not unexpected that many of the cardiovascular effects originally ascribed to 5-HT_{2A} receptors could be due to activation of the 5-HT_{2C} or 5-HT_{2B} receptors, because the 5-HT_2 receptor subtypes are to a great extent homologous, with pharmacological similarities (Giorgetti and Tecott 2004). Of the 5-HT_2 receptor agonists, DOI and mCPP have been frequently used in both experimental and clinical studies to investigate the functional role of 5-HT_{2A} and 5-HT_{2C} receptors, although they are not selective.

The 5-HT_{2C} receptors in the brain are widely distributed throughout the CNS (Clement et al. 2000), and activation of 5-HT_2 receptors located in prosencephalic areas is associated with a sympathoexcitatory and hypertensive response, whereas the opposite effect is seen after activation of the 5-HT_2 receptors located in the hindbrain. Third ventricle injections of 5-HT_{2C} receptor agonist, mCPP in rats produces an increase BP attenuated by the antagonist SDZ-SER 082. Also, third ventricle injections of SDZ-SER 082 significantly blunted stress-induced hypertension without modifying the increase in HR induced by the stress (Ferreira et al. 2005). Because the administration of the SDZ-SER 082 alone in nonstressed rats failed to induce any significant change in BP, it is worthwhile to suggest that 5-HT_{2C} receptors appear not to exert endogenous, tonic, modulatory role in the control of blood pressure, under nonstress conditions. On the other hand, under stress conditions, the activation of 5-HT_{2C} receptors became relevant to cardiovascular coping-stress response.

The increase in BP produced by mCPP was sustained for several minutes and accompanied by an initial sharp decrease followed by a sustained increase in HR (Ferreira et al. 2005). This probably means that the initial phase of mCPP-induced hypertensive response triggers baroreflex-mediated bradycardia. On the other hand, the coexistence of hypertension and tachycardia in the subsequent phase indicates a hypertensive, sympathoexcitatory drive, in which baroreflex inhibition of HR is suppressed. It is possible that such drive overcomes the baroreflex-mediated sympathoinhibition associated with AVP, which is also released in response to third ventricular injection of mCPP (Jorgensen et al. 2003; Knowles and Ramage 2000; McCall and Clement 1994).

Third ventricle injections of the selective 5-HT_3 receptor agonist, *m*-CPBG, in rats cause a significant decrease in BP without any change in HR. The opposed effects are observed with 5-HT_3 receptor blockade by third ventricle injections of the selective 5-HT$_3$ receptor antagonist, ondansetron. These data seem to indicate that the stimulation of brain 5-HT$_3$ receptors induces a fall in BP through a decrease in sympathetic activity. Indeed, central activation of the 5-HT$_3$ receptor inhibits the baroreflex-mediated tachycardia; conversely, baroreflex-mediated bradycardia is maintained, indicating normal parasympathetic activity in these animals. These data suggest that central serotonin acting on 5-HT$_3$ receptors promotes a tonic inhibitory drive on blood pressure.

The pharmacological activation of central 5-HT$_3$ receptors by *m*-CPBG was able to impair stress-induced hypertensive response and inhibit tachycardic response in a dose-dependent manner (Ferrcira et al. 2004). This allows us to conclude that central 5-HT$_3$ receptor activation may efficiently exert an inhibitory drive on acute stress-induced changes in cardiovascular function. On the other hand, in stressed animals receiving a third ventricle injection of ondansetron, there was an increase in BP that was not significantly different from that found in saline-treated, stressed controls (Ferreira et al. 2004). This may mean that the endogenous inhibitory drive in blood pressure regulation exerted by serotonin via central 5-HT$_3$ receptors as observed in nonstressed rats is somehow suppressed during stress. Alternatively, during stress, the sum of the many integrated hypertensive drives exerted by different neurochemical components largely exceeds the 5-HT$_3$ receptor–dependent inhibitory effect on blood pressure. A similar inhibitory effect on hypertensive response to stress is observed after injection of *m*-CPBG into the medial septum/vertical limb of diagonal band (MS/vDB), but it failed to cause any significant change in resting blood pressure (Urzedo-Rodrigues et al. 2011). This suggests that, in this particular region, 5-HT$_3$ receptors generating a tonic inhibitory drive on blood pressure are already fully activated. Therefore, the inhibitory 5-HT$_3$ receptor–dependent drive exerted by the serotonergic pathways located in the MS/vDB plays a crucial role in maintaining blood pressure within its physiological range (Urzedo-Rodrigues et al. 2011). Typical behavior referred to as a defense reaction is triggered by different stressful conditions and the dorsomedial nucleus of the hypothalamus and the dorsal part of the periaqueductal gray appear to be the key sites in the brain involved in this response (for reviews, see Depaulis et al. 1994; DiMicco et al. 2002). 5-HT$_3$ receptors located at the NTS are essential for the bradycardic reflex observed during the defense reaction (Comet et al. 2004, 2005; Sévoz-Couche et al. 2003).

It appears that an increase in the sympathetic drive may explain the rise in blood pressure observed after blockade of the 5-HT$_3$ receptors in the MS/vDB by ondansetron. Indeed, administration of prazosin, an alpha1-adrenoceptor blocker, reverses the hypertensive response induced by the injection of ondansetron into the MS/vDB. An additional mechanism by which the blockade of 5-HT$_3$ receptors located in the MS/vDB induces a hypertensive response may be related to the increase in brain angiotensinergic activity. Injection of losartan, an AT1 receptor antagonist, into the MS/vDB blunted the previously detected hypertensive response induced by ondansetron (Urzedo-Rodrigues et al. 2011). Taken together, these data suggest that 5-HT$_3$ receptors at this brain level exert a tonic inhibition of local release of AII and sympathoinhibition.

The hypotension induced by central 5-HT$_3$ receptor activation may result from an interaction with different brain neurotransmitter pathways such as GABAergic, glutamatergic, and opiatergic (Chameau and Van Hooft 2006; Chu et al. 2009). Lateral ventricle injection of mu, kappa, or delta opioid receptor antagonists impairs the hypotensive response to central 5-HT_3 receptor stimulation (Fregoneze et al. 2011). These opioid receptors have been identified in the cell body, as well as in axon terminals and at synaptic terminals. Their activation may change the spike duration controlling Ca^{2+} influx, thus inhibiting neurotransmitter release (Chandra and Chandra 1993; Schoffelmeer et al. 1992a,b). It is possible that the release of opioid peptides induced by 5-HT_3 receptor activation inhibits neurotransmitter release that controls sympathetic tonus and normal blood pressure. However, the intrinsic cellular mechanism by which 5-HT_3 and opioid receptors interact remains to be established.

Another serotonergic receptor that has more recently been studied in cardiovascular function is the 5-HT_7 receptor. However, there is no selective 5-HT_7 receptor agonist available and the antagonist (SB269970) that has been used also binds to the 5-HT_{5A} receptors, alpha-adrenergic receptors, and dopaminergic receptors (Foong and Bornstein 2009; Lovell et al. 2000). The exact physiological role of the 5-HT_7 receptor remains to be clarified. It has been observed that central administration of SB269970 blocks the bradycardic reflex induced by baroreceptor, cardiopulmonary, and chemoreceptor stimulation (Damaso et al. 2007; Kellett et al. 2005). 5-HT_7 receptors may also exert a functional effect on sleep and thermoregulation (Hedlund and Sutcliffe 2004; Hedlund et al. 2003), as well as on antidepressant-like activity in the forced swim test (Guscott et al. 2003; Hedlund et al. 2005). Acute restraint stress up-regulates 5-HT_7 receptor messenger RNA in the rat hippocampus (Yau et al. 2001). In addition, a possible role of 5-HT_7 receptors at the NTS in cardiovascular control during stressful conditions has been suggested, because injection of SB269970 into the cisterna magna increased BP without altering HR (Ramage and Villalón 2008).

10.4.1 Conclusion

The physiological role of brain serotonin receptors on cardiovascular function has been studied over the past decade; however, the participation of all the different subtypes of serotonin receptors on this function remains to be clarified. Usually, systemic serotonin activity seems to induce a triphasic response characterized by initial hypotension followed by an increase in BP and then a phase of stable hypotension. The brain action of serotonin on cardiovascular function appears to depend on the sites of the brain and on the receptor subtype activated. Stimulation of central 5-HT_{1A} receptors by injection of an agonist into the lateral ventricle, dorsal raphe, raphe magnus and pallidus, and the rostral ventrolateral medulla decreases blood pressure, while promoting an increase in blood pressure by injection into the raphe obscurus and the preoptic area. Central activation of 5-HT_2 receptors located in prosencephalic areas is linked to a sympathoexcitatory, hypertensive response, whereas the opposite effect is seen after activation of the 5-HT_2 receptors located in the hindbrain. These receptors also seem to participate in cardiovascular control during stressful conditions, because the blockade of central 5-HT_2 receptors blunted stress-induced hypertension. Pharmacological activation of central 5-HT_3 receptors at MS/vDB has a tonic sympathoinhibitory effect, while causing sympathoexcitation at the NTS. Additionally, the sympathoinhibitory effect of 5-HT_3 receptors at the MS/vDB appears to involve an inhibition of the release of angiotensin in this area. Furthermore, the interaction between serotonergic and opiatergic pathways may be functionally important in the tonic inhibitory drive on blood pressure exerted by 5-HT_3 receptors at prosencephalic areas. The role of central 5-HT_7 receptors in cardiovascular control remains unclear, with clarification dependent on the development of new appropriate pharmacological tools.

10.5 CONCLUDING REMARKS

Several studies have shown that central serotonergic pathways are important for the maintenance of hydroelectrolytic balance and cardiovascular function. The functional role of serotonin in these parameters seems to be dependent on the brain area stimulated and subtype of receptor activated. Furthermore, the physiological responses induced by serotonin may involve interaction with different neurotransmitters. Although many studies have been conducted in an attempt to clarify the role of serotonin and its receptors in cardiovascular function and hydroelectrolytic balance, several questions remain to be answered. The main problem concerns the lack of appropriate pharmacological tools for each subtypes of serotonin receptor and the use of different experimental protocols that make it very difficult to draw a clear picture of the role of serotonin receptors in the different brain areas on salt appetite, drinking behavior, and cardiovascular control. Although there are several limitations in the techniques used so far, important results have been obtained with serotonergic drugs selective for different receptors on the cardiovascular function and hydroelectrolytic balance.

REFERENCES

Abdelaal, A. E., P. F. Mercer, and G. J. Mogenson. 1976. Plasma angiotensin II levels and water intake following β-adrenergic stimulation, hypovolemia, cellular dehydration and water deprivation. *Pharmacol Biochem Behav* 4:317–21.

Alex, K. D., and E. A. Pehek. 2007. Pharmacologic mechanisms of serotonergic regulation of dopamine neurotransmission. *Pharmacol Ther* 113:296–20.

Anderson, I. K., G. R. Martin, and A. G. Ramage. 1992. Central administration of 5-HT activates 5-HT1A receptors to cause sympathoexcitation and 5-HT2/5-HT1C receptors to release vasopressin in anaesthetized rats. *Br J Pharmacol* 107:1020–8.

Anderson, I. K., G. R. Martin, and A. G. Ramage. 1995. Evidence that activation of 5-HT2 receptors in the forebrain of anaesthetized cats causes sympathoexcitation. *Br J Pharmacol* 116:1751–6.

Andrade-Franzé, G. M., C. A. Andrade, L. A. De Luca Jr., P. M. De Paula, and J. V. Menani. 2010. Lateral parabrachial nucleus and central amygdala in the control of sodium intake. *Neuroscience* 165:633–41.

Antunes-Rodrigues, J., M. De Castro, L. L. Elias, M. M. Valença, and S. M. McCann. 2004. Neuroendocrine control of body fluid metabolism. *Physiol Rev* 84:169–208.

Bagdy, G., K. T. Kalogeras, and K. Szemeredi. 1992. Effect of 5-HT1C and 5-HT2 receptor stimulation on excessive grooming, penile erection and plasma oxytocin concentrations. *Eur J Pharmacol* 229:9–14.

Bakish, D., Y. D. Lapierre, R. Weinstein et al. 1993. Ritanserin, imipramine, and placebo in the treatment of dysthymic disorder. *J Clin Psychopharmacol* 13:409–14.

Barnes, J. M., N. M. Barnes, N. M. Costall, R. J. Naylon, and M. B. Tyers. 1989. 5-HT3 receptors mediate inhibition of acetylcholine release in cortical tissue. *Nature* 338:762–3.

Barnes, N. M., and T. Sharp. 1999. A review of central 5-HT receptors and their function. *Neuropharmacology* 38:1083–152.

Blier, P., P. J. Monroe, C. Bouchard, D. L. Smith, and D. J. Smith. 1993. 5-HT3 receptors which modulate [^{3}H] 5-HT release in the guinea-pig hypothalamus are not autoreceptors. *Synapse* 15:143–8.

Bockaert, J., and J. P. Pin. 1998. Use of a G-protein-coupled receptor to communicate: An evolutionary success. *C R Acad Sci III* 321:529–51.

Boess, F. G., and I. L. Martin. 1994. Molecular biology of 5-HT receptors. *Neuropharmacology* 33:275–317.

Bolaños-Jiménez, F., R. Manhaes de Castro, and G. Fillon. 1994. Effect of chronic antidepressant treatment on 5-HT1B presynaptic hetero-receptors inhibiting acetylcholine release. *Neuropharmacology* 33:77–81.

Brownfield, M. S. J., Greathouse, S. A. Lorens, J., Armstrong, J. H., Urban, and L. D. Van de Kar. 1988. Neuropharmacological characterization of serotoninergic stimulation of vasopressin secretion in conscious rats. *Neuroendocrinology* 47:277–83.

Bruinvels, A. T., B. Landwehrmeyer, E. L. Gustafson et al. 1994. Localization of 5-HT1B, 5-HT1D alpha, 5-HT1E and 5-HT1F receptor messenger RNA in rodent and primate brain. *Neuropharmacology* 33:367–86.

Buisson-Defferier, S., and M. Van den Buuse. 1992. Cardiovascular effects of the 5-HT1A receptor ligand, MDL 73005EF, in conscious spontaneously hypertensive rats. *Eur J Pharmacol* 223:133–41.

Cao, C., J. H. Han, S. Z. Kim, K.W. Cho, and S. H. Kim. 2003. Diverse regulation of atrial natriuretic peptide secretion by serotonin receptor subtypes. *Cardiovasc Res* 59:360–8.

Carson, M. J., E. A. Thomas, P. E. Danielson, and J. G. Sutcliffe. 1996. The 5HT5A serotonin receptor is expressed predominantly by astrocytes in which it inhibits cAMP accumulation: A mechanism for neuronal suppression of reactive astrocytes. *Glia* 17:317–26.

Castro, L., R. Athanazio, M. Barbetta, A. C. Ramos et al. 2003. Central 5-HT2B/2C and 5-HT3 receptor stimulation decreases salt intake in sodium-depleted rats. *Brain Res* 981:151–9.

Castro, L., E. De Castro-e-Silva, C. P. Luz et al. 2000. Central 5-HT4 receptors and drinking behavior. *Pharmacol Biochem Behav* 66:443–8.

Castro, L., I. Maldonado, I. Campos et al. 2002a. Central administration of *m*-CPP, a serotonin 5-HT2B/2C agonist, decreases water intake in rats. *Pharmacol Biochem Behav* 72:891–8.

Castro, L., B. Varjão, I. Maldonado et al. 2002b. Central 5-HT3 receptors and water intake in rats. *Physiol Behav* 77:349–59.

Castro, L., B. Varjão, I. Silva et al. 2001. Effect of intracerebroventricular administration of GR113808, a selective 5HT4 antagonist, on water intake during hyperosmolarity and hypovolemia. *Brazilian J Med Biol Res* 34:791–796.

Cavalcante-Lima, H. R., H. R. Lima, R. H. Costa-e-Sousa et al. 2005. Dipsogenic stimulation in ibotenic DRN-lesioned rats induces concomitant sodium appetite. *Neurosci Lett* 374:5–10.

Chameau, P., and J. A. Van Hooft. 2006. Serotonin 5HT3 receptors in the central nervous system. *Cell Tissues Res* 326:573–81.

Chandra, M., and N. Chandra. 1993. Serotonergic mechanisms in hypertension. *Int J Cardiol* 42:189–96.

Chu, L. F., D. Y. Liang, X. Li et al. 2009. From mouse to man: The 5-HT3 receptor modulates physical dependence on opioid narcotics. *Pharmacogenet Genomics* 19:193–205.

Clement, D. A., T. Punhani, D. Y. Duxon, T. P. Blackburn, and K. C. F. Fone. 2000. Immunohistochemical localization of the 5-HT2C receptor protein in the rat CNS. *Neuropharmacology* 39:123–32.

Clement, M. E., and R. B. Mccall 1990. Studies on the site and the mechanism of the sympatholytic action of 8-OH-DPAT. *Brain Res* 525:232–41.

Comet, M. A., R. Laguzzi, M. Hamon, and C. Sévoz-Couche. 2005. Functional interaction between nucleus tractus solitarius NK1 and 5-HT3 receptors in the inhibition of baroreflex in rats. *Cardiovasc Res* 65:930–9.

Comet, M. A., C. Sévoz-Couche, N. Hanoun, M. Hamon, and R. Laguzzi. 2004. 5-HT-mediated inhibition of cardiovagal baroreceptor reflex response during defense reaction in the rat. *Am J Physiol Heart Circ Physiol* 287:H1641–9.

Connor, H. E., and G. E. Higgins. 1990. Cardiovascular effects of 5-HT1A receptor agonists injected into the dorsal raphe nucleus of conscious rats. *Eur J Pharmacol* 182:63–72.

Côté, F., C., Fligny, Y. Fromes, J. Mallet, and G. Vodjdani. 2004. Recent advances in understanding serotonin regulation of cardiovascular function. *Trends Mol Med* 10:233–8.

Crespi, D., M. Gobbi, and T. Mennini. 1997. 5-HT3 serotonin heteroreceptors inhibit [^{3}H] acetylcholine release in rat cortical synaptosomes. *Pharmacol Res* 35:351–4.

Dabiré, H., C. Cherqui, B. Fournier, and H. Schmitt. 1987. Comparison of effects of some 5-HT1 agonists on blood pressure and heart rate of normotensive anaesthetized rats. *Eur J Pharmacol* 140:259–66.

Dabiré, H., C. Cherqui, M. Safar, and H. Schmitt. 1990. Haemodynamic aspects and serotonin. *Clin Physiol Biochem* 3:56–63.

Damaso, E. L., L. G. Bonagamba, D. O. Kellett, D. Jordan, A. G. Ramage, and B. H. Machado. 2007. Involvement of central 5-HT7 receptors in modulation of cardiovascular reflexes in awake rats. *Brain Res* 1144:82–90.

Dampney, R. A., M. J. Coleman, M. A. Fontes et al. 2002. Central mechanisms underlying short- and long-term regulation of the cardiovascular system. *Clin Exp Pharmacol Physiol* 29:261–8.

Davern, P. J., and M. J. McKinley. 2010. Forebrain regions affected by lateral parabrachial nucleus serotonergic mechanisms that influence sodium appetite. *Brain Res* 1339:41–8.

De Arruda Camargo, G. M., L. A. A. De Arruda Camargo, and W. A. Saad. 2010a. Role of serotonergic 5-HT1A and oxytocinergic receptors of the lateral septal area in sodium intake regulation. *Behav Brain Res* 209:260–6.

De Arruda Camargo, G. M. P., L. A. A. de Arruda Camargo, and W. A. Saad. 2010b. On a possible dual role for the lateral septal area 5-HT1A receptor system in the regulation of water intake and urinary excretion. *Behav Brain Res* 215:122–8.

De Castro-e-Silva, E., C. Sarmento, T. A. Nascimento et al. 1997. Effect of third ventricle administration of L-694,247, a selective 5-HT1D receptor agonist, on water intake in rats. *Pharmacol Biochem Behav* 57:749–54.

De Gobbi, J. I., S. P. Barbosa, L. A. De Luca Jr., R. L. Thunhorst, A. K. Johnson, and J. V. Menani. 2005. Activation of serotonergic 5-HT1A receptors in the lateral parabrachial nucleus increases NaCl intake. *Brain Res* 1066:1–9.

De Gobbi, J. I., G. Martinez, S. P. Barbosa et al. 2007. 5-HT2 and 5-HT3 receptors in the lateral parabrachial nucleus mediate opposite effects on sodium intake. *Neuroscience* 146: 1453–61.

De Gobbi, J. I., J. V. Menani, T. G. Beltz, R. F. Johnson, R. L. Thunhorst, and A. K. Johnson. 2008. Right atrial stretch alters fore- and hind-brain expression of c-fos and inhibits the rapid onset of salt appetite. *J Physiol* 586:3719–29.

De Mello Cruz, A. P., G. Pinheiro, S. H. Alves, G. Ferreira, M. Mendes, and C. E. Macedo, 2005. Behavioral effects of systematically administered MK-212 are prevented by ritanserin microinfusion into the basolateral amygdala of rats exposed to the elevated plus-maze. *Psychopharmacology* 182:345–54.

De Souza Villa, P., G. M. De Arruda Camargo, L. A. A. De Arruda Camargo, and W. A. Saad. 2007. Activation of paraventricular nucleus of hypothalamus 5-HT1A receptor on sodium intake. *Regul Pept* 140:142–7.

De Souza Villa, P., J. V. Menani, G. M. de Arruda Camargo, L. A. A. de Arruda Camargo, and W. A. Saad. 2008. Activation of the serotonergic 5-HT1A receptor in the paraventricular nucleus of the hypothalamus inhibits water intake and increases urinary excretion in water-deprived rats. *Regul Pept* 150:14–20.

De Vries, P., J. P. Heiligers, C. M. Villalón, and P. R. Saxena. 1996. Blockade of porcine carotid vascular response to sumatriptan by GR 127935, a selective 5-HT1D receptor antagonist. *Br J Pharmacol* 118:85–92.

Depaulis, A., K. A. Keay, and R. Bandler. 1994. Longitudinal organization of defensive reactions in the midbrain periaqueductal gray region of the rat. *Exp Brain Res* 90:307–18.

Derkach, V., A. Surprenant, and R.A. North. 1989. 5-HT3 receptors are membrane ion channels. *Nature* 339:706–9.

DiMicco, J. A., B. C. Samuels, M. V. Zaretskaia, and D. V. Zaretsky. 2002. The dorsomedial hypothalamus and the response to stress: Part renaissance, part revolution. *Pharmacol Biochem Behav* 71:469–80.

Dreteler, G. H., W. Wouters, P. R. Saxena, and A. G. Ramage. 1991. Pressor effects of microinjection of 5-HT1A agonists into the raphe obscurus of the anaesthetized rat. *Br J Pharmacol* 102:317–22.

Duxon, M. S., T. P. Flanigan, A. C. Reavley, G. S. Baxter, T. P. Blackburn, and K. F. C. Fone. 1997. Evidence for expression of the 5-hydroxytryptamine 2B receptor protein in the rat central nervous system. *Neuroscience* 76:323–9.

Faerber, L., S. Drechsler, S. Ladenburger, H. Gschaidemeier, and W. Fischer. 2007. The neuronal 5-HT3 receptor network after 20 years of research: Evolving concepts in management of pain and inflammation. *Eur J Pharmacol* 560:1–8.

Ferreira, H. S., E. De Castro-e-Silva, C. Cointeiro, E. Oliveira, T. N. Faustino, and J. B. Fregoneze. 2004. Role of central 5-HT3 receptors in the control of blood pressure in stressed and non-stressed rats. *Brain Res* 1028:48–58.

Ferreira, H. S., E. Oliveira, T. N. Faustino, E. De Castro-e-Silva, and J. B. Fregoneze. 2005. Effect of the activation of central 5-HT2C receptors by the 5-HT2C agonist mCPP on blood pressure and heart rate in rats. *Brain Res* 1040:64–72.

Fitts, D. A., S. N. Thornton, A. A. Ruhf, D. K. Zierath, A. K. Johnson, and R. L. Thunhorst. 2003. Effects of central oxytocin receptor blockade on water and saline intake, mean arterial pressure, and c-fos expression in rats. *Am J Physiol Regul Integr Comp Physiol* 285:1331–9.

Fitzsimons, J. T. 1972. Thirst. *Physiol Rev* 52:468–561.

Fitzsimons, J. T. 1998. Angiotensin, thirst, and sodium appetite. *Physiol Rev* 78:583–686.

Fonseca, F. V., A. S. Mecawi, I. G. Araujo et al. 2009. Role of the 5-HT1A somatodendritic autoreceptor in the dorsal raphe nucleus on salt satiety signaling in rats. *Exp Neurol* 217:353–60.

Foong, J. P., and J. C. Bornstein. 2009. 5-HT antagonists NAN-190 and SB 269970 block alpha2-adrenoceptors in the guinea pig. *Neuroreport* 20:325–30.

Foster-Schubert, K. E., and D. E. Cummings. 2006. Emerging therapeutic strategies for obesity. *Endocrine Rev* 27:779–93.

Fozard, J. R., A. K. Mir, and D. N. Middlemiss. 1987. Cardiovascular response to 8-hydroxy-2-di-*n*-propylaminotetralin 8-OH-DPAT in the rat: Site of action and pharmacological analysis. *J Cardiovasc Pharmacol* 9:328–47.

Fregoneze, J. B., E. F. Oliveira, V. F. Ribeiro, H. S. Ferreira, and E. De Castro-e-Silva. 2011. Multiple opioid receptors mediate the hypotensive response induced by central 5-HT3 receptor stimulation. *Neuropeptides* 45:219–27.

Fujii, K., and Takeda, N. 1988. Phylogenetic detection of serotonin immunoreactive cells in the central nervous system of invertebrates *Comp Biochem Physiol* 89C:233–39.

Galeotti, N., C. Ghelardini, and A. Bartolini. 1998. Role of 5-HT4 receptors in the mouse passive avoidance test. *J Pharmacol Exp Ther* 286:1115–21.

Gaster, L. M., and F. D. King. 1997. Serotonin 5-HT3 and 5-HT4 receptor antagonists. *Med Res Rev* 17:163–214.

Gentili, L., A. Saija, G. Luchetti, and M. Massi. 1991. Effect of the 5-HT2 antagonist ketanserin on salt appetite in the rat. *Pharmacol Biochem Behav* 39:171–6.

Giorgetti, M., and L. H. Tecott. 2004. Contributions of 5-HT(2C) receptors to multiple actions of central serotonin systems. *Eur J Pharmacol* 488:1–9.

Glass, J. D., G. H. Grossman, L. Farnbauch, and L. DiNardo. 2003. Midbrain raphe modulation of nonphotic circadian clock resetting and 5-HT release in the mammalian suprachiasmatic nucleus. *J Neurosc* 23:7451–60.

Godino, A., L. A. De Luca Jr., J. Antunes-Rodrigues, and L. Vivas. 2007. Oxytocinergic and serotonergic systems involvement in sodium intake regulation, satiety or hypertonicity markers? *Am J Physiol Regul Integr Comp Physiol* 293:R1027–36.

Gradin, K., A. Pettersson, T. Hedner, and B. Persson. 1985. Acute administration of 8-hydroxy-2-di-*n*-propyl-amino tetralin 8-OH-DPAT, a selective 5-HT-receptor agonist, causes a biphasic blood pressure response and bradycardia in the normotensive Sprague–Dawley rat and in the spontaneously hypertensive rat. *J Neural Transm* 62:302–19.

Green, A. R. 2006. Neuropharmacology of 5-hydroxytryptamine. *Br J Pharmacol* 147:S145–52.

Guscott, M. R., E. Egan, G. P. Cook et al. 2003.The hypothermic effect of 5-CT in mice is mediated through the 5-HT7 receptor. *Neuropharmacology* 44:1031–7.

Hackler, E. A., G. H. Turner, P. J. Gresch et al. 2006. 5-HT2C receptor contribution to *m*-chlorophenylpierazine and *N*-methyl-[beta]-carboline-3-carboxamide-induced anxiety-like behavior and limbic brain activation. *J Pharmacol Exp Ther* 320:1023–9.

Harel-Dupas, C., I. Cloez, and G. Fillion. 1991. The inhibitory effect of trifluoromethylphenylpiperazine on [3H]acetylcholine release in guinea-pig hippocampal synaptosomes is mediated by a 5-hydroxytryptamine1 receptor distinct from 1A, 1B and 1C subtypes. *J Neurochem* 56:221–7.

Harsing, L. G. Jr., I. Prauda, J. Barkoczy, P. Matyus, and Z. Juranyi. 2004. A 5-HT7 heteroreceptor-mediated inhibition of [3H] serotonin release in raphe nuclei slices of rat: Evidence for a serotonergic–glutamatergic interaction. *Neurochem Res* 29:1487–97.

Hayes, D. J., and A. J. Greenshaw. 2011. 5-HT receptors and reward-related behavior: A review. *Neurosci Biobehav Rev* 35:1419–49.

Hayrapetyan, V., M. Jenschke, G. H. Dillon, and T. K. Machu. 2005. Co-expression of 5-HT3B subunit with the 5-HT3A receptor reduces alcohol sensitivity. *Brain Res Mol Brain Res* 142:146–50.

Hedlund, P. B., and J. G. Sutcliffe. 2004. Functional, molecular and pharmacological advances in 5-HT7 receptor research. *Trends Pharmacol Sci* 25:481–6.

Hedlund, P. B., P. E. Danielson, E. A. Thomas, K. Slanina, M. J. Carson, and J. G. Sutcliffe. 2003. No hypothermic response to serotonin in 5-HT7 receptor knockout mice. *Proc Natl Acad Sci USA* 100:1375–80.

Hedlund, P. B., S. Huitron-Resendiz, S. J. Henriksen, and J. G. Sutcliffe. 2005. 5-HT7 receptor inhibition and inactivation induce antidepressantlike behavior and sleep pattern. *Biol Psychiatry* 58:831–7.

Hill, D. L., and C. M. Mistretta. 1990. Developmental neurobiology of salt taste sensation. *Trends Neurosci* 13:188–95.

Hoyer, D., D. E. Clarke, J. R. Fozard et al. 1994. VII International union of pharmacology classification of receptors for 5-hydroxytryptamine serotonin. *Pharmacol Rev* 46:157–203.

Hoyer, D., J. P. Hannon, and G. R. Martin. 2002. Molecular, pharmacological and functional diversity of 5-HT receptors. *Pharmacol Biochem Behav* 71:533–54.

Jacobs, B. L., and E. C. Azmitia. 1992. Structure and function of the brain serotonin system. *Physiol Rev* 2:165–229.

Jacobs, K. M., G. P. Mark, and T. R. Scott. 1988. Taste responses in the nucleus tractus solitarius of sodium deprived rats. *J Physiol* 406:393–410.

Jhamandas, J. H., T. Petrov, K. H. Harris, T. Vu, and T. L. Krukoff. 1996. Parabrachial nucleus projection to the amygdala in the rat, electrophysiological and anatomical observations. *Brain Res Bull* 39:115–26.

Johnson, A. K. 2007. The sensory psychobiology of thirst and salt appetite. *Med Sci Sports Exerc* 39:1388–400.

Johnson, A. K., and R. L. Thunhorst. 1997. The neuroendocrinology of thirst and salt appetite, visceral sensory signals and mechanisms of central integration. *Front Neuroendocrinol* 18:292–353.

Johnson, B. A., D. R. Jasinski, G. P. Galloway et al. 1996. Ritanserin in the treatment of alcohol dependence—a multi-center clinical trial. Ritanserin Study Group. *Psychopharmacology* 128:206–15.

Jorgensen, H., M. Riis, U. Knigge, A. Kjaer, and J. Warberg. 2003. Serotonin receptors involved in vasopressin and oxytocin secretion. *J Neuroendocrinol* 15:242–9.

Joshi, P. R., A. Suryanarayanan, E. Hazai, M. K. Schulte, G. Maksai, and Z. Bikádi. 2006. Interactions of granisetron with an agonist-free 5-HT3A receptor model. *Biochemistry* 45:1099–105.

Kellett, D. O., S. C. Stanford, B. H. Machado, D. Jordan, and A. G. Ramage. 2005. Effect of 5-HT depletion on cardiovascular vagal reflex sensitivity in awake and anesthetized rats. *Brain Res* 1054:61–72.

Khateb, A., P. Fort, A. Alonso, B. E. Jones, and M. Mühlethaler. 1993. Pharmacological and immunohistochemical evidence for serotonergic modulation of cholinergic nucleus basalis neurons. *Eur J Neurosci* 5:541–7.

Knight, A. R., A. Misra, K. Quirk et al. 2004. Pharmacological characterisation of the agonist radioligand binding site of 5-HT2A, 5-HT2B and 5-HT2C receptors. *Naunyn Schmiedebergs Arch Pharmacol* 370:114–23.

Knowles, I. D., and A. G. Ramage. 2000. Evidence that activation of central 5-HT(2B) receptors causes renal sympathoexcitation in anaesthetized rats. *Br J Pharmacol* 129:177–83.

Koshimizu, T. A., Y. Nasa, A. Tanoue et al. 2006. V1a vasopressin receptors maintain normal blood pressure by regulating circulating blood volume and baroreflex sensitivity. *Proc Natl Acad Sci USA* 10320:7807–12.

Krobert, K. A., T. Bach, T. Syversveen, A. M. Kvingedal, and F. O. Levy. 2001. The cloned human 5-HT7 receptor splice variants: A comparative characterization of their pharmacology, function and distribution. *Naunyn Schmiedebergs Arch Pharmacol* 363:620–32.

Kuhn, D. M., W. A. Wolf, and W. Lovenberg. 1980. Review of the role of the central serotonergic neuronal system in blood pressure regulation. *Hypertension* 2:243–55.

Laer, S., F. Remmers, H. Scholz, B. Stein, F. U. Muller, and J. Neumann. 1998. Receptor mechanisms involved in the 5-HT-induced inotropic action in the rat isolated atrium. *Br J Pharmacol* 123:1182–8.

Landry, E. S., and P. A. Guertin. 2004. Differential effects of 5-HT1 and 5-HT2 receptor agonists on hindlimb movements in paraplegic mice. *Prog Neuropsychopharmacol Biol Psychiatry* 28:1053–60.

Lanfumey, L., and M. Hamon. 2004. 5-HT1 receptors. *Curr Drug Targets CNS Neurol Disord* 3:1–10.

Leysen, J. E. 2004. 5-HT2 receptors. *Curr Drug Targets CNS Neurol Disord* 3:11–26.

Lovell, P. J., S. M. Bromidge, S. Dabbs et al. 2000. A novel, potent, and selective 5-HT(7) antagonist: (*R*)-3-(2-(2-(4-methylpiperidin-1-yl)ethyl)pyrrolidine-1-sulfonyl) phenol (SB-269970). *J Med Chem* 43:342–5.

Lu, C. C., C. J. Tseng, F. J. Wan, T. H. Yin, and C. S. Tung. 1992. Role of locus coeruleus and serotonergic drug actions on schedule-induced polydipsia. *Pharmacol Biochem Behav* 43:255–61.

Luz, C., A. Souza, R. Reis, J. B. Fregoneze, and E. De Castro e Silva. 2006. Role of 5-HT3 and 5-HT2C receptors located within the medial amygdala in the control of salt intake in sodium-depleted rats. *Brain Res* 1099:121–32.

Luz, C. P., A. Souza, R. Reis et al. 2007. The central amygdala regulates sodium intake in sodium-depleted rats: Role of 5-HT3 and 5-HT2C receptors. *Brain Res* 1139:178–94.

Margatho, L. O., S. P. Barbosa, L. A. De Luca Jr., and J. V. Menani 2002. Central serotonergic and adrenergic/imidazoline inhibitory mechanisms on sodium and water intake. *Brain Res* 956:103–9.

Margatho, L. O., A. Giusti-Paiva, J. V. Menani, L. L. Elias, L. M. Vivas, and J. Antunes-Rodrigues. 2007. Serotonergic mechanisms of the lateral parabrachial nucleus in renal and hormonal responses to isotonic blood volume expansion. *Am J Physiol Regul Integr Comp Physiol* 292:R1190–7.

Maricq, A. V., A. S. Peterson, A. J. Brake, R. M. Myers, and D. Julius. 1991. Primary structure and functional expression of the 5-HT3 receptor, a serotonin-gated ion channel. *Science* 254:432–7.

Martin, G. R. 1994. Vascular receptors for 5-hydroxytryptamine: Distribution, function and classification. *Pharmacol Ther* 62:283–324.

Matthes, H., U. Boschert, N. Amlaiky et al. 1993. Mouse 5-hydroxytryptamine and 5-hydroxytryptamine receptors define a new family of serotonin receptors: Cloning, functional expression, and chromosomal localization. *Mol Pharmacol* 43:313–9.

Maura, G., G. C. Andrioli, P. Cavazzani, and M. Raiteri. 1992. 5-Hydroxytryptamine 3 receptors sided on cholinergic axon terminals of human cerebral cortex mediate inhibition of acetylcholine release. *J Neurochem* 58:2334–7.

McCall, R. B., and M. E. Clement. 1994. Role of serotonin1A and serotonin2 receptors in the central regulation of the cardiovascular system. *Pharmacol Rev* 46:231–43.

McCall, R. B., B. N. Patel, and L. T. Harris. 1987. Effects of serotonin1 and serotonin2 receptor agonists and antagonists on blood pressure, heart rate and sympathetic nerve activity. *J Pharmacol Exp Ther* 242:1152–9.

McKinley, M. J., and A. K. Johnson. 2004. The physiological regulation of thirst and fluid intake. *News Physiol Sci* 19:1–6.

Melis, C., P. L. Chau, K. L. Price, S. C. Lummis, and C. Molteni. 2006. Exploring the binding of serotonin to the 5-HT3 receptor by density functional theory. *J Phys Chem B* 110:26313–9.

Meller, S. T., S. J. Lewis, M. J. Brody, and G. F. Gebhart. 1992. Vagal afferent-mediated inhibition of a nociceptive reflex by i.v. serotonin in the rat. Role of 5-HT receptor subtypes. *Brain Res* 585:71–86.

Menani, J. V., and A. K. Johnson. 1995. Lateral parabrachial serotonergic mechanisms: Angiotensin-induced pressor and drinking responses. *Am J Physiol Regul Integr Comp Physiol* 269:R1044–9.

Mertens, M. J. F., M. Pfaggendorf, and P. A. van Zwieten. 1993. Impaired vasodilator and chronotropic responses to 5-hydroxytryptamine in two models of hypertension-associated cardiac hypertrophy. *Blood Pressure* 1:254–9.

Mitchell, E. S., and J. F. Neumaier. 2005. 5-HT6 receptors: A novel target for cognitive enhancement. *Pharmacol Ther* 108:320–33.

Molliver, M. E. 1987. Serotonergic neuronal systems: What their anatomic organization tells us about function. *J Clin Psychopharmacol* 7:3S–23S.

Montes, R., and A. K. Johnson. 1990. Efferent mechanisms mediating renal sodium and water excretion induced by centrally administered serotonin. *Am J Physiol Regul Integr Comp Physiol* 259:R1267–73.

Morales, M., E. Battenberg, and F. E. Bloom. 1998. Distribution of neurons expressing immunoreactivity for the 5HT3 receptor subtype in the rat brain and spinal cord. *J Comp Neurol* 402:385–401.

Morilak, D. A., S. J. Garlow, and R. D. Ciaranello. 1993. Immunocytochemical localization and description of neurons expressing serotonin2 receptors in the rat brain. *Neuroscience* 54:701–17.

Muntzel, M. S., S. J. Lewis, and A. K. Johnson. 1996. Anteroventral third ventricle lesions attenuate pressor responses to serotonin in anesthetized rats. *Brain Res* 714:104–10.

Neill, J. C., and S. J. Cooper. 1989. Selective reduction by serotonergic agents of hypertonic saline consumption in rats, evidence for possible 5-HT1C receptor mediation. *Psychopharmacology (Berl)* 99:196–201.

Nelson, D. L. 2004. 5-HT5 receptors. *Curr Drug Targets CNS Neurol Disord.* 3:53–8.

Niesler, B., B. Frank, J. Kapeller, and G. A. Rappold. 2003. Cloning, physical mapping and expression analysis of the human 5-HT3 serotonin receptor-like genes HTR3C, HTR3D and HTR3E. *Gene* 310:101–11.

Nonogaki, K., K. Nozue, and Y. Oka. 2006. Increased hypothalamic 5-HT2A receptor gene expression and effects of pharmacologic 5-HT2A receptor in obese AY mice. *Biochem Biophys Res Commun* 351:1078–82.

Norgren, R. 1995. Gustatory system. In: *The Rat Nervous System*, 2nd ed., ed. G. Paxinos, 751–771. San Diego: Academic Press.

Oikawa, R., Y. Nasa, R. Ishii et al. 2007. Vasopressin V1A receptor enhances baroreflex via the central component of the reflex arc. *Eur J Pharmacol* 558:144–50.

Oliver, K. R., A. M. Kinsey, A. Wainwright, and D. J. S. Sirinath-Singhji. 2000. Localization of 5-HT5A receptor-like immunoreactivity in the rat brain. *Brain Res* 867:131–42.

Pasqualetti, M., M. Ori, I. Nardi, M. Castagna, G. B. Cassano, and D. Marazziti. 1998. Distribution of the 5-HT5A serotonin receptor mRNA in the human brain. *Mol Brain Res* 56:1–8.

Patel, S., J. Roberts, J. Moorman, and C. Reavill. 1995. Localization of serotonin-4 receptors in the striatonigral pathway in rat brain. *Neuroscience* 69:1159–67.

Pauwels, P. J. 1997. 5-HT 1B/D receptor antagonists. *Gen Pharmacol* 29:293–303.

Pazos, A., and J. M. Palacios. 1985. Quantitative autoradiographic mapping of serotonin receptors in the rat brain: I. Serotonin-1 receptors. *Brain Res* 346:205–30.

Peroutka, S. J. 1994. 5-Hydroxytryptamine receptors in vertebrates and invertebrates: Why are there so many? *Neurochem Int* 25:533–6.

Peters, J. A., S. C. R. Lummis, N. M. Barnes, T. G. Hales, and B. Niesler. 2011. 5-HT_3 receptors. Last modified on 2011-08-29. IUPHAR database IUPHAR-DB, http://www.iuphar-db.org/DATABASE/FamilyMenuForward?familyId = 68. (Accessed on 2011-11-06).

Pratt, G. D., N. G. Bowery, G. J. Kilpatrick et al. 1990. Consensus meeting agrees distribution of 5-HT3 receptors in mammalian hindbrain. *Trends Pharmacol Sci* 11:135–7.

Ramage, A. G. 2001. Central cardiovascular regulation and 5-hydroxytryptamine receptors. *Brain Res. Bull* 56:425–39.

Ramage, A. G., and C. M. Villalón. 2008. 5-Hydroxytryptamine and cardiovascular regulation. *Trends Pharmacol Sci* 29:472–81.

Reeves, D. C., and S. C. R. Lummis. 2006. Detection of human and rodent 5-HT3B receptors subunits by anti-peptide polyclonal antibodies. *BMC Neurosci* 27:1–8.

Reis, L. C., M. J. Ramalho, A. L. Favaretto, J. Gutkowska, S. M. McCann, and J. Antunes-Rodrigues. 1994. Participation of the ascending serotonergic system in the stimulation of atrial natriuretic peptide release. *Proc Natl Acad Sci USA* 91:12022–6.

Reis, L. C., M. J. P. Ramalho, and J. Antunes-Rodrigues. 1992. Brain serotonergic stimulation reduces the water intake induced by systemic and central beta-adrenergic administration. *Braz J Med Biol Res* 25:529–36.

Reis, L. C., M. J. P. Ramalho, and J. Antunes-Rodrigues. 1990a. Central serotonergic modulation of drinking behavior induced by angiotensin II and carbachol in normally hydrated rats, effect of intracerebroventricular injection of MK-212. *Braz J Med Biol Res.* 23:1339–42.

Reis, L. C., M. J. P. Ramalho, and J. Antunes-Rodrigues. 1990b. Central serotonergic modulation of drinking behavior induced by water deprivation, Effect of a serotonergic agonist MK-212 administered intracerebroventricularly. *Braz J Med Biol Res.* 23:1335–8.

Reyntjens, A., Y. G. Gelders, M. L. J. A. Hoppenbrouwers, and G. Vanden Bussche. 1986. Thymosthenic effects of ritanserin R 55667, a centrally acting serotonin S2 receptor blocker. *Drug Dev Res* 8:205–11.

Rouah-Rosilio, M., M. Orosco, and S. Nicolaidis. 1994. Serotoninergic modulation of sodium appetite in the rat. *Physiol Behav* 55:811–6.

Rowland, N. E., B.-H. Li, M. J. Fregly, and G. C. Smith. 1994. Involvement of angiotensin in water intake induced by peripheral administration of a serotonin agonist, 5-carboxyamidotryptamine. *Brain Res* 664:148–154.

Saavedra, J. M. 1992. Brain and pituitary angiotensin. *Endocr Rev* 13:329–80.

Saxena, P. R. 1995. Serotonin receptors: Subtypes, functional response and therapeutic relevance. *Pharmacol Ther* 66:339–68.

Saxena, P. R., and C. M. Villal6n. 1991. 5-Hydroxytryptamine: A chameleon in the heart. *Trends in Pharmacol Sci* 12:223–7.

Schoffelmeer, A. N., T. J. De Vries, F. Hogenboom, V. J. Hruby, P. S. Portoghese, and A. H. Mulder. 1992a. Opioid receptor antagonists discriminate between presynaptic mu and delta receptors and the adenylate cyclase-coupled opioid receptor complex in the brain. *J Pharmacol Exp Ther* 263:20–4.

Schoffelmeer, A. N., B. J. Van Vliet, T. J. De Vries, M. H. Heijna, and A. H. Mulder. 1992b. Regulation of brain neurotransmitter release and of adenylate cyclase activity by opioid receptors. *Biochem Soc Trans* 20:449–53.

Schulkin, J. 1982. Behavior of sodium deficient rats, the search for a salty taste. *J Comp Physiol Psychol* 96:628–34.

Sévoz-Couche, C., M. A. Comet, M. Hamon, and R. Laguzzi. 2003. Role of nucleus tractus solitarius 5-HT3 receptors in the defense reaction-induced inhibition of the aortic baroreflex in rats. *J Neurophysiol* 90:2521–30.

Simansky, K. J. 1995. Peripheral 5-carboxamidotryptamine 5-CT elicits drinking by stimulating 5-HT1-like serotonergic receptors in rats. *Pharmacol Biochem Behav* 38:459–62.

Siniscalchi, A., I. Badini, L. Beani, and C. Bianchi. 1999. 5-HT4 receptor modulation of acetylcholine outflow in guinea pig brain slices. *Neuroreport* 10:547–51.

Stachenfeld, N. S. 2008. Acute effects of sodium ingestion on thirst and cardiovascular function. *Curr Sports Med Rep* 7(4 Suppl):S7–13.

Steardo, L., and M. Iovino. 1986. Vasopressin released after enhanced serotonergic transmission is not due to activation of the peripheral renin–angiotensin system. *Brain Res* 382:145–8.

Strauss, W. H., and E. Klieser. 1991. Psychotropic effects of ritanserin, a selective S2 antagonist, an open study. *Eur Neuropsychopharmacol* 1:101–5.

Stricker, E. M., and A. F. Sved. 2000. Thirst. *Nutrition* 16:821–6.

Szabo, A., B. L. Butz, and R. H. Alper. 1998. Further characterization of forebrain serotonin receptors mediating tachycardia in conscious rats. *Brain Res Bull* 45:583–8.

Takeuchi, Y., J. H. McLean, and D. A. Hopkins. 1982. Reciprocal connections between the amygdala and parabrachial nuclei, ultrastructural demonstration by degeneration and axonal transport of horseradish peroxidase in the cat. *Brain Res* 239:583–8.

Tangaprégasson, M. J., A. M. Tangaprégasson, and A. Soulairac. 1974. Effets des lesions de la region du raphé mesencephalique sur le comportment de soif et de la neurosécrétion hypothalamique antérieure du rat. *Ann Endocrinol (Paris)* 35:667–8.

Twarog, B. M., and I. H. Page. 1953. Serotonin content of some mammalian tissues and urine and a method for its determination. *Am J Physiol* 175:157–61.

Thomas, D. R., and J. J. Hagan. 2004. 5-HT7 receptors. *Curr Drug Targets CNS Neurol Disord* 3:81–90.

Uphouse, L. 1997. Multiple serotonin receptors: Too many, not enough, or just the right number? *Neurosc Biobehav Rev* 21:679–98.

Urzedo-Rodrigues, L. S., H. S. Ferreira, D. O. Almeida et al. 2011. Blockade of 5-HT3 receptors at septal area increase blood pressure in unanaesthetized rats. *Auton Neurosci* 159:51–61.

Valenta, B., and E. A. Singer. 1990. Hypotensive effects of 8-hydroxy-2-di-*n*-propylamino tetralin and 5-methylurapidil following stereotaxic microinjection into the ventral medulla of the rat. *Br J Pharmacol* 99:713–6.

Van de Kar, L. D., A. Javed, Y. Zhang, F. Serres, D. K. Raap, and T. S. Gray. 2001. 5-HT2A receptors stimulate ACTH, corticosterone, oxytocin, renin, and prolactin release and activate hypothalamic CRF and oxytocin-expressing cells. *J Neurosci* 21:3572–9.

Van Oekelen, D., W. H. M. L. Luyten, and J. E. Leysen. 2003. 5-HT2A and 5-HT2C receptors and their atypical regulation properties. *Life Sci* 72:2429–49.

Verbalis, J. G. 2003. Disorders of body water homeostasis. *Best Pract Res Clin Endocrinol Metab* 17:471–503.

Vilaró, M. T., R. Cortes, and G. Mengod. 2005. Serotonin 5-HT4 receptors and their mRNAs in rat and guinea pig brain: Distribution and effects of neurotoxic lesions. *J Comp Neurol* 484:418–39.

Villalón, C. M., D. Centurion, M. Lujan-Estrada, J. A. Terron, and A. Sanchez-Lopez. 1997a. Mediation of 5-HT-induced external carotid vasodilatation in GR 127935-pretreated vagosympathectomized dogs by the putative 5-HT7 receptor. *Br J Pharmacol* 120:1319–27.

Villalón, C. M., P. De Vries, and P. R. Saxena. 1997b. Serotonin receptors as cardiovascular targets. *Drug Disc Today* 2:294–300.

Villalón, C. M., A. Sanchez-Lopez, and D. Centurion. 1996. Operational characteristics of the 5-HT1-like receptors mediating external carotid vasoconstriction in vagosympathectomized dogs. Close resemblance to the 5-HT1D receptor subtype. *Naunyn Schmiedebergs Arch Pharmacol* 354:550–6.

Waeber, C., M. Sebben, J. Bockaert, and A. Dumuis. 1996. Regional distribution and ontogeny of 5-HT4 binding sites in rat brain. *Behav Brain Res* 73:259–62.

Wesolowska, A. 2002. In the search for selective ligands of 5-HT5, 5-HT6 and 5-HT7 serotonin receptors. *Pol J Pharmacol* 54:327–41.

Wood, M. D. 2003. Therapeutic potential of 5-HT2C receptor antagonists in the treatment of anxiety disorders. *Curr Drug Targets CNS Neurol Disord* 2:383–7.

Yakel, J. L., and M. B. Jackson. 1988. 5-HT3 receptors mediate rapid responses in cultured hippocampal and clonal cell line. *Neuron* 1:615–21.

Yau, J. L., J. Noble, J. R. Seckl. 2001. Acute restraint stress increases 5-HT7 receptor mRNA expression in the rat hippocampus. *Neurosci Lett* 309:141–4.

Yusuf, S., N. Al-Saady, and A. J. Camm. 2003. 5-Hydroxytryptamine and atrial fibrillation: How significant is this piece in the puzzle? *J. Cardiovasc Electrophysiol* 14:209–14.

Zhang, Y., T. S. Gray, D. N. D'Souza et al. 2004. Desensitization of 5-HT1A receptors by 5-HT2A receptors in neuroendocrine neurons in vivo. *J Pharmacol Exp Ther* 310:59–66.

11 Serotonergic Autoinhibition within Dorsal Raphe Nucleus Modulates Sodium Appetite

André S. Mecawi, Fabricia V. Fonseca, Iracema G. de Araujo, and Luis C. Reis

CONTENTS

11.1 ORGANIZATION OF MIDBRAIN DORSAL RAPHE NUCLEUS (DRN): CONTROL OF SEROTONERGIC NEUROTRANSMISSION

Serotonin [5-hydroxytryptamine (5-HT)] is a widespread monoaminergic neurotransmitter in the brain. Its neuronal population is among the first to differentiate in the mammalian brain (Gaspar et al. 2003), playing an important role in neurogenesis (Lauder and Krebs 1978) and in several normal physiological functions and pathological disorders throughout the mammalian life. There are 235,000 serotonergic neurons in the human midbrain raphe and 10,000–12,000 multipolar cell in rats (Abrams et al. 2004; Adell et al. 2002). The 5-HT neurons exist in a widespread distributed system in the mammalian brain (Jacobs and Azmitia 1992), with a peculiar elevated density of serotonergic perikarya dispersed on or near the brainstem's midline, where they are primarily located in the median (MRN) and dorsal raphe (DRN) nuclei (Azmitia and Segal 1978; Parent et al. 1981). The DRN is located in the ventral part of the midbrain periaqueductal gray matter and extends caudally, bordering the anterior portion of the pons. The DRN's serotonergic neurons are organized into a cluster arranged in several topographical subdivisions. Rostral projections of the DRN are arranged in a conspicuous manner, connected to an extensive collaborative distribution in the terminal fields (Azmitia 1987, 2001). According to Azmitia, different subsets of serotonergic neurons take part in the integrated coordination of neural control systems (Azmitia 1987, 2001; Jacobs and Azmitia 1992). This idea has been revisited by Abrams et al. (2004) through morphological and functional approaches.

Observations concerning fibers projecting from the rostral ventromedial subregion toward forebrain sites have led to a hypothesis that the subpopulation of topographically arranged serotonergic neurons has functional properties correlated with the modulation of specific forebrain systems, particularly those related to neuroendocrine, hydroelectrolytic, and cardiovascular control (Azmitia and Segal 1978; Bosler and Descarries 1988). This concept is supported by Abrams et al. (2004) and Clark et al. (2006), according to whom distinctive clusters of serotonergic neurons within the DRN may comprise a restricted circuitry. Each one may be related to the control of a unique motivated behavior (and corresponding neuroendocrine and autonomic-related responses) as well as cognitive and psychoemotional and anxiety status control.

Serotonergic neurons produce serotonin from the tryptophan (Boadle-Biber 1993; Tyce 1990), and serotonin's transmission takes place in a rhythmic pattern that occurs spontaneously in a complex way, with electrophysiological differences among raphe nuclei clusters in firing rate frequency, period, and the interval between each burst (Allers and Sharp 2003; Hajós et al. 1995; Marinelli et al. 2004). In 2007, Hajós et al. used an *in vivo* juxtacellular labeling method to strongly suggest that the neurons from DRN and MRN that develop a pattern of stereotyped burst firing are 5-HT- and/or tryptophan hydroxylase-containing neurons (Hajós et al. 2007). Although 5-HT neuronal activity may be modulated by several factors (Berridge and Waterhouse 2003; Cedraz-Mercez et al. 2007; Lind 1986; Petrov et al. 1992), there is a well-established modulation of this neuronal population by a somatodendritic autofeedback mechanism mediated through the activation of 5-HT_{1A} somatodendritic autoreceptors (Aghajanian et al. 1987; Azmitia 1987, 2001; Hajós et al. 1995).

Briefly, there are at least 16 different types of identified 5-HT receptors: (a) the 5-HT_1 receptors (A, B, D, E, and F), which are classically coupled to inhibitory G proteins; (b) the 5-HT_2 (A, B, and C) receptors, which are coupled to G_q proteins; (c) the 5-HT_3 receptors (A, B, and C), which are ionotropic receptors; (d) the 5-HT_4,

5-HT_6, and 5-HT_7 receptors, which are coupled to stimulatory G proteins; and (e) the 5-HT_5 (A and B) receptors, which, despite have not been well described, some evidences suggests that it may be linked to inhibitory G proteins (Bockaert et al. 2006; Polter and Li 2010). The 5-HT_{1A} receptor is the most well-characterized serotonin receptor because of its early discovery and its translational implications for several physiological and pathological conditions.

The 5-HT_{1A} receptors may be located in two different sites in the mammalian brain: presynaptically (on dendrites and soma) of serotonergic neurons from the raphe nuclei or, postsynaptically, on non-serotonergic neurons, mainly from the limbic areas (Lanfumey and Hamon 2004). Briefly, the activation of neuronal 5-HT_{1A} receptors activates G-protein–coupled inwardly rectifying potassium channels, which hyperpolarizes neurons and decreases their firing rate (Andrade and Nicol 1987; Tada et al. 1999). These receptors also reduce calcium currents and the evoked calcium influx (Bayliss et al. 1995; Cheng et al. 1998). Overall, serotonin signaling via 5-HT_{1A} receptors therefore induces a rapid and effective inhibition of neuronal transmission. Thus, postsynaptic 5-HT_{1A} receptors (also called heteroreceptors) signal the efferent inputs of raphe nuclei serotonergic neurons and may also play a role in indirect negative feedback, inhibiting pyramidal cortical neurons that project to raphe nuclei serotonergic neurons (Celada et al. 2004). Additionally, the activation of 5-HT_{1A} autoreceptors, which are located presynaptically on serotonergic neurons, decreases the activity of the firing neuron, leading to a decrease in its own serotonergic neurotransmission (Azmitia 2001).

The presence of autoreceptors on serotonergic neurons from the raphe nuclei was previously suggested when a 5-HT_{1A} agonist receptor was found to reduce serotonin release, whereas an unspecific agonist or 5-HT itself also reduced serotonin release (Aghajanian et al. 1987). Soon afterward, electrophysiological studies showed that this negative feedback directly affected serotonergic neurons themselves (Aghajanian 1982), mediated by 5-HT_{1A} receptors (Liu et al. 2005; Sotelo et al. 1990; Sprouse and Aghajanian 1987). It was also demonstrated that ablation of 5-HT_{1A} autoreceptors increased serotonergic neurotransmission, whereas their overexpression decreased 5-HT neurotransmission (Bortolozzi et al. 2004; Richardson-Jones et al. 2010); these findings strongly implicate the importance of 5-HT_{1A} autoreceptors in the serotonergic system.

There is an elevated density of 5-HT_{1A} autoreceptors in the MRN and DRN (Chalmers and Watson 1991; Kia et al. 1996), the majority of which are from serotonergic neurons in the DRN and, to a lesser extent, in the MRN (Sotelo et al. 1990). Serotonin's action on these receptors produces a reduction in the neuronal firing rate and, subsequently, a decrease in the synthesis and turnover of activity-dependent serotonin, which is directly dependent on extracellular serotonin concentration. For example, as serotonin content increases, 5-HT_{1A} signaling mediates a decrease in serotonergic neuronal activity and, as a consequence, reduces serotonin's release in the forebrain terminals (Hutson et al. 1989; Sprouse and Aghajanian 1987). Thus, the activation of somatodendritic 5-HT_{1A} autoreceptors reduces the activity of the entire serotonergic system.

11.2 MIDBRAIN DRN MODULATES SALT INTAKE AND RENAL SODIUM EXCRETION: INTERACTION WITH LAMINA TERMINALIS

The involvement of the midbrain serotonergic system in hydroelectrolytic balance has been shown in responses such as renal sodium excretion, atrial natriuretic peptide (ANP) release, and sodium intake. Intracerebroventricular (icv) injections of serotonin and serotonin 5-HT_{2A} or 5-HT_{2C} agonists increased urinary sodium excretion in hydrated rats, whereas natriuresis was decreased by 5-HT_{1A} agonists or brain serotonin depletion induced by icv injection of *para*-chloro-phenyl-alanine (Reis et al. 1991, 1994), which irreversibly inhibits tryptophan-hydroxylase, the limiting enzyme for serotonin synthesis (Tyce 1990). A similar response was observed in rats submitted to electrolytic lesion of the DRN, which led to a significant decrease in renal sodium excretion in hydrated rats (Reis et al. 1994). Both DRN lesioning and brain serotonin depletion drastically reduced basal and, stimulated ANP release after blood volume expansion (Antunes-Rodrigues et al. 2004; Reis et al. 1994). Briefly, some reports have suggested that the brain's serotonergic system could modulate natriuresis directly, mediating sympathetic renal nerve activity (Montes and Johnson 1990), or indirectly via oxytocin signaling, modulating ANPergic neuronal activity from the anteroventral third ventricle region (AV3V) or ANP's release from the heart atria (Haanwinckel et al. 1995; Reis et al. 1994). The possible concurrences between ANP release and natriuresis versus extracellular volume regulation are discussed elsewhere (Reis 2007), and this result most likely occurs in parallel with the control of the salt intake response.

It is well established that salt intake is essential to the homeostatic processes for adjusting body fluid volume and for maintaining the concentration of the body's main ionic component (Na^+) in ranges compatible with life (Antunes-Rodrigues et al. 2004; Fitzsimons 1998; Johnson and Thunhorst 1997; Reis 2007; Thunhorst et al. 1994). Thus, several studies have investigated connections between the forebrain and brainstem systems that may be related to sodium appetite modulation (Godino et al. 2007, 2010; Menani et al. 1996, 1998a,b; Stricker and Verbalis 1996).

Munaro and Chiaraviglio (1981) were the first to demonstrate that sodium-depleted rats with free access to hypertonic saline showed an increase in the hypothalamic concentration of serotonin and its metabolite, 5-hydroxyindoleacetic acid (5-HIAA), which is now known to be related to sodium satiety. Apart from behavioral studies involving DRN lesions, there were few studies before 1990 regarding the recruitment of serotonergic neurons involved in sodium satiety.

The inhibitory influence of the serotonergic system on sodium intake, both induced (sodium appetite) and spontaneous (under need-free conditions), was initially demonstrated in rats submitted to a peripheral injection of serotonergic compounds [e.g., serotonin releasers and agonists or specific serotonin presynaptic uptake inhibitors (SSRI)]. Acute systemic administration of 5-HT_{1C} receptor (currently classified as 5-HT_{2C}) agonists or serotonin presynaptic inhibitors in water-deprived rats selectively reduced the consumption of hypertonic NaCl compared to isotonic NaCl solution or water (Neill and Cooper 1989). Additionally, the administration of serotonin releaser drastically reduced hypertonic NaCl ingestion in sodium-depleted and hydrated rats, with no effect on water ingestion (Rouah-Rosilio et al. 1994). We have also shown that acute systemic treatments with a brain serotonin releaser (fenfluramine), an SSRI (fluoxetine), or 5-HT_{2C} receptor agonists (MK212 and mCPP) significantly reduces hypertonic saline ingestion in both fluid- and food-deprived rats as well as in sodium-depleted rats (Badauê-Passos et al. 2003). Hypertonic saline ingestion induced by sodium depletion was also inhibited by icv administration of 5-HT_{2C} and 5-HT_3 receptor agonists (mCPP and m-CPBG, respectively), as described by Castro et al. (2003). In contrast, serotonin 5-HT_{1A} agonists (8-OH-DPAT and gepirone) and a 5-HT_2 antagonist (metergoline) increased hypertonic saline ingestion in hydrated, rehydrating, (Cooper et al. 1988) and sodium-depleted rats (Rouah-Rosilio et al. 1994), with no effect on water ingestion, most likely through the reduction of serotonergic signaling. Further investigation of these issues with new methodological approaches has made it possible to identify serotonergic neurons of the DRN that participate in the behavioral expression of sodium satiety (Franchini et al. 2002). In this study, the authors used an immunohistochemistry technique to detect c-Fos, a protein expressed by immediate early genes. Their results suggested the presence of serotonin in c-Fos–marked neurons, which confirmed the hypothesis that serotonin neurons are recruited during sodium ingestion when sodium depletion was induced by peritoneal dialysis.

A possible neural substrate for serotonin control of sodium intake was then demonstrated by our laboratory through electrolytic (Olivares et al. 2003) and ibotenic (Cavalcante-Lima et al. 2005a) DRN lesions producing exaggerated sodium appetite responses in rats under basal or sodium-depleted conditions. Furthermore, rats submitted to DRN ibotenic lesions developed a sodium appetite under thirst-induced protocols (Cavalcante-Lima et al. 2005b). Thus, it has been hypothesized that the DRN serotonergic system participates in sodium appetite regulation (1) in basal conditions, inhibiting need-free salt ingestion, and (2) after sodium depletion, inhibiting need-induced sodium ingestion. Increased serotonergic activity from clusters of ascending midbrain neurons therefore seems to be associated with sodium appetite control, promoting body fluid homeostasis in parallel with renal sodium excretion. Moreover, it is likely that the DRN is a marker of satiety for sodium appetite, which could operate through bidirectional cross-talk with forebrain structures that facilitate sodium intake (such as the nuclei from the lamina terminalis) and through another key inhibitor of sodium appetite, the lateral parabrachial nucleus (LPBN) (Badauê-Passos et al. 2007; Franchini et al. 2002; Margatho et al. 2008; Reis 2007).

The most intriguing question is how DRN neurons participate in the control of hydroelectrolytic balance, with special attention to sodium appetite. In this context, the most important anatomical evidence of DRN serotonergic modulation of hydroelectrolytic balance and sodium appetite is derived from a report by Azmitia and Segal (1978), which identified, by an autoradiographic analysis, DRN projections to the lamina terminalis, which contains important nuclei implicated in hydroelectrolytic regulation, such as the organum vasculosum of lamina terminalis (OVLT), median preoptic nucleus (MnPO), and subfornical organ (SFO). Subsequently, this hypothesis was readdressed by Lind (1986) with evidence of a reciprocal interaction, or bidirectional cross-talk, with the forebrain, with specific attention to the SFO. The SFO sends modulatory projections to the DRN, thus tuning its serotonergic ascending drive. On this circumstance, Tanaka et al. (1998) showed that SFO neurons exhibiting AT1 receptors, which monitor actions of circulating angiotensin II and convey information about cardiovascular and possibly behavioral (sodium appetite) status. After sodium and secondarily water ingestion, increased volume expansion activates the DRN, which in turn releases serotonin in the LPBN. Inhibitory inputs from both structures are sent to the lamina terminalis, particularly the SFO.

Regarding water- or sodium-deprived models, several reports have demonstrated that the angiotensin II excitatory effects on sodium appetite involves reciprocal interactions with inhibitory serotonergic systems in male rats (Badauê-Passos et al. 2007; Colombari et al. 1996; Franchini et al. 2002; Menani et al. 1998a). Recall that the lamina terminalis (particularly the SFO, which expresses c-Fos protein after salt consumption in response to sodium depletion in rats) projects fibers toward the DRN, as shown by iontophoretic administration of fluorogold (FG), a retrograde tracer, into the DRN (Badauê-Passos et al. 2007). Moreover, subpopulations of SFO neurons expressing angiotensin II AT1 receptors have been identified projecting to the DRN (Tanaka et al. 1998), and a pharmacological and electrophysiological study showed an excitatory (mediated by 5-HT_{2A} and 5-HT_{2C} receptors) and inhibitory (mediated by 5-HT_{1A} receptors) serotonin role in SFO neurons, which were also sensitive to angiotensin II (Scrogin et al. 1998). Thus, this body of evidence would suggest that reduced serotonergic activity within the DRN, possibly mediated by increased synaptic angiotensin II (from SFO or even from glial cells, as proposed by Grobe et al. 2008), could culminate in the removal of inhibitory influences on neural systems excited by angiotensin II in the lamina terminalis, which would stimulate natriorexigenic responses (Badauê-Passos et al. 2007; Cavalcante-Lima et al. 2005a; Reis et al. 2007). After water or sodium deprivation, angiotensin II mediation of the sodium appetite trigger may then involve both the activation of angiotensinergic pathways, which stimulate water and sodium appetite, and the inhibition of serotonergic systems, which tonically curb sodium appetite.

These findings also lead to the idea that DRN serotonergic neurons receive information by monitoring plasma angiotensin II oscillations (related to variations in blood pressure, plasma volume, and plasma Na^+ concentration) processed by the SFO. Taken together, the observations above constitute circumstantial evidence of the existence of an SFO–DRN–SFO circuit that may be implicated in the regulation of hydroelectrolytic and cardiovascular balance, as previously proposed (Cavalcante-Lima et al. 2005a,b; Lind 1986; Reis et al. 1994; Scrogin et al. 1998; Tanaka et al. 1998). In addition to this possibility, other reciprocal interactions between forebrain areas and the midbrain raphe must not be ruled out, particularly among the OVLT, MnPO and lateral hypothalamic area, and DRN. These speculations from the experimental observations of Tanaka et al. (1998) raise the likelihood that other humoral signals (e.g., plasma levels of ANP and oxytocin) may be monitored by the lamina terminalis, which sends processed information to the DRN regarding hydroelectrolytic and cardiovascular status.

Several independent data have raised the possibility that there is a physiological integration of LPBN and DRN serotonergic neurons (Franchini et al. 2002; Margatho et al. 2008; Menani et al. 1996, 1998a,b, 2002; Tanaka et al. 2004). Consequently, after this integration, both structures most likely convey information to the forebrain to coordinate sodium excretion adjustments and the salt satiety response. In light of this concept, Margatho et al. (2008) demonstrated that the DRN displayed a significant contingent of retrogradely labeled c-Fos–immunoreactive-FG neurons after blood volume expansion in rats subjected to FG injection into the LPBN 7 days beforehand. This result reinforces the concept that the identified neurons in the DRN project to the LPBN. Furthermore, Petrov et al. (1992) showed that the LPBN sends peptidergic (neurotensinergic, galaninergic and CGRPergic) pathways toward the DRN, from which serotonergic efferent fibers project to the pontine nucleus. Later, Jolas and Aghajanian (1997) showed that neurotensin induces a concentration-dependent increase in the firing rate of a serotonergic neuron subpopulation. These observations strengthened the hypothesis that one of the mechanisms responsible for triggering sodium satiety in response to salt ingestion and subsequent blood volume increase results from an interaction between the DRN and the LPBN (Margatho et al. 2008; Reis 2007).

11.2.1 Does Activation of Serotonin Somatodendritic 5-HT$_{1A}$ Receptors at DRN Control Sodium Appetite?

We may assume that excitatory signals (satiety signals) reach the specific cluster of serotonergic neurons in the DRN that induce sodium satiety, mediated by serotonin release into the forebrain natriorexigenic circuits at the lamina terminalis, particularly in the SFO. Serotonin is released by axonal terminals, and it is recognized (see below) that it is also released into extracellular somatodendritic loci within the DRN, where it acts on its autoreceptors. Therefore, serotonin itself might modulate the ascending serotonergic activity by activation of serotonin somatodendritic 5-HT$_{1A}$ autoreceptors. The ensuing autoinhibition possibly constitutes a negative feedback loop related to diminished ascending serotonergic activity and a subsequent gradually increasing natriorexigenic status.

This understanding of autoinhibition is supported by neurochemical and electrophysiological studies showing that serotonin 5-HT$_{1A}$ somatodendritic autoreceptor activation in the DRN reduces spontaneous firing rate (Hjorth and Magnusson 1988; Liu et al. 2005; Sprouse and Aghajanian 1987) and serotonin release from serotonin-containing neurons, in a G$_i$ protein–coupled manner, which reduces the Ca^{+2} inflow (Blier and de Mongtiny 1990; Blier et al. 1998). Therefore, considering the aforementioned ascending serotonergic projections from the DRN to forebrain structures, the acute systemic or intraraphe administration of serotonin 5-HT$_{1A}$ receptor agonists would decrease the propagation of action potentials along the ascending serotonergic pathway, self-adjusting and optimizing DRN serotonergic neurons to several environmental contexts. These adjustments would occur through oscillatory activation of serotonin 5-HT$_{1A}$ somatodendritic autoreceptors, owing to variation in extracellular serotonin concentration and excitatory versus inhibitory signals acting on other receptors. In this context, Liu et al. (2005) suggested that after the increase in firing rate in clusters of serotonergic neurons occurs (e.g., by alpha-1 adrenergic stimulation and possibly other excitatory inputs), subsequent increases in the inhibition rate occur, mediated by 5-HT$_{1A}$ autoreceptors.

In a behavioral approach to studying this paradigm, we hypothesized that the activation of somatodendritic autoreceptors by a serotonin 5-HT_{1A} agonist within the DRN would decrease the ascending transmission, which would result in an increased natriorexigenic response in sodium-depleted rats. We conjectured, meanwhile, that 5-HT_{1A} receptor blockade by peripheral or intra-DRN administration of a serotonin 5-HT_{1A} antagonist could produce an opposite effect by removing somatodendritic autoinhibition.

11.2.1.1 Acute Systemic or Intra-DRN Injection of a 5-HT_{1A} Agonist Increases Sodium Intake, Whereas Its Repeated Administration Reduces Intake under Basal Conditions and after Sodium Depletion

We reported for the first time the direct participation of DRN 5-HT_{1A} somatodendritic autoreceptors in sodium appetite modulation (Fonseca et al. 2009). Acute systemic administration of the serotonin 5-HT_{1A} agonist, 8-OH-DPAT [8-hydroxy-2-(*n*-dipropylamino)tetralin-HBr], was performed in separate groups of normally sodium-sated male rats. The rats were submitted to a free choice paradigm between water and hypertonic saline simultaneously. The systemically administered 8-OH-DPAT promoted a significant increase in hypertonic saline intake, and the clear sodium preference compared to sodium-sated control group was sustained 24 h after administration. Similar results were obtained with acute administration of 8-OH-DPAT into the DRN of male rats, which led to an intense increase in salt preference compared to control group under basal conditions when the free-choice paradigm was used.

An acute intra-DRN administration of 8-OH-DPAT was also performed in sodium-depleted rats. Sodium depletion was induced by both injection of the diuretic/natriuretic furosemide and the removal of sodium from the diet for 24 h. Compared to the respective control, the microinjection of 8-OH-DPAT evoked an additional and significant increase in cumulative sodium intake from 2 to 5 h afterward when both water and hypertonic saline were offered to sodium-depleted rats (Fonseca et al. 2009). It is possible that this experimental maneuver mimics a physiological moment in which extracellular serotonin is enhanced, as previously described (Beer et al. 1990). This enhancement of extracellular serotonin could decrease the firing rate associated with subsequent reduced serotonergic activity and forebrain serotonin release, which could trigger natriorexigenic behavior. Garratt et al. (1991) have demonstrated that treatment with 8-OH-DPAT reduces the serotonergic neuron firing rate to 50% of control values in anesthetized rats. Such a change would start a novel cycle of increased natriorexigenic drive (see Figure 11.1).

In contrast to the increased sodium intake produced by a single acute injection, we observed that repeated 8-OH-DPAT intraperitoneal or intra-DRN injections over 6 days resulted in a long-lasting reduction in cumulative sodium intake in sodium-depleted rats (Fonseca et al. 2009). Chronic treatment with a serotonin 5-HT_{1A} agonist may have induced desensitization of the serotonin somatodendritic autoreceptors, as previously described (Albert et al. 1996; Assié et al. 2006; Hjorth et al. 2000). This desensitization would possibly be associated with a reduction of the receptor density

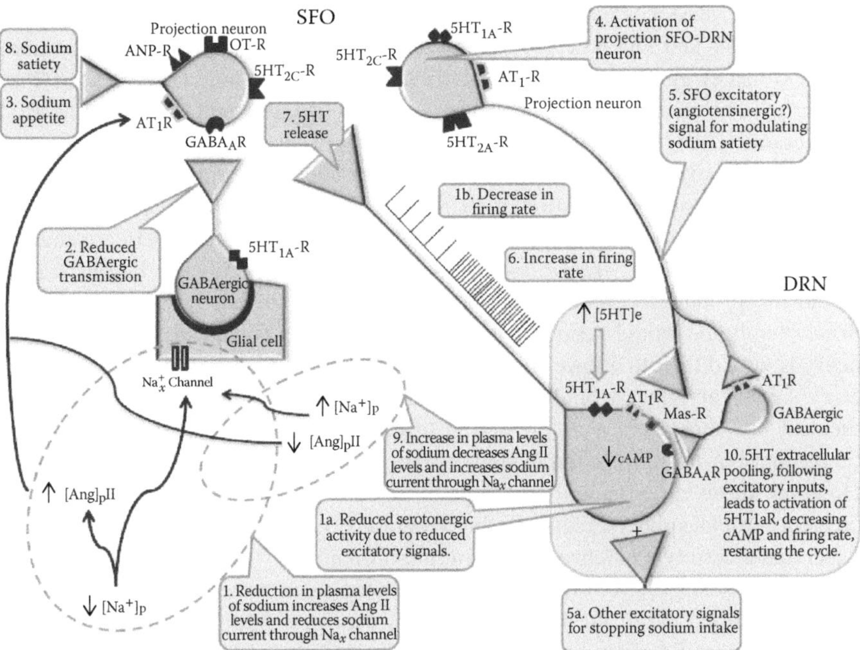

FIGURE 11.1 Hypothetical model for control of sodium intake through connections between dorsal raphe nucleus (DRN) and subfornical organ (SFO): excitatory and inhibitory inputs to serotonin $5\text{-}HT_{1A}$ somatodendritic autoreceptors of DRN neurons control sodium satiety signaling. Thirteen reports among others comprise part of this illustration: Lind (1986) has demonstrated a bidirectional connection between SFO and DRN. Hosono et al. (1999) raised the hypothesis that oxytocin (OT) is released in synapses on OT-sensitive neurons of SFO. Lenkei et al. (1995), Ray and Saavedra (1997), and Saavedra (1986) identified angiotensin II AT_1 receptors and atrial natriuretic peptide receptors in SFO. Tanaka et al. (1998) evidenced that SFO neurons expressing angiotensin II AT_1 receptors project toward DRN. Scrogin et al. (1998) showed that part of SFO neurons expressing serotonin $5\text{-}HT_{1A}$, $5\text{-}HT_{2A}$, and $5\text{-}HT_{2C}$ receptors also expresses angiotensin II AT_1 receptors. Hiyama et al. (2004) and Noda (2007) showed a close communication between Na_x channels on ependimal glial cell and GABAergic neurons representing basis for a sodium sensor at SFO. Cavalcante-Lima et al. (2005a,b) showed that damage of DRN serotonergic neurons increases sodium appetite raised by natriorexigenic paradigms that enhance angiotensin II plasma levels. Franchini et al. (2002) and Badauê-Passos et al. (2007) confirmed the bidirectional connection between SFO and DRN serotonergic neurons. Numbers display hypothetical sequential physiological steps implied on sodium appetite vs. sodium satiety responses. Other information can be viewed in the text. ANP-R, ANP receptor; AT_1R, angiotensin II AT_1 receptor; Ang II, angiotensin II; [Ang II]p, Ang II plasma concentration; 5-HT, serotonin; $5\text{-}HT_{1A}$-R, serotonin $5\text{-}HT_{1A}$ receptor; $5\text{-}HT_{2A}$-R, serotonin $5\text{-}HT_{2A}$ receptor; $5\text{-}HT_{2C}$-R, serotonin $5\text{-}HT_{2C}$ receptor; $GABA_AR$, $GABA_A$ receptor; OT-R, oxytocin receptor; $[Na^+]p$, Na^+ plasma concentration; [5-HT]e, extracellular 5-HT concentration; Na_x^+, Na^+ channel sensitive to $[Na^+]$. (Figure digitized by Dr. Luciano Gonçalves Fernandes.)

(Fanelli and McMonagle-Strucko 1992) or with changes in receptor G-protein coupling (Hensler and Durgam 2001), resembling antidepressant and anxiolytic treatments (Jolas et al. 1995; Picazo et al. 1995). However, it may involve a different cluster of serotonergic neurons because, according to the contextual reading from the Abrams (2004) report, topographically organized subpopulations of serotonergic neurons may have specific specialties of different behaviors and physiological properties. Thus, as the concentration of serotonin in the extracellular medium increases, 5-HT_{1A} autoreceptor desensitization would occur, allowing greater effectiveness of ascending serotonergic transmission to create sodium satiety. The desensitization would also produce a new electrophysiological timing in which the DRN somatodendritic membrane would become more excitable. In this condition, the serotonergic neurons would tend to a lower threshold of excitability. Consequently, satiety excitatory signals arriving at the DRN would create a facilitatory effect for the recruitment of inhibitory pathways projecting to the forebrain, particularly to structures of the lamina terminalis implicated in the genesis of sodium appetite behavior. However, these experimental and theoretical observations are not yet sufficient to create a complete understanding of the physiology behind homeostatic adjustment of water and electrolyte balance of the internal milieu.

Angiotensinergic (from the SFO), noradrenergic (from the locus coeruleus), and neurotensinergic (from the LPBN) pathways can modulate the activity of serotonergic neurons (Berridge and Waterhouse 2003; Cedraz-Mercez et al. 2007; Lind 1986; Petrov et al. 1992). Identification of angiotensin II AT_1 receptors in the DRN provides support for the existence of a possible feedback loop between the SFO and midbrain raphe (Moulik et al. 2002), as previously discussed. However, there are no reports on whether DRN neurons expressing AT_1 receptors are in fact serotonergic. Alternatively, serotonergic and GABAergic neurons within the DRN could be endowed with angiotensin II AT_1 and AT_2 receptors and angiotensin 1–7 Mas receptors. The nature of these receptors and the transduction of the cellular signal would allow variable cellular responses after stimulation by each peptide or by a spatial sum of their stimulation. In light of the physiological relevance of GABAergic neurons (Aghajanian et al. 1987; Liu et al. 2002; Sharkey et al. 2008) in the internal circuit of the DRN, there are multiple possibilities for altering extracellular serotonin concentration and serotonin 5-HT_{1A} autoreceptor activation after the arrival of excitatory and inhibitory inputs (Figure 11.1).

11.2.1.2 Acute Systemic and Intra-DRN Administration of a Serotonin 5-HT_{1A} Antagonist Decreases Sodium Intake in Sodium-Sated and Sodium-Deficient Rats

Previous reports in rats have shown that systemic administration of 5-HT_{1A} receptor antagonists increases the activity of serotonergic cells in the DRN (Arborelius et al. 1995). Thus, one may predict that a similar treatment would reduce sodium intake. We have tested this hypothesis using WAY100135 [*N-tert*-butyl-3-(4)-(2-methoxyphenyl) piperazin-1-yl)-2-phenylpro-pionamide di-hydrochloride], which is a 5-HT_{1A} autoreceptor antagonist (Fletcher et al. 1993; Routledge et al. 1993). Acute subcutaneous (s) administration of WAY100135 was performed in sodium-sated and sodium-depleted rats through furosemide injection 24 h before the experiment. In both situations, the central 5-HT_{1A} blockade induced a long-lasting decrease in sodium intake (Fonseca et al., in preparation).

Acute intra-DRN WAY100135 administration was also carried out in three other paradigms that induce bodily sodium deficiency: (1) sodium depletion induced by an acute injection of furosemide (sc), followed by a low-sodium diet for 24 h (Furosemide protocol); (2) acute sodium depletion produced by the interaction between furosemide and captopril (sc) 1 h before the experiment (Furocap protocol), in which the animals did not receive water or food (because captopril, which is an angiotensin-converting enzyme inhibitor, when associated with a diuretic/natriuretic substance, produces significant sodium and volume depletion); and (3) feeding only sodium-deficient food for 4 days followed by the Furocap protocol (modified Furocap protocol).

Intraraphe injection of WAY100135 produced long-term sodium intake inhibition in all sodium-induced deficiency models. Each of these three coexisting conditions (hyponatremia, hypovolemia, and arterial hypotension) is enough to induce sodium appetite; when all are present, their signaling is processed in concert to induce a more powerful craving for salt, resulting in a consummatory phase (Reis 2007; Thunhorst et al. 1990; Weisinger et al. 1997). Because each stimulus (hyponatremia, hypovolemia, and arterial hypotension) reached the lamina terminalis by different endocrine and neural afferent pathways (directly through circumventricular organs or sodium-sensitive channels or, indirectly, by the nucleus of solitary tract), our results suggest that the serotonergic system acts through differential projecting pathways and receptor mechanisms to induce sodium satiety. Sodium intake induced by triple stimulus is mediated by (1) the barosensitive mechanism (Thornton et al. 1995); (2) reduced aversion to salt during hyponatremia, as deduced from the hypothesis of Na_x^+ channels in the SFO (Hiyama et al. 2004; Noda 2007); and (3) elevated plasma angiotensin II levels, sensitizing the SFO, OVLT (both from the *lamina terminalis*), and area postrema, which are circumventricular organs (Fitzsimons 1998).

Thus, the results from acute systemic and intra-DRN 5-HT_{1A} antagonist administration raise an interesting conclusion: tonic serotonergic activity seems to modulate sodium consumption in both basal and sodium-depleted conditions, which reduces raphe serotonergic neuron activity as described by Franchini et al. (2002). Thus, these observations corroborate data from Franchini et al. (2002) and Badauê-Passos et al. (2007), who have demonstrated that serotonin-producing neurons in the DRN are excited (by c-Fos expression according to immunohistochemistry) after sodium consumption under a basal condition (sodium-sated condition) or after sodium depletion.

Additionally, the blockade of serotonin somatodendritic 5-HT_{1A} autoreceptors appears to facilitate the recruitment of excitatory signals to the serotonergic system, which then leads to increased serotonergic transmission and the manifestation of the sodium satiety status. It remains to be elucidated which excitatory inputs reach the serotonergic DRN neurons and through which pathway they come. Thus, taking into account the intense and long-lasting sodium satiety effect obtained after the blockade of the serotonin somatodendritic 5-HT_{1A} autoreceptors and the exaggerated sodium intake response in ibotenic DRN-damaged rats submitted to different experimental sodium-deficiency protocols associated with increased angiotensin II plasma levels (Cavalcante-Lima et al. 2005a,b; Lima et al. 2004; Olivares et al. 2003), it is plausible to suggest that the natriorexigenic stimulus is affected by changes in serotonergic ascending activity.

11.3 CONCLUDING REMARKS

According to the advanced morphofunctional approaches to understanding the DRN envisaged by Abrams et al. (2004), topographically arranged subpopulations of serotonergic neurons consist of exclusive clusters that project to distinct forebrain loci. From this point of view, it is therefore conceivable that homeostatic behaviors work concomitantly or serially with autonomic and neuroendocrine responses, both triggered to adjust the bodily fluid balance as they do other physiological systems.

The current data support the proposal that fluctuations in the activation of 5-HT_{1A} somatodendritic autoreceptors may affect the ascending serotonergic responsiveness related to sodium intake control. The data obtained from chronic treatment with a 5-HT_{1A} agonist led us to speculate that the increased firing rate of serotonergic neurons and the ensuing increase in extracellular serotonin levels may desensitize 5-HT_{1A} somatodendritic autoreceptors, thus allowing greater efficiency in serotonergic signaling of homeostatic sodium satiety during the consummatory phase.

Additional studies are needed to characterize the immunohistochemical and electrophysiological correlates underlying the activity of 5-HT_{1A} somatodendritic autoreceptors and the reciprocal morphological and physiological connections between the DRN and the lamina terminalis structures.

ACKNOWLEDGMENTS

This research was supported in part by grants from Conselho Nacional de Desenvolvimento Científico e Tecnológico (CNPq) and Fundação de Amparo à Pesquisa Carlos Chagas Filho do Rio de Janeiro (FAPERJ). We are indebted to Dr. Luciano Gonçalves Fernandes (Department of Physiological Sciences, Federal Rural University of Rio de Janeiro) for preparing Figure 11.1.

REFERENCES

Abrams, J. K., P. L. Johnson, J. H. Hollis et al. 2004. Anatomic and functional topography of the dorsal raphe nucleus. *Ann N Y Acad Sci* 1018:46–57.

Adell, A., P. Celada, M. T. Abellán et al. 2002. Origin and functional role of the extracellular serotonin in the midbrain raphe nuclei. *Brain Res Rev* 39:154–80.

Aghajanian, G. K., J. S. Sprouse, and K. Rasmussen. 1987. Physiology of the midbrain serotonin system. In: *Psychopharmacology: The Third Generation of Progress*, ed. H. Y. Meltzer, 141–9. New York: Raven Press.

Aghajanian, G. K. 1982. Regulation of serotonergic neuronal activity: Autoreceptors and pacemaker potentials. *Adv Biochem Psychopharmacol* 34:173–81.

Albert, P. R., P. Lembo, J. M. Storring et al. 1996. The 5-HT1A receptor: Signaling, desensitization, and gene transcription. *Neuropsychopharmacology* 14:19–25.

Allers, K. A., and T. Sharp. 2003. Neurochemical and anatomical identification of fast- and slow-firing neurones in the rat dorsal raphe nucleus using juxtacellular labelling methods in vivo. *Neuroscience* 122:193–204.

Andrade, R., and R. A. Nicoll. 1987. Pharmacologically distinct actions of serotonin on single pyramidal neurones of the rat hippocampus recorded in vitro. *J Physiol* 394:99–124.

Antunes-Rodrigues, J., M. De-Castro, L. L. Elias et al. 2004. Neuroendocrine control of body fluid metabolism. *Physiol Rev* 84:169–208.

Arborelius, L., G. G. Nomikos, P. Grillner et al. 1995. 5-HT1A receptor antagonists increase the activity of serotonergic cells in the dorsal raphe nucleus in rats treated acutely or chronically with citalopram. *Naunyn Schmiedebergs Arch Pharmacol* 352:157–65.

Assié, M. B., H. Lomenech, V. Ravailhe et al. 2006. Rapid desensitization of somatodendritic 5-HT1A receptors by chronic administration of the high-efficacy 5-HT1A agonist, F13714: A microdialysis study in the rat. *Br J Pharmacol* 149:170–8.

Azmitia, E. C. 2001. Modern views on an ancient chemical: Serotonin effects on cells proliferation, maturation, and apoptosis. *Brain Res Bull* 56:413–24.

Azmitia, E. C. 1987. The CNS serotonergic system: Progression toward a collaborative organization. In: *Psychopharmacology: The Third Generation of Progress*, ed. H. Y. Meltzer, 61–73. New York: Raven Press.

Azmitia, E. C., and M. Segal. 1978. An autoradiographic analysis of the differential ascending projections of the dorsal and median raphe nuclei in the rat. *J Comp Neurol* 179:641–68.

Badauê-Passos, D. Jr., A. Godino, A. K. Johnson et al. 2007. Dorsal raphe nuclei integrate allostatic information evoked by depletion-induced sodium ingestion. *Exp Neurol* 206:86–97.

Badauê-Passos, D. Jr., R. R. Ventura, L. F. S. Silva et al. 2003. Effect of brain serotoninergic stimulation on sodium appetite of euthyroid and hypothyroid rats. *Exp Physiol* 88:251–60.

Bayliss, D. A., M. Umemiya, and A. J. Berger. 1995. Inhibition of N- and P-type calcium currents and the after-hyperpolarization in rat motoneurones byserotonin. *J Physiol* 485:635–47.

Beer, M., G. A. Kennett, and G. Curzon. 1990. A single dose of 8-OH-DPAT reduces raphe binding of [^{3}H]8-OH-DPAT and increases the effect of raphe stimulation on 5-HT metabolism. *Eur J Pharmacol* 178:179–87.

Berridge, C. W., and B. D. Waterhouse. 2003. The locus coeruleus–noradrenergic system: Modulation of behavioral state and state-dependent cognitive processes. *Brain Res Rev* 42:33–84.

Blier, P., G. Piñeyro, M. Mansari et al. 1998. Role of somatodendritic 5-HT autoreceptors in modulating 5-HT neurotransmission. *Ann N Y Acad Sci* 861:204–16.

Blier, P., and C. de Montigny. 1990. Electrophysiological investigation of the adaptive response of the 5-HT system to the administration of 5-HT1A receptor agonists. *J Cardiovasc Pharmacol* 7:S42–8.

Boadle-Biber, M. C. 1993. Regulation of serotonin synthesis. *Prog Biophys Mol Biol* 60:1–15.

Bockaert, J., S. Claeysen, C. Bécamel et al. 2006. Neuronal 5-HT metabotropic receptors: Fine-tuning of their structure, signaling, and roles in synaptic modulation. *Cell Tissue Res* 326:553–72.

Bortolozzi, A., M. Amargós-Bosch, M. Toth et al. 2004. In vivo efflux of serotonin in the dorsal raphe nucleus of 5-HT1A receptor knockout mice. *J Neurochem* 88:1373–9.

Bosler, O., and L. Descarries. 1988. Monoamine innervation of the organum vasculosum laminae terminalis (OVLT): A high resolution radioautographic study in the rat. *J Comp Neurol* 272:545–61.

Castro, L., R. Athanazio, M. Barbetta et al. 2003. Central 5-HT2B/2C and 5-HT3 receptor stimulation decreases salt intake in sodium-depleted rats. *Brain Res* 981:151–8.

Cavalcante-Lima, H. R., H. R. Lima, R. H. Costa-e-Sousa et al. 2005a. Dipsogenic stimulation in ibotenic DRN-lesioned rats induces concomitant sodium appetite. *Neurosci Lett* 374:5–10.

Cavalcante-Lima, H. R., D. Badauê-Passos Jr., W. de-Lucca Jr. et al. 2005b. Chronic excitotoxic lesion of the dorsal raphe nucleus induces sodium appetite. *Braz J Med Biol Res* 38:1669–75.

Cedraz-Mercez, P. L., A. S. Mecawi, A. Lepletier et al. 2007. Noradrenergic stimulation within midbrain raphe increases electrolyte excretion in rats. *Exp Physiol* 92:923–31.

Celada, P., M. Puig, M Amargós-Bosch et al. 2004. The therapeutic role of 5-HT1A and 5-HT2A receptors in depression. *J Psychiatry Neurosci* 29:252–65.

Chalmers, D. T., and S. J. Watson. 1991. Comparative anatomical distribution of 5-HT receptor mRNA and 5-HT binding in rat brain—a combined in situ hybridisation/in vitro receptor autoradiographic study. *Brain Res* 561:51–60.

Cheng, L. L., S. J. Wang, and P. W. Gean. 1998. Serotonin depresses excitatory synaptic transmission and depolarization-evoked Ca^{2+} influx in rat basolateral amygdala via 5-HT1A receptors. *Eur J Neurosci* 10:2163–72.

Clark, M. S., R. A. McDevit, and J. F. Neumaier. 2006. Quantitative mapping of tryptophan hydroxylase-2, 5-HT1A, 5-HT1B, and serotonin transporter expression across the anteroposterior axis of the rat dorsal and median raphe nuclei. *J Comp Neurol* 498:611–23.

Colombari, D. S., J. V. Menani, and A. K. Johnson. 1996. Forebrain angiotensin type 1 receptors and parabrachial serotonin in the control of NaCl and water intake. *Am J Physiol Regul Integr Comp Physiol* 271:R1470–6.

Cooper, S. J., M. J. Fryer, and J. C. Neill. 1988. Specific effect of putative 5-HT1A, agonists, 8-OH-DPAT and gepirone, to increase hypertonic saline consumption in the rat. Evidence against a general hyperdipsic action. *Physiol Behav* 43:533–7.

Fanelli, R. J., and K. McMonagle-Strucko. 1992. Alteration of 5-HT1A receptor binding sites following chronic treatment with ipsapirone measured by quantitative autoradiography. *Synapse* 1:75–81.

Fitzsimons, J. T. 1998. Angiotensin, thirst, and sodium appetite. *Physiol Rev* 78:583–686.

Fletcher, A., D. J. Bill, S. J. Bill et al. 1993. WAY100135: A novel, selective antagonist at presynaptic and postsynaptic 5-HT1A receptors. *Eur J Pharmacol* 24:283–91.

Fonseca, F. V., A. S. Mecawi, I. G. Araujo et al. 2009. Role of the 5-HT1A somatodendritic autoreceptor in the dorsal raphe nucleus on salt satiety signaling in rats. *Exp Neurol* 217:253–60.

Franchini, L. F., A. K. Johnson, J. de Olmos et al. 2002. Sodium appetite and Fos activation in serotonergic neurons. *Am J Physiol Regul Integr Comp Physiol* 282:R235–43.

Garratt, J. C., F. Crespi, R. Mason et al. 1991. Effect of idazoxan on DRN 5-HT neuronal function. *Eur J Pharmacol* 193:87–93.

Gaspar, P., O. Cases, and L. Maroteaux. 2003. The developmental role of serotonin: News from mouse molecular genetics. *Nat Rev Neurosci* 12:1002–12.

Godino, A., L. O. Margatho, X. E. Caeiro et al. 2010. Activation of lateral parabrachial afferent pathways and endocrine responses during sodium appetite regulation. *Exp Neurol* 221:275–84.

Godino, A., L. A. De Luca Jr., J. Antunes-Rodrigues et al. 2007. Oxytocinergic and serotonergic systems involvement in sodium intake regulation: Satiety or hypertonicity markers? *Am J Physiol Regul Integr Comp Physiol* 293:R1027–36.

Grobe, L. J., D. Xu, and D. C. Sigmund. 2008. An intracellular renin–angiotensin system in neurons: Fact, hypothesis, or Fantasy. *Physiology* 23:187–93.

Haanwinckel, M. A., L. K. Elias, A. L. Favaretto et al. 1995. Oxytocin mediates atrial natriuretic peptide release and natriuresis after volume expansion in the rat. *Proc Natl Acad Sci USA* 92:7902–6.

Hajós, M., S. E. Gartside, A. E. Villa et al. 1995. Evidence for a repetitive (burst) firing pattern in a sub-population of 5-hydroxytryptamine neurons in the dorsal and median raphe nuclei of the rat. *Neuroscience* 69:189–97.

Hajós, M., K. A. Allers, K. Jennings et al. 2007. Neurochemical identification of stereotypic burst-firing neurons in the rat dorsal raphe nucleus using juxtacellular labelling methods. *Eur J Neurosci* 25:119–26.

Hensler, J., and H. Durgam. 2001. Regulation of 5-HT(1A) receptor-stimulated [^{35}S]-GTPgammaS binding as measured by quantitative autoradiography following chronic agonist administration. *Br J Pharmacol* 132:605–11.

Hiyama, T. Y., E. Watanabe, H. Okado et al. 2004. The subfornical organ is the primary locus of sodium-level sensing by Nax sodium channels for the control of salt intake behavior. *J Neurosci* 24:9276–81.

Hjorth, S., and T. Magnusson. 1988. The 5-HT 1A receptor agonist, 8-OH-DPAT, preferentially activates cell body 5-HT autoreceptors in rat brain in vivo. *Naunyn Schmiedebergs Arch Pharmacol* 338:463–71.

Hjorth, S., H. J. Bengtsson, A. Kullberg et al. 2000. Serotonin autoreceptor function and antidepressant drug action. *J Psychopharmacol* 14:177–85.

Hosono, T., H. A. Schmid, E. Kanosue et al. 1999. Neuronal actions of oxytocin on the subfornical organ of male rats. *Am J Physiol Endocrinol Metab* 276:E1004–8.

Hutson, P. H., G. S. Sarna, M. T. O'Connel et al. 1989. Hippocampal 5-HT synthesis and release in vivo is decreased by infusion of 8-OHDPAT into the nucleus raphe dorsalis. *Neurosci Lett* 100:276–80.

Jacobs, B. L., and E. C. Azmitia. 1992. Structure and function of the brain serotonin system. *Physiol Rev* 72:165–231.

Johnson, A. K., and R. L. Thunhorst. 1997. The neuroendocrinology of the thirst and salt appetite: Visceral sensory signals and mechanisms of central integration. *Front Neuroendocrinol* 18:292–353.

Jolas, T., R. Schreiber, A. M. Laporte et al. 1995. Are postsynaptic 5-HT1A receptors involved in the anxiolytic effects of 5-HT1A receptor agonists and in their inhibitory effects on the firing of serotonergic neurons in the rat? *J Pharmacol Exp Ther* 272:920–9.

Jolas, T., and G. K. Aghajanian. 1997. Neurotensin and the serotonergic system. *Prog Neurobiol* 52:455–68.

Kia, H. K., M. C. Miquel, M. J. Brisorgueil et al. 1996. Immunocytochemical localization of serotonin receptors in the rat central nervous system. *J Comp Neurol* 365:289–305.

Lanfumey, L., and M. Hamon. 2004. 5-HT1 receptors. *Curr Drug Targets CNS Neurol Disord* 3:1–10.

Lauder, J. M., and H. Krebs. 1978. Serotonin as a differentiation signal in early neurogenesis. *Dev Neurosci* 1:15–30.

Lenkei, Z., P. Corvol, and C. Llorens-Cortes. 1995. The angiotensin receptor subtype AT1A predominates in rat forebrain areas involved in blood pressure, body fluid homeostasis and neuroendocrine control. *Mol Brain Res* 30:53–60.

Lima, H. R. C., H. R. Cavalcante-Lima, P. L. Cedraz-Mercez et al. 2004. Brain serotonin depletion enhances the sodium appetite induced by sodium depletion or beta-adrenergic stimulation. *An Acad Bras Cienc* 76:85–92.

Lind, R. W. 1986. Bi-directional, chemically specified neural connections between the subfornical organ and the midbrain raphe system. *Brain Res* 384:250–61.

Liu, R.-J., E. K. Lambe, and G. K. Aghajanian. 2005. Somatodendritic autoreceptor regulation of serotonergic neurons: Dependence on l-tryptophan and tryptophan hydroxylase-activating kinases. *Eur J Neurosci* 21:945–58.

Liu, R. J., A. N. van den Pol, and G. K. Aghajanian. 2002. Hypocretins (orexins) regulate serotonin neurons in the dorsal raphe nucleus by excitatory direct and inhibitory indirect actions. *J Neurosci* 22:9453–64.

Margatho, L. O., A. Godino, F. R. Oliveira et al. 2008. Lateral parabrachial afferent areas and serotonin mechanisms activated by volume expansion. *J Neurosci Res* 86:3613–21.

Marinelli, S., S. A. Schnell, S. P. Hack et al. 2004. Serotonergic and nonserotonergic dorsal raphe neurons are pharmacologically and electrophysiologically heterogeneous. *J Neurophysiol* 92:3532–7.

Menani, J. V., S. P. Barbosa, L. A. De Luca Jr. et al. 2002. Serotonergic mechanisms of the lateral parabrachial nucleus and cholinergic-induced sodium appetite. *Am J Physiol Regul Integr Comp Physiol* 282:R837–41.

Menani, J. V., D. S. Colombari, T. G. Beltz et al. 1998a. Salt appetite: Interaction of forebrain angiotensinergic and hindbrain serotonergic mechanisms. *Brain Res* 801:29–35.

Menani, J. V., L. A. De Luca Jr., and A. K. Johnson. 1998b. Lateral parabrachial nucleus serotonergic mechanisms and salt appetite induced by sodium depletion. *Am J Physiol Regul Integr Comp Physiol* 274:R555–60.

Menani, J. V., R. L. Thunhorst, and A. K. Johnson. 1996. Lateral parabrachial nucleus and serotonergic mechanisms in the control of salt appetite in rats. *Am J Physiol Regul Integr Comp Physiol* 270:R162–8.

Montes, R., and A. K. Johnson. 1990. Efferent mechanisms mediating renal sodium and water excretion induced by centrally administered serotonin. *Am J Physiol Regul Integr Comp Physiol* 259:R1267–73.

Moulik, S., R. C. Speth, B. B. Turner et al. 2002. Angiotensin II receptor subtype distribution in the rabbit brain. *Exp Brain Res* 142:275–83.

Munaro, N., and E. Chiaraviglio. 1981. Hypothalamic levels and utilization of noradrenaline and 5-hydroxytryptamine in the sodium-depleted rat. *Pharmacol Biochem Behav* 15:1–5.

Neill, J. C., and S. J. Cooper. 1989. Selective reduction by serotonergic agents of hypertonic saline consumption in rats. Evidence for 5-HT1C receptor mediation. *Psychopharmacology* 99:196–201.

Noda, M. 2007. Hydromineral neuroendocrinology: Mechanism of sensing sodium levels in the mammalian brain. *Exp Physiol* 92:513–22.

Olivares, E. L., R. H. Costa-e-Sousa, H. R. Cavalcante-Lima et al. 2003. Effect of electrolytic lesion of the dorsal raphe nucleus on water intake and sodium appetite. *Braz J Med Biol Res* 36:1709–16.

Parent, A., L. Descarries, and A. Beaudet. 1981. Organization of ascending serotonin systems in the adult rat brain. A radioautographic study after intraventricular administration of [3H]5-hydroxytryptamine. *Neuroscience* 6:115–38.

Petrov, T., J. H. Jhamandas, and T. L. Krukoff. 1992. Characterization of peptidergic efferents from the lateral parabrachial nucleus to identified neurons in the rat dorsal raphe nucleus. *J Chem Neuroanat* 5:367–73.

Picazo, O., C. López-Rubalcava, and A. Fernández-Guasti. 1995. Anxiolytic effect of the 5-HT1A compounds 8-hydroxy-2-(di-*n*-propylamino) tetralin and ipsapirone in the social interaction paradigm: Evidence of a presynaptic action. *Brain Res Bull* 37:169–75.

Polter, A. M., and X. Li. 2010. 5-HT1A receptor-regulated signal transduction pathways in brain. *Cell Signal* 22:1406–12.

Ray, P. E., and J. M. Saavedra. 1997. Selective chronic sodium or chloride depletion specifically modulates subfornical organ atrial natriuretic peptide receptor number in young rats. *Cell Mol Neurobiol* 17:455–70.

Reis, L. C. 2007. Role of the serotoninergic system in the sodium appetite control. *An Acad Bras Cienc* 79:261–83.

Reis, L. C., M. J. Ramalho, A. L. Favaretto et al. 1994. Participation of the ascending serotonergic system in the stimulation of atrial natriuretic peptide release. *Proc Nat Acad Sci USA* 91:12022–26.

Reis, L. C., M. J. Ramalho, and J. Antunes-Rodrigues. 1991. Effect of central administration of serotoninergic agonists on electrolyte excretion control. *Braz J M Biol Res* 24:633–41.

Richardson-Jones, J. W., C. P. Craige, B. P. Guiard et al. 2010. 5-HT1A autoreceptor levels determine vulnerability to stress and response to antidepressants. *Neuron* 65:40–52.

Rouah-Rosilio, M., M. Orosco, and S. Nicolaidis. 1994. Serotoninergic modulation of sodium appetite in the rat. *Physiol Behav* 55:811–6.

Routledge, C., J. Gurling, I. K. Wright et al. 1993. Neurochemical profile of the selective and silent 5-HT1A receptor antagonist WAY100135: An in vivo microdialysis study. *Eur J Pharmacol* 239:195–202.

Saavedra, J. M. 1986. Atrial natriuretic peptide (6–33) binding sites: Decreased number and affinity in the subfornical organ of spontaneously hypertensive rats. *J Hypertens* Suppl 4:S313–6.

Scrogin, K. E., A. K. Johnson, and H. A. Schmid. 1998. Multiple receptor subtypes mediates the effect of serotonin on rat subfornical organ neurons. *Am J Physiol Regul Integr Comp Physiol* 275:R2035–42.

Sharkey, L. M., S. G. Madamba, G. R. Siggins et al. 2008. Galanin alters GABAergic neurotransmission in the dorsal raphe nucleus. *Neurochem Res* 33:285–91.

Sotelo, C., B. Cholley, S. el Mestikawy et al. 1990. Direct immunohistochemical evidence of the existence of 5-HT1A autoreceptors on serotoninergic neurons in the midbrain raphe nuclei. *Eur J Neurosci* 2:1144–54.

Sprouse, J. S., and G. K. Aghajanian. 1987. Electrophysiological responses of serotoninergic dorsal raphe neurons to 5-HT1A and 5-HT1B agonists. *Synapse* 1:3–9.

Stricker, E. M., and J. G. Verbalis. 1996. Central inhibition of salt appetite by oxytocin in rats. *Regul Pept* 66:83–5.

Tada, K., K. Kasamo, N. Ueda et al. 1999. Anxiolytic 5-hydroxytryptamine1A agonists suppress firing activity of dorsal hippocampus CA1 pyramidal neurons through a postsynaptic mechanism: Single-unit study in unanesthetized, unrestrained rats. *J Pharmacol Exp Ther* 288:843–8.

Tanaka, J., A. Ushigome, K. Hori et al. 1998. Responses of raphe nucleus projecting subfornical organ neurons to angiotensin II in rats. *Brain Res Bull* 45:315–8.

Tanaka, J., Y. Hayashi, K. Yamato et al. 2004. Involvement of serotonergic systems in the lateral parabrachial nucleus in sodium and water intake: A microdialysis study in the rat. *Neurosci Lett* 26:41–4.

Thornton, S. N., A. Sanchez, and S. Nicolaïdis. 1995. An angiotensin-independent, hypotension-induced, sodium appetite in the rat. *Physiol Behav* 57:555–61.

Thunhorst, R. L., M. Morris, and A. K. Johnson. 1994. Endocrine changes associated with a rapidly developing sodium appetite in rats. *Am J Physiol Regul Integr Comp Physiol* 267:R1168–73.

Thunhorst, R. L., K. J. Ehrlich, and J. B. Simpson. 1990. Subfornical organ participates in salt appetite. *Behav Neurosci* 104:637–42.

Tyce, G. M. 1990. Origin and metabolism of serotonin. *J Cardiovasc Pharmacol* 16:S1–S7.

Weisinger, R. S., J. R. Blair-West, P. Burns et al. 1997. Role of brain angiotensin in thirst and sodium appetite of rats. *Peptides* 18:977–84.

12 A Classic Innate Behavior, Sodium Appetite, Is Driven by Hypothalamic Gene-Regulatory Programs Previously Linked to Addiction and Reward

Wolfgang Liedtke

CONTENTS

12.1 OUR RECENT RESULTS AGAINST CLASSIC CONCEPTS OF INSTINCTIVE BEHAVIOR

Instinctive behavior is intricately linked to organismal homeostasis and species survival (Denton et al. 1996, 2009; Lehrman 1953; Richter 1956). Instincts are embodied by the brain, in particular, the hypothalamus. The classic instinct of sodium appetite is commonly evoked by depletion, that is, lack of appropriate intake in the face of sodium loss, further by adrenocorticotropic hormone (ACTH) acting on the suprarenal cortex to produce gluco- and mineralocorticoids, which act on the brain, and associated with reproduction in pregnancy and lactation (Blair-West et al. 1995; De Luca et al. 2010; Denton et al. 1999; Johnson and Thunhorst 1997; Fitzsimons

1998; Morris et al. 2008; Na et al. 2007; Roitman et al. 1997; Wolf 1964). In a recent publication (Liedtke et al. 2011), we have elucidated gene-regulatory programs in the hypothalamus as underlying sodium appetite. Furthermore, we were able to attenuate sodium intake behavior of sodium appetite by application of antagonists of dopamine receptor-1 (DRD1) both systemically as well as by microinjection into the lateral hypothalamus. This chapter will elucidate these results and discuss them in a number of contexts not covered in the original publication. In the original report, the term "gratification" is used throughout. This refers to the subjective sentiment of feeling gratified after satiation of an instinctive craving—in this context, sodium appetite and thirst for water—but in a general sense applicable to any instinct. Gratification, it can be argued, is difficult to verify in a setting of animal experimentation. However, its use to describe animal behavior by veteran researchers in the field is noteworthy and should be respected.

12.2 MORE DETAILED REVISITING OF OUR RESULTS

In our study (Liedtke et al. 2011), using microarrays, we found that genes related to monoaminergic signaling and reward pathways were regulated in the mammalian hypothalamus. This prompted dedicated follow-up with gene set enrichment analysis (GSEA) (Subramanian et al. 2005) in order to address whether there is coordinated gene expression following previously established mechanisms of concerted gene-regulatory programs (Figure 12.1). Surprisingly, gene sets previously found to be enriched in addiction to opioids and cocaine were significantly associated with sodium appetite (Liedtke et al. 2011). We then focused on DARPP32, a critical intraneuronal signaling molecule for addiction and reward (Bateup et al. 2008; Nairn et al. 2004; Ouimet et al. 1984; Stipanovich et al. 2008; Svenningsson et al. 2003), and detected that it was regulated in lateral hypothalamic neurons that also expressed orexin. These neurons have previously been shown to be part of an addiction–reward circuit (Aston-Jones et al. 2006, 2009; Boutrel et al. 2005; Carter et al. 2009; Harris and Aston-Jones 2006; Harris et al. 2005; Rao et al. 2008; Sharf et al. 2008). Next, we inhibited dopamine receptors and mGlu5. All inhibitors attenuated sodium appetite. Antagonists of DRD1 did not influence thirst for water, which the other inhibitors did. Therefore, we established the specificity of the DRD1 blocking compound by verifying the absence of an effect in *Drd1*$^{-/-}$ mice, and then microinjected the DRD1 antagonist topically into the lateral hypothalamus on both sides. This led to a striking reduction of sodium appetite.

12.3 PARTICULAR FEATURES OF GSEA

Based on these findings, it is worth noting that we obtained the metrical parameters of GSEA that were hinting at gene set enrichment, yet that left final uncertainty, when we were analyzing sodium appetite evoked either by depletion or by ACTH. By combining a need-free and a depletion-based strategy to evoke sodium appetite, we used an experimental approach so that the two different experimental paradigms shared the induction of sodium appetite, but not the associated stresses and secondary effects of either method of induction. This led to an increase of GSEA metrical parameters indicative of enrichment. Increased significance of gene set enrichment, in turn, can only be explained by sodium appetite being causally linked to this particular instance of concerted gene regulation, and it appears very unlikely that this could be an epiphenomenon. In other words, the default for combining gene expression data from two conditions is a "watering down" of the tightness of concerted gene regulation, and the more genes that are coregulated in one condition, the more this will become sensitive to "watering down" when a second condition is added, which is not sharing a similar mode of concerted gene regulation. Finally, in regard to the concerted regulation of genes previously associated with alcohol addiction, we note that GSEA metrics fall short of significance levels by a very small degree in combined datasets. Yet when looking at significance levels upon combining data from need-free and depletion-induced sodium appetite, the alcohol gene set has the steepest increase of all sets. We believe that this indicates that alcoholism shares many significant features with addiction to opioids and cocaine, but that there are also pathways and gene regulation involved in alcoholism that are significantly different from its "dry" counterparts.

These results were surprising. In view of our experience with sodium appetite and addiction-related genes, we do recommend the concept of combining two stimuli that evoke the same physiological or pathological condition/state. Then gene expression in an organ or group of cells can be determined by microarray (with genome-wide coverage) or RNA-seq methodology, followed by GSEA. By analogy, there is no reason why this approach should not also work with proteomics data. Gene sets that will be enriched more significantly upon combination of the datasets are likely to embody a gene-regulatory correlate to the condition under study.

12.4 SODIUM APPETITE AND ITS DEPENDENCE ON DOPAMINE RECEPTOR 1

Another aspect deserving comment is the strict dependence of sodium appetite on DRD1 signaling. This is shown by systemic application of the compound SCH23990, a known DRD1 antagonist, which we found to be very powerful, in a dose-dependent manner. It has no off-target effects as shown by its lack of effects in *Drd1*$^{-/-}$ mice, and it showed no significant effects on thirst for water (antagonism of DRD2 does). Not too surprisingly, *Drd1*$^{-/-}$ mice showed a sodium appetite, implying genetic compensatory mechanisms, but they were completely unaffected by SCH23990. Moreover, bilateral microinjection of SCH23990 into the rat lateral hypothalamus, targeted to the area of the lateral hypothalamus (LH) that harbors orexin-expressing neurons, completely eliminated (!) sodium appetite in two rats with the more lateral targeting, and substantially attenuated it in two rats with more medial targeting (Liedtke et al. 2011). It is attractive to postulate direct expression of DRD1 receptors on LH orexin-expressing neurons given the previously established role of these neurons in addictive circuitry (Harris and Aston-Jones 2006; Harris et al. 2005). However, we remain well aware of an alternative explanation, namely, DRD1 expression on projections that are located in this area. These projections would extend into known or as yet unknown circuitry that is critical for sodium appetite.

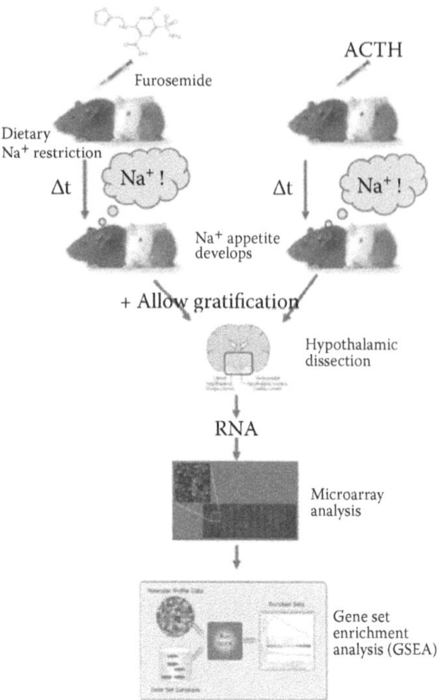

Identification of enriched gene sets that have previously been found associated with addictions to opioids and cocaine

Gratification immediately loosens lateral hypothalamus gene regulation (!)

Lateral hypothalamus neurons that express orexin, critical players in addiction brain circuitry, upregulate DARPP32 and ARC

Antagonism of DRD1, both systemically and by targeted microinjection into the lateral hypothalamus, greatly attenuates sodium appetite

FIGURE 12.1 **(See color insert.)** Left-hand illustration depicts experimental workflow; right-hand side statements reiterate critical results obtained.

The latter possibility—and more so established work on the role of the nucleus accumbens in sodium appetite (Roitman et al. 1999, 2002, 2005; Na et al. 2007)—Suggests that orexin-expressing LH neurons are unlikely to be the only location in which blocking DRD1 will attenuate sodium appetite.

12.5 SATIETY/GRATIFICATION GENE REGULATION IN VIRTUAL ABSENCE OF PHYSIOLOGICAL RE-EQUILIBRATION

The above explanations have obvious and logical extensions that will be pursued further in future studies. However, one more result from our study deserves highlighting here. We found that the tightness of gene regulation does indeed "water down" upon gratification of the specific need. Many genes that were regulated in one direction reversed upon gratification. This finding has to be viewed against the fact that for the satiation/gratification group, the hypothalamic RNA was sampled directly after the animals stopped drinking 0.3 M NaCl. We therefore view this finding as no less than a fascinating molecular correlate of satiety. Because the majority of mature and protein-encoding mRNAs is not sufficiently unstable to undergo natural degradation/decay within this time frame, we strongly suspect the process to be caused by small noncoding RNA molecules that target mature mRNAs for subsequent rapid decay. Micro-RNAs are one such type of small noncoding RNAs (Hwang et al.

2004; Seoane et al. 2003; Sethi and Lukiw 2009; Tapocik et al. 2009; Trivedi and Ramakrishna 2009; Yoo et al. 2004). We thus hypothesize the existence of "satiety miRNAs." Knowledge of their molecular identity would indeed be of considerable relevance, and questions could be asked as to the molecular specificity of "satiety miRNAs" for related—yet specifically distinct—instincts, such as thirst for water and hunger, in relation to sodium appetite.

12.6 SUM-UP

In sum, for the classic instinct of sodium appetite, we have established that groups of genes previously shown to be associated with drug addiction were significantly regulated in the hypothalamus of mice rendered hungry for salt. Administration of dopamine- and metabotropic glutamate-5 receptor antagonists attenuated satiation of sodium appetite in mice and rats. We were able to attenuate satiation/gratification behavior of sodium appetite by application of antagonists of DRD1 both systemically as well as by microinjection into the LH. In this chapter, several particular and previously underappreciated aspects of GSEA and of the gene changes observed with satiation/gratification were discussed in the context of our own results, observations of others, and the prevailing concepts about instinctive and addiction-related behavior.

REFERENCES

Aston-Jones, G., R. J. Smith, D. E. Moorman and K. A. Richardson. 2009. Role of lateral hypothalamic orexin neurons in reward processing and addiction. *Neuropharmacology* 56:112–21.

Aston-Jones, G., R. J. Smith, G. C. Sartor et al. 2006. Lateral hypothalamic orexin/hypocretin neurons: A role in reward-seeking and addiction. *Brain Res* 1314:74–90.

Bateup, H. S., P. Svenningsson, M. Kuroiwa et al. 2008. Cell type-specific regulation of DARPP-32 phosphorylation by psychostimulant and antipsychotic drugs. *Nat Neurosci* 11:932–9.

Blair-West, J. R., D. A. Denton, M. McBurnie, E. Tarjan and R. S. Weisinger. 1995. Influence of adrenal steroid hormones on sodium appetite of Balb/c mice. *Appetite* 24:11–24.

Boutrel, B., P. J. Kenny, S. E. Specio et al. 2005. Role for hypocretin in mediating stress-induced reinstatement of cocaine-seeking behavior. *Proc Natl Acad Sci USA* 102:19168–73.

Carter, M. E., J. S. Borg and L. de Lecea. 2009. The brain hypocretins and their receptors: Mediators of allostatic arousal. *Curr Opin Pharmacol* 9:39–45.

De Luca Jr., L. A., D. T. Pereira-Derderian, R. C. Vendramini, R. B. David and J. V. Menani. 2010. Water deprivation-induced sodium appetite. *Physiol Behav* 100:535–44.

Denton, D. A., J. R. Blair-West, M. I. McBurnie, J. A. Miller, R. S. Weisinger and R. M. Williams. 1999. Effect of adrenocorticotrophic hormone on sodium appetite in mice. *Am J Physiol Regul Integr Comp Physiol* 277:R1033–40.

Denton, D. A., M. J. McKinley, M. Farrell and G. F. Egan. 2009. The role of primordial emotions in the evolutionary origin of consciousness. *Conscious Cogn* 18:500–14.

Denton, D. A., M. J. McKinley and R. S. Weisinger. 1996. Hypothalamic integration of body fluid regulation. *Proc Natl Acad Sci USA* 93:7397–404.

Fitzsimons, J. T. 1998. Angiotensin, thirst, and sodium appetite. *Physiol Rev* 78:583–686.

Harris, G. C. and G. Aston-Jones. 2006. Arousal and reward: A dichotomy in orexin function. *Trends Neurosci* 29:571–7.

Harris, G. C., M. Wimmer and G. Aston-Jones. 2005. A role for lateral hypothalamic orexin neurons in reward seeking. *Nature* 437:556–9.

Hwang, C. K., C. S. Kim, H. S. Choi, S. R. McKercher and H. H. Loh. 2004. Transcriptional regulation of mouse mu opioid receptor gene by PU.1. *J Biol Chem* 279:19764–74.

Johnson, A. K. and R. L. Thunhorst. 1997. The neuroendocrinology of thirst and salt appetite: Visceral sensory signals and mechanisms of central integration. *Front Neuroendocrinol* 18:292–353.

Lehrman, D. S. 1953. A critique of Konrad Lorenz's theory of instinctive behavior. *Q Rev Biol* 28:337–78.

Liedtke, W. B., M. J. McKinley, L. L. Walker et al. 2011. Relation of addiction genes to hypothalamic gene changes subserving genesis and gratification of a classic instinct, sodium appetite. *Proc Natl Acad Sci USA* 108:12509–14.

Morris, M. J., E. S. Na and A. K. Johnson. 2008. Salt craving: The psychobiology of pathogenic sodium intake. *Physiol Behav* 94:709–21.

Na, E. S., M. J. Morris, R. F. Johnson, T. G. Beltz and A. K. Johnson. 2007. The neural substrates of enhanced salt appetite after repeated sodium depletions. *Brain Res* 1171:104–10.

Nairn, A. C., P. Svenningsson, A. Nishi, G. Fisone, J. A. Girault and P. Greengard. 2004. The role of DARPP-32 in the actions of drugs of abuse. *Neuropharmacology* 47:S14–23.

Ouimet, C. C., P. E. Miller, H. C. Hemmings, Jr., S. I. Walaas and P. Greengard. 1984. DARPP-32, a dopamine- and adenosine 3′:5′-monophosphate-regulated phosphoprotein enriched in dopamine-innervated brain regions: III. Immunocytochemical localization. *J Neurosci* 4:111–24.

Rao, Y., M. Lu, F. Ge et al. 2008. Regulation of synaptic efficacy in hypocretin/orexin-containing neurons by melanin concentrating hormone in the lateral hypothalamus. *J Neurosci* 28:9101–10.

Richter, C. 1956. Salt appetite of mammals: Its dependence on instinct and metabolism. *L'Instinct dans le comportement des animaux et de l'homme*. Paris: Masson et Cie.: 577–633.

Roitman, M. F., E. Na, G. Anderson, T. A. Jones and I. L. Bernstein. 2002. Induction of a salt appetite alters dendritic morphology in nucleus accumbens and sensitizes rats to amphetamine. *J Neurosci* 22:RC225.

Roitman, M. F., T. A. Patterson, R. R. Sakai, I. L. Bernstein and D. P. Figlewicz. 1999. Sodium depletion and aldosterone decrease dopamine transporter activity in nucleus accumbens but not striatum. *Am J Physiol Regul Integr Comp Physiol* 276:R1339–45.

Roitman, M. F., G. E. Schafe, T. E. Thiele and I. L. Bernstein. 1997. Dopamine and sodium appetite: Antagonists suppress sham drinking of NaCl solutions in the rat. *Behav Neurosci* 111:606–11.

Roitman, M. F., R. A. Wheeler and R. M. Carelli. 2005. Nucleus accumbens neurons are innately tuned for rewarding and aversive taste stimuli, encode their predictors, and are linked to motor output. *Neuron* 45:587–97.

Seoane, L. M., M. Lopez, S. Tovar, F. F. Casanueva, R. Senaris and C. Dieguez. 2003. Agouti-related peptide, neuropeptide Y, and somatostatin-producing neurons are targets for ghrelin actions in the rat hypothalamus. *Endocrinology* 144:544–51.

Sethi, P. and W. J. Lukiw. 2009. Micro-RNA abundance and stability in human brain: Specific alterations in Alzheimer's disease temporal lobe neocortex. *Neurosci Lett* 459:100–4.

Sharf, R., M. Sarhan and R. J. Dileone. 2008. Orexin mediates the expression of precipitated morphine withdrawal and concurrent activation of the nucleus accumbens shell. *Biol Psychiatry* 64:175–83.

Stipanovich, A., E. Valjent, M. Matamales et al. 2008. A phosphatase cascade by which rewarding stimuli control nucleosomal response. *Nature* 453:879–84.

Subramanian, A., P. Tamayo, V. K. Mootha et al. 2005. Gene set enrichment analysis: A knowledge-based approach for interpreting genome-wide expression profiles. *Proc Natl Acad Sci USA* 102:15545–50.

Svenningsson, P., E. T. Tzavara, R. Carruthers et al. 2003. Diverse psychotomimetics act through a common signaling pathway. *Science* 302:1412–5.

Tapocik, J. D., N. Letwin, C. L. Mayo, B. Frank, T. Luu, O. Achinike, C. House, R. Williams, G. I. Elmer and N. H. Lee. 2009. Identification of candidate genes and gene networks specifically associated with analgesic tolerance to morphine. *J Neurosci* 29:5295–307.

Trivedi, S. and G. Ramakrishna. 2009. miRNA and neurons. *Int J Neurosci* 119:1995–2016.

Wolf, G. 1964. Effect of dorsolateral hypothalamic lesions on sodium appetite elicited by desoxycorticosterone and by acute hyponatremia. *J Comp Physiol Psychol* 58:396–402.

Yoo, J. H., S. Y. Lee, H. H. Loh, I. K. Ho and C. G. Jang. 2004. Altered emotional behaviors and the expression of 5-HT1A and M1 muscarinic receptors in micro-opioid receptor knockout mice. *Synapse* 54:72–82.

13 Development of Local RAS in Cardiovascular/Body Fluid Regulatory Systems and Hypertension in Fetal Origins

Caiping Mao, Lijun Shi, Na Li, Feichao Xu, and Zhice Xu

CONTENTS

13.1 INTRODUCTION

The renin–angiotensin system (RAS) plays important roles in physiological control of cardiovascular systems and body fluid homeostasis in adults, and also in the pathogenesis of various cardiovascular (e.g., hypertension, stroke, ischemic heart disease), metabolic (e.g., diabetes mellitus, metabolic syndrome), and renal (e.g., glomerulitis, renal fibrosis) diseases (Barker 2002; Bichu et al. 2009; Carey and Siragy 2003; Nilsson et al. 2005; Re 2004a; Seckl and Holmes 2007). Since the theory of adult diseases in fetal origins was introduced, the development of the RAS in normal and abnormal patterns before birth has attracted considerable attention. A variety of evidence has demonstrated that the prenatal RAS is important, as an endocrine and paracrine system, in the control of body fluid homeostasis and neuroendocrine functions. This system is subject to a number of conditions or environmental insults during pregnancy and, through plasticity phenomena, could cause alterations either in physiological functions or morphological phenotypes related to blood pressure and body fluid homeostasis.

The classic RAS means systemic or circulatory RAS, but growing evidence has demonstrated that almost all tissues and organs, and even single cells (e.g., cardiomyocyte, neuron), possess all components of RAS (Kumar et al. 2007; Re 2004b). This chapter focuses on the functional development of local RAS related to control of fetal cardiovascular and body fluid homeostasis, and impact of alternations of RAS during development *in utero* on "programming" of cardiovascular diseases. The ontogeny of local RAS in the fetal brain, cardiovascular, and renal systems is reviewed, then we discuss the functions of local RAS in regulation of organ development during prenatal period, and finally we consider the possible role or involvement of an altered RAS during critical developmental periods in health or diseases in later life.

13.2 OVERVIEW OF RAS

Since the discovery of renin by Tigerstedt more than 100 years ago (Tigerstedt and Bergman 1898), the elements of RAS family have been significantly expanded over decades, including substrates, enzymes, bioactive products, and receptors. This section briefly summarizes the metabolism of bioactive angiotensin peptides and their functional receptors important in prenatal life.

13.2.1 Catalytic Formation of Bioactive Angiotensin Peptides

Classical theory stated that the sequential enzymatic cascades of RAS are involved in production of angiotensin I (ANG I) from angiotensinogen (ATG) catalyzed by renin, followed by the formation of angiotensin II (ANG II) from ANG I generated by angiotensin-converting enzyme (ACE) (Peach 1977). The past three decades of research has provided evidence, indicating that metabolic pathways generating the bioactive angiotensin peptides are much more complicated, as discussed below.

13.2.1.1 Generation of ANG II and ANG III

At least three biological pathways have been investigated in generation of octapeptide ANG II, and two of them entail the formation of decapeptide ANG I. ANG I is mainly produced from hydrolyzing ATG by renin bond of Leu–Leu in rodents, and Leu–Val in human (member of A1 family of aspartyl proteases, EC 3.4.23.15). In addition, another two aspartyl proteases, cathepsin D (EC 3.4.23.5) and cathepsin E (EC 3.4.23.34), and two serine proteases of S1 family, elastase (EC 3.4.21.37) and proteinase 3 (EC 3.4.21.76), could also generate ANG I from ATG (Hackenthal et al. 1978; Kageyama et al. 1995; Ramaha and Patston 2002) (Figure 13.1).

The first classical ANG II formation pathway is a direct hydrolysis reaction from ANG I. It entails ACE (member of M2 family of zinc metallopeptidases, EC 3.4.15.1) to hydrolyze the Phe–His peptide bond in ANG I with its dipeptidyl carboxypeptidase activity (Lew 2004; Rice et al. 2004). In addition, chymase α (member of serine proteases of S1 family, EC 3.4.21.39) possesses chymotrypsin-like serine protease activity that is capable of cleaving the Phe–His peptide bond in ANG I to form ANG II. Recently, several lines of evidence showed that chymase α plays an important role in ACE-independent synthesis of ANG II (Bacani and Frishman 2006). Furthermore, four serine proteases of the S1 family, including tonin (EC 3.4.21.35), cathepsin G (EC 3.4.21.20), elastase and proteinase 3, cysteine protease of C1 family, and cathepsin B (EC 3.4.22.1), can convert ANG I into ANG II in a manner similar to ACE (Azaryan et al. 1985; Cardoso et al. 2010; Ramaha and Patston 2002). The second ANG II formation pathway is an indirect hydrolysis reaction from ANG I associated with catalysis by cathepsin A (a member of serine proteases of S10 family, EC 3.4.16.5). With its carboxypeptidase activity, cathepsin A could target

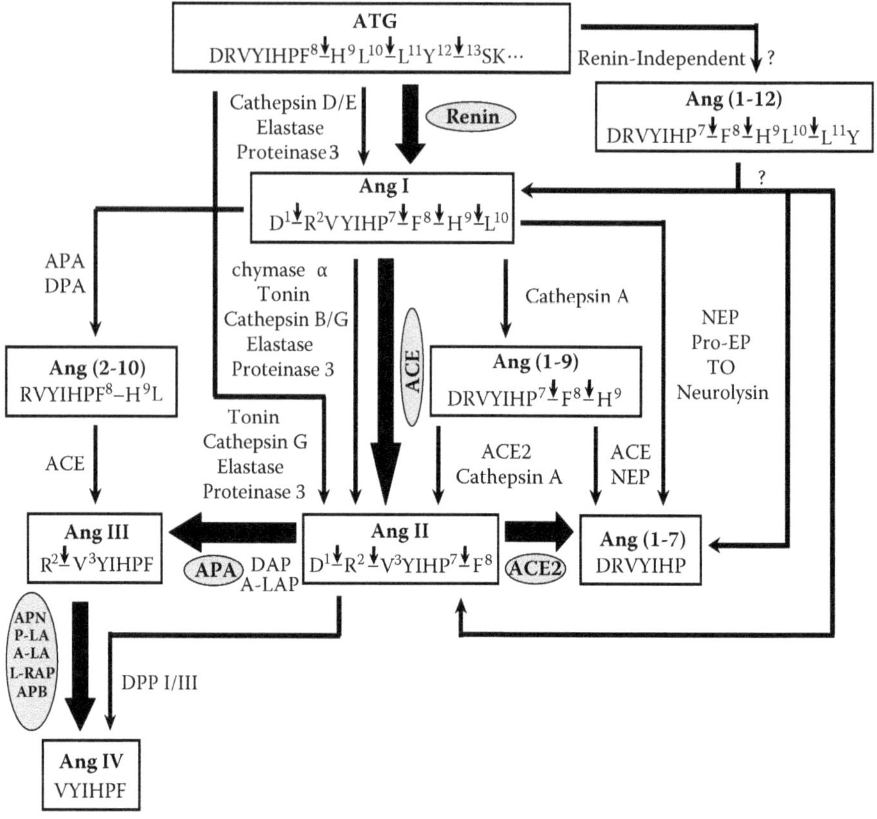

FIGURE 13.1 Enzymatic cascade in formation of bioactive angiotensins.

ANG I to remove carboxyl terminal Leu residue in generating a nonapeptide angiotensin (1–9) [Ang (1–9)], which also could be catalyzed by ACE2 (discussed below). Subsequently, the carboxyl terminal His residue in Ang (1–9) is further cleaved out by cathepsin A to form ANG II (Hiraiwa 1999). The third pathway of ANG II formation does not need ANG II as an intermediate. Accumulated evidence showed that at least four enzymes such as tonin, cathepsin G, elastase, and proteinase 3 could break the Phe–His peptide bond in ATG to produce ANG II directly (Cardoso et al. 2010; Ramaha and Patston 2002) (Figure 13.1).

The heptapeptide angiotensin (2–8) (ANG III), sharing some functions with ANG II, could be either generated directly from ANG II or indirectly from ANG I. In the predominant way, aminopeptidase A (APA; member of zinc metallopeptidases of M1 family), known as the most important protease responsible for ANG III generation from ANG II, with its glutamyl/aspartyl aminopeptidase activity, could remove amino terminal Asp residue in ANG II to form ANG III directly (Reaux et al. 2000). Other two zinc metallopeptidases, aspartyl aminopeptidase (DAP, member of M18 family, EC 3.4.11.21) and adipocyte-derived leucine aminopeptidase (A-LAP, member of M1 family), also could convert ANG II to ANG III in a similar manner as

APA[21,22]. In another way, using monoaminopeptidase activity of APA and DAP, ANG I could be first converted into nonapeptide angiotensin (2–10) [Ang (2–10)], and then using carboxyl terminal dipeptide cleavage activity, ACE could target Ang (2–10) as a substrate to generate ANG III (Lew 2004; Matsumoto et al. 2000; Tsujimoto and Hattori 2005).

13.2.1.2 Generation of Ang (1–7)

To generate the heptapeptide angiotensin (1–7) [Ang (1–7)], three angiotensin peptides—ANG I, ANG II, and Ang (1–9)—could be candidates. In general, three enzymatic pathways are incorporated into the formation of Ang (1–7) (Figure 13.1). The most important pathway generating Ang (1–7) is the catalysis by ACE2 (EC 3.4.17.23) under specific conditions such as hypertension (Ferrario 2011; Shi et al. 2010a). ACE2 is a zinc metallopeptidases of M2 family. Unlike ACE, ACE2 merely could remove single carboxyl terminal residues of the substrates with its monocarboxypeptidase activity. Although both ANG II and ANG I are substrates of ACE2, the catalytic efficiency of ANG II is much higher (about 400-folds) (Trask et al. 2007; Vickers et al. 2002). The first pathway is the catalysis removing Phe residue in the carboxyl terminal of ANG II to form Ang (1–7), where executive enzymes are ACE2, prolyl endopeptidase (Pro-EP, member of serine proteases of S9A family, EC 3.4.21.26), and prolyl carboxypeptidase (Pro-CP, member serine proteases of S28 family, EC 3.4.21.26) (Tan et al. 1993; Welches et al. 1991). The second pathway for Ang (1–7) formation entails ACE and another zinc metallopeptidase of M13 family referred to as neutral endopeptidase (NEP; neprilysin, EC 3.4.24.11). The dipeptidyl carboxypeptidase activity of ACE and peptide cleavage activity (preferential between hydrophobic residues, particularly Pro–Phe/Tyr) of NEP could release Phe–His dipeptide from Ang (1–9) to generate Ang (1–7) (Rice et al. 2004). The third pathway generating Ang (1–7) is directly from hydrolyzing ANG I by NEP, and Pro-EP breaking the Pro–Phe peptide bond as described above. Additionally, the other two zinc metallopeptidases of the M3 family—thimet oligopeptidase (TO, EC 3.4.24.15) and neurolysin (EC 3.4.24.16)—may also be able to convert ANG I directly into Ang (1–7) by cleaving the Pro–Phe peptide bond (Karamyan and Speth 2007; Rice et al. 2004; Serizawa et al. 1995; Welches et al. 1991).

13.2.1.3 Generation of ANG IV

The hexapeptide angiotensin (3–8) (ANG IV) is primarily produced from ANG III by removing the amino terminal Arg residue via catalysis by five zinc metallopeptidases of M1 family, including placental leucine aminopeptidase (P-LAP; EC 3.4.11.3), A-LAP, leukocyte-derived arginine aminopeptidase (L-RAP), aminopeptidase N (APN; EC 3.4.11.2), and aminopeptidase B (APB; EC 3.4.11.6) (Reaux et al. 1999; Tsujimoto and Hattori 2005; Foulon et al. 1999). In addition, using diaminopeptidase activity of dipeptidyl peptidase I (DPP I; member of cysteine proteases of C1 family, EC 3.4.14.1) and dipeptidyl peptidase III (DPP III; member of zinc metallopeptidases of M49 family, EC 3.4.14.4), ANG IV could directly be produced from ANG II by removing its amino terminal Asp–Arg residues (Abramić et al. 2000; Davis and Pieringer 1987).

13.2.1.4 New Story of Ang (1–12)

As discussed above, ANG I seems to be a sole precursor next to ATG in generating the downstream angiotensin peptides, including ANG II, Ang (1–7), ANG III, and ANG IV. Notably, there is a lack of metabolites of ATG in the upstream of ANG I *in vivo*. Nagata et al. in 2006 found a new *in situ* ATG-derived dodecapeptide proangiotensin-12 [Ang (1–12)] from rodent tissues (Nagata et al. 2006). Subsequent studies further showed that Ang (1–12) could be converted into ANG I, ANG II, and Ang (1–7) via catalysis by tissue-specific ACE, chymase, and NEP. For instance, ANG II formation in cardiomyocytes is chymase-dependent, and in the brain it is ACE-dependent (Arnold et al. 2010; Prosser et al. 2009; Trask et al. 2008). Ang (1–12) formation is renin-independent, which is important in both physiological regulation and pathogenesis of certain diseases (Ahmad et al. 2011; Bujak-Gizycka et al. 2010; Ferrario et al. 2009).

13.2.2 Functional Receptors of RAS

Following formation of angiotensin peptides described above, the bioactive oligopeptides may bind to and activate their respective specific receptors. Consequently, specific intracellular signals could be triggered so as to further elicit various physiological or pathophysiological responses.

13.2.2.1 AT1 and AT2 Receptors

The AT1 receptor belongs to seven-transmembrane G protein–coupled receptor (GPCR) superfamily that is predominantly linked to pertussis toxin–insensitive G proteins. ANG II mainly activates AT1 receptor for its various physiological actions and pathological roles in cardiovascular, renal, endocrine, and body fluid systems (Fitzsimons 1998; Xu et al. 2009a,b). Unlike humans and other mammals, there exist two subtypes of AT1 receptors—AT1a and AT1b—in rodents, which exhibit the homology of about 95% and 92% identical at amino acid and nucleotide levels, respectively. These two AT1 subtypes share the resemblance of ligand binding and intracellular signaling, but differ in tissue distribution and molecular features (e.g., chromosomal localization, genomic structure, and transcriptional regulation). The emergence of these two subtypes in rodents was suggested to be the consequence of gene duplication during evolution (Clauser et al. 1996; Guo and Inagami 1994). In prenatal life in both human and other mammals such as rodents and sheep, AT1 receptor appears early in the embryonic period, and is widely distributed throughout renal, neuronal, cardiovascular, endocrine, and hepatic tissues and organs. Furthermore, the AT1 receptor mediates various arrays of developmental events and functional processes before birth (Mao et al. 2009a).

The AT2 receptor also belongs to the seven-transmembrane GPCR superfamily without homology with the AT1 receptor. Although the AT2 receptor displays typical motifs and signature residues of classical GPCR, its intracellular signaling exhibited little similarity with that of the AT1 receptor and other classical GPCR members (Ichiki et al. 1994; Mukoyama et al. 1993; Nouet and Nahmias 2000). The physiological roles of AT2 receptor remain unclear, although its possible counteracting effects, including vasodilation and natriuresis, against AT1 receptor in cardiovascular and renal systems have been suggested (Carey et al. 2000). In another facet, the AT2 receptor may act as a culprit together with AT1 receptor involved in the pathogenesis of diseases such as hypertension, atherosclerosis, myocardial hypertrophy, cardiac fibrosis, and glomerular fibrosis (Kobori et al. 2007; Savoia et al. 2011; Steckelings et al. 2005). The expression of the AT2 receptor could be earlier and higher than that of the AT1 receptor in prenatal period, and its distribution pattern was somewhat parallel with that of the AT1 receptor. Growing evidence indicated that AT2 receptor could promote tissue differentiation during early development and exhibit proapoptotic and antiproliferative effects regulating fetal growth *in utero*. Moreover, AT2 has been considered in controlling fetal functional maturation related to cardiovascular and nervous systems (Cui et al. 2001; Mao et al. 2009a; Zhang and Pratt 1996).

It is well known that ANG II is the most potent ligand in binding and activating AT1 and AT2 receptors. Recent findings have suggested that ANG III may be more potent than ANG II responsible for AT1 receptor activation, especially in central regulation of blood pressure and release of arginine vasopressin (AVP) (Reaux et al. 2001; Wright et al. 2003). In addition, ANG III could elicit critical natriuresis responses mediated by activation of AT2 receptor in the kidney (Padia et al. 2008, 2010). However, little is known about distribution and localization of ANG III and its executive enzymes (e.g., APA) during the prenatal period, which should be investigated sooner or later. Furthermore, whether ANG III acts as a functional ligand on AT1 and AT2 receptor before birth also deserves further study.

13.2.2.2 Mas Receptor

The seven-transmembrane receptor Mas was originally discovered as a protooncogene, and served as an endogenous specific receptor for Ang (1–7) of the GPCR superfamily (Santos et al. 2003; Young et al. 1986, 1988). In adults, Mas receptor is widely distributed throughout the body, and involved in the regulation of vascular tone, renal glomerular, and tubular functions, and neural oxidative stress, mainly mediated by phosphatidylinositol 3 kinase (PI3K) protein kinase B (PKB, Akt)-nitric oxide synthase (NOS) signaling pathway (Handa et al. 1996; Paula et al. 1999; Santos et al. 2003; Sampaio et al. 2007). Furthermore, there is no apparent structural change in Mas receptor knockout mice (Alenina et al. 2008; Walther et al. 1998). Although some cardiovascular and behavioral functions can be altered in the absence of the Mas receptor, it is largely unknown if and how Ang (1–7) and its receptors exert functional control over embryonic and fetal development (Walther et al. 1998, 2000; Xu et al. 2008). To date, the explicit expression of Mas and its distributions are rarely studied in the prenatal period.

13.2.2.3 AT4 Receptor—Dichotomy of Insulin-Regulated Aminopeptidase and c-Met

Previous studies showed that the binding site for ANG IV, termed as AT4 receptor, was distinctive from AT1 and AT2 receptors (Harding et al. 1992; Swanson et al. 1992). In 2001, Albiston et al. (2001) showed that the binding site for ANG IV might be the insulin-regulated aminopeptidase (IRAP), a membrane-associated aminopeptidase coupled with glucose transporter 4 (GLUT4). In adults, the AT4 receptor was found mainly distributed in the central nervous system in control of learning, memory, and cognition. Although the AT4 receptor could also be detected in peripheral tissues, such as in the adrenal gland, vasculature, and kidney (Swanson et al. 1992; Wright et al. 2008). New studies have indicated that the AT4 receptor does not belong to the seven-transmembrane GPCR superfamily (Zhang et al. 1999), and suggested that ANG IV–induced biological events resulted from competitively inhibiting enzymatic activity of IRAP for catabolizing endogenous bioactive peptides (Lew et al. 2003). Accumulating evidence has also demonstrated that ANG IV together with AT4 receptor is indispensable for the regulation of neuronal plasticity concerning learning and memory (Wright and Harding 2004).

Scattered data have indicated that ANG IV could augment the process of neurite outgrowth; however, the expression of AT4 receptor or ANG IV binding during neurogenesis and fetal brain development is still unclear (Wright et al. 2008). Notably, based on the homology search, researchers found that ANG IV shared partial match to the hepatocyte growth factor (HGF). Thus, the receptor for HGF—c-Met—might be the candidate for AT4 receptor (Yamamoto et al. 2010). A large body of evidence has shown that c-Met is a member of type I tyrosine kinase receptor family, which could elicit proliferation and differentiation for stem cells and promotion of angiogenesis (Forte et al. 2006). From this point of view, the ANG IV/AT4 receptor might play possible roles in prenatal development. This intriguing field remains unknown and deserves to be further investigated.

13.2.2.4 (Pro)Renin Receptors—Bypass of Classical RAS Pathway

The aspartyl protease renin is an active enzymatic product of its inactive precursor prorenin. Classical thought ascertains that renin is a key rate-limiting factor for metabolism of a great array of angiotensin peptides derived from ATG. Interestingly, there exists a single transmembrane receptor (pro)renin receptor (PRR) for either renin or prorenin binding, which is fairly conserved among humans and other mammals (Burcklé and Bader 2006). Activation of PRR stimulated by renin or prorenin would in turn trigger intracellular signaling such as extracellular signal-regulated protein kinases 1 and 2 (ERK1/2), mitogen-activated protein (MAP) kinase (MAPK) p38 and PI3K pathways (Feldt et al. 2008; Sakoda et al. 2007; Schefe et al. 2006). Consequently, the stimulated intracellular signaling could in turn lead to up-regulation of growth factors and extracellular matrix (e.g., TGFβ 1, collagens, fibronectin) related to the pathogenesis of cardiovascular and renal diseases (Huang et al. 2006; Saris et al. 2006). Notably, nuclear translocation of promyelocytic leukemia zinc finger transcription factor (PLZF), which was the upstream of PI3K pathway, could reciprocally inhibit expression of PRR, and a short negative feedback loop of PRR was established (Schefe et al. 2006). Furthermore, PRR could explicitly magnify enzymatic activities of both renin and prorenin, which are fundamentally important for intracellular RAS especially in the brain (Grobe et al. 2008; Kumar et al. 2007). To date, the distribution and localization of PRR in the prenatal period are undetermined. However, emerging data suggest that activation of PRR might be involved in neural development concerning neuritogenesis, axonal targeting, and electrophysiological remodeling as discussed below.

As has been pointed out throughout this section, almost all physiological and pathological actions of RAS are mediated by stimulation of membrane-bond receptors induced by extracellular angiotensin peptides. However, growing evidence suggest that perhaps all bioactive angiotensin peptides could be generated intracellularly, at least in some types of cells (Kumar and Boim 2009; Zhuo and Li 2011). Some cellular proliferative and apoptotic events could be regulated by intracellular RAS in pathogenesis of diseases (Kumar and Boim 2009; Zhuo and Li 2011). We would like to propose that intracellular RAS may play a role in the regulation of development before birth, especially in early embryonic period, because (1) PRR serves as an indispensible factor for early development in zebrafish and xenopus (Amsterdam et al. 2004; Cruciat et al. 2010), and (2) in early developmental periods, especially before the barrier of uteroplacental system is established, and the embryo is fragile to various environmental insults. In order to survive environmental insults, secretion processes of RAS components might be inhibited and shielded in certain intracellular compartments to regulate differentiation and growth together with other key molecular systems during early development. To test such hypothesis, developing new techniques such as combination of single cell experiments with genetic manipulations and induced differentiation approaches may be key.

13.3 EFFECTS OF RAS IN EARLY DEVELOPMENT

In humans and other mammals, ATG, renin, and ACE are generated and secreted from the liver, kidney, and lung, respectively. These proteins form the foundation for the circulatory RAS. Almost all components of RAS are widely distributed and localized throughout embryonic tissues and organs, including in the uteroplacental unit, independently from the circulatory RAS. The localized RAS controls growth and differentiation processes in organogenesis, even in peri-implantation period before embryonic blood circulation is established (Kon et al. 1989; Schütz et al. 1996). As RAS-mediated cardiovascular and body fluid control are a main topic of this chapter, we will now focus on the local RAS in the brain, and cardiovascular and renal systems.

13.3.1 RAS BEFORE ORGANOGENESIS

After fertilization, the zygote undergoes cleavage to form blastocyst and implant into the uterus before organogenesis. Recent evidence show that ANG II elicits proliferation of mouse embryonic stem cells mediated by AT1 activation via calcium/protein kinase C (Ca^{2+}/PKC) signaling pathway. Moreover, high glucose-induced proliferation of mouse embryonic stem cells could be augmented by ANG II via AT1 receptor (Han et al. 2007; Kim et al. 2010). In addition, ANG II could promote growth in both somitic and yolk sac/allantoic development mediated by AT2 receptor (Tebbs et al. 1999). Those *in vitro* experiments suggest that ANG II–induced activation of AT1 and AT2 receptors could synergistically elicit potent growth signals for early embryonic development before organogenesis. It remains to be determined whether similar roles also apply for *in vivo* studies.

13.3.2 RAS and Development of Brain

13.3.2.1 Expression and Localization of RAS in Developing Brain

In the human prenatal brain, expression and distribution of components of RAS are still relatively unclear. In rodents, mRNA and protein of ATG could be detected on E18 in the choroid plexus and ependymal cells lining the third ventricle and increased until birth (Mungall et al. 1995). In the fetal brain, ATG predominantly locates in the hypothalamic and thalamic nuclei, and in cerebellar and cortical neurons. Those areas are associated with regulation of prenatal fluid and electrolyte balance, sensorimotor development, and brain maturation (Mungall et al. 1995; Sernia et al. 1997). Renin activity could be detected in the fetal rodent brain accompanied with ATG from E19 (Sood et al. 1987, 1989, 1990). With the immunostaining technique, ACE was detected in the choroid plexus, subfornical organ (SFO), and posterior pituitary on E19 in the rodent brain (Tsutsumi et al. 1993). In ovine fetuses, the brain ACE is present and relatively intact in its functions, at least at 70% gestation (Shi et al. 2010a,b). Other enzymes for production of angiotensin peptides, such as tonin, chymase, and cathepsin D/G, were found functional in the adult brain.

By quantitative autoradiographic analysis, AT1 and AT2 receptors was detected on E18 in the fetal rodent brain (Tsutsumi et al. 1991). Later, using radiolabeled cRNA probes for *in situ* hybridization, appearance of AT1a mRNA was found on E19 in many regions of the fetal brain, and that of AT2 mRNA was on E13 located in the differentiating lateral hypothalamus (Nuyt et al. 1999, 2001). Notably, unlike that of adults, the AT2 receptor was more abundant than AT1 in the developing brain during prenatal periods. In the fetal rat brain, the AT1 receptor was mainly detected in the SFO, paraventricular nucleus (PVN), nucleus of the solitary tract (NTS), choroid plexus, and anterior pituitary. Unlike that of AT1 receptor, AT2 receptor expression appeared earlier but existed transiently (von Bohlen und Halbach et al. 2001). AT2 mRNA was first detected in the lateral hypothalamic neuroepithelium on E13. Later, the AT2 receptor was found in (1) subthalamic and hypoglossus nuclei on E15; (2) the motor facial nucleus, pedunculopontine nucleus, cerebellum, and the inferior olivary complex on E17; (3) the thalamus, interstitial nucleus of Cajal, bed nucleus of the supraoptic decussation, nuclei of the lateral lemniscus, locus coeruleus, and supragenual nucleus on E19; and (4) the lateral septal and medial amygdaloid nuclei, medial geniculate body, and the superior colliculus on E21 (Mao et al. 2009a). In ovine fetuses, both AT1 and AT2 receptors were detected in the developing brain related to central pathways in cardiovascular and body fluid regulation at 70% of gestational age (Hu et al. 2004). It should be noted that other functional RAS receptors such as AT4 receptor, Mas receptor, and PRR are all functional in the adult brain, whereas their distribution and possible functions are still unclear in the prenatal brain.

13.3.2.2 Effects of RAS on Brain Development

Previous research showed that antisense to ATG mRNA inhibited growth of neuroblastoma cells *in vitro* (Sernia et al. 1997), suggesting that RAS may play an important role in brain development. As discussed above, AT2 could be neuron-specific expressed and detected very early in brain development. In addition, AT2 receptor knockout mice had more neural cells in barn regions, including the cortex and limbic system (von Bohlen und Halbach et al. 2001). Those studies suggested that AT2 receptors may mediate cellular differentiation and apoptosis during neuronal development.

In cell lines of neural origin (e.g., NG108-5, PC12W), ANG II could stimulate the AT2 receptor to increase the number and length of neurite. This effect of neurite outgrowth was accompanied with an increase of polymerized β-tubulin and mitogen-associated protein 2 (MAP2) (Laflamme et al. 1996; Meffert et al. 1996). One intracellular signaling of neurite outgrowth induced by AT2 activation was involved in delayed and sustained phosphorylation of $p42/p44^{mapk}$ (Gendron et al. 1999). The activation of $p42/p44^{mapk}$ could be associated with inhibition of p21ras and stimulation of Rap1/B-Raf (Gendron et al. 1999, 2003). In addition, neurite outgrowth stimulated by AT2 receptor was related to activation of neuronal NO synthase (nNOS) and subsequent NO/cGMP/PKG pathway (Gendron et al. 2002; Yamazaki et al. 2005). Furthermore, recent results show that ANG II activates AT2 receptor to trigger its carboxyl-terminal binding protein—AT2 receptor-interacting protein (ATIP)—to form a complex with Src homology 2 domain-containing protein-tyrosine phosphatase 1 (SHP-1). In turn, the ATIP/SHP-1 complex could translocate to nucleus to increase mRNA expression of methyl methanesulfonate sensitive 2 (MMS2), which belongs to a family of ubiquitin-conjugating enzyme variants important for neurite outgrowth (Li et al. 2007).

AT1 receptor activation also has the capability to regulate neuritogenesis processes (Yang et al. 2001, 2002). In one way, stimulated PI3K/Akt pathway was triggered by AT1 receptor activation, in turn, to influence transcriptional activity of GAP-43 (growth-associated protein-43) gene and activate certain cytoskeletal molecules such as FAK (focal adhesion kinase) and paxillin in neurite (Brunet et al. 1999; Casamassima and Rozengurt 1998; Kops et al. 1999). In another way, activated AT1 receptor was shown to stimulate the PI3K/p70 ribosomal S6 kinase (p70S6K) pathway to enhance expression of a potent neurotropic factor termed plasminogen activator inhibitor-1 (PAI-1) (Zelezna et al. 1992). Very recently, the ANG II–independent PRR activation by (pro)renin was observed to be capable of up-regulating the p85α subunit of PI3K mediated by stimulation of PLZF, and in turn to trigger the PI3K/GAP-43 and PI3K/PAI-1 signaling involved in regulation of neurite outgrowth (Lazartigues 2009). The ERK1/2 signaling triggered by PRR may also be involved in this process (Contrepas et al. 2009). Taken together, it is rational to propose that ANG II–dependent AT1/AT2 receptors and ANG II–independent PRR may work cooperatively in controlling processes of neuritogenesis synergistically (Figure 13.2).

It is well accepted that ion homeostasis plays remarkable roles in neuronal differentiation. In neuron origin cell lines, AT2 stimulation could inhibit the activity of phosphotyrosine phosphatase (PTPase), and in turn attenuate T-type calcium channel current (I_{Ca}) (Buisson et al. 1992, 1995). In addition, ANG II was shown to activate the AT2 receptor to stimulate a delayed-rectifier potassium channel current (I_K) and a transient potassium channel current (I_A) by a mechanism involved in stimulation of Ser/Thr phosphatase (PP2A) mediated by phospholipase A2 (PLA2) activation and arachidonic acid derivatives release (Zhu et al. 1998). In general, AT1 receptor

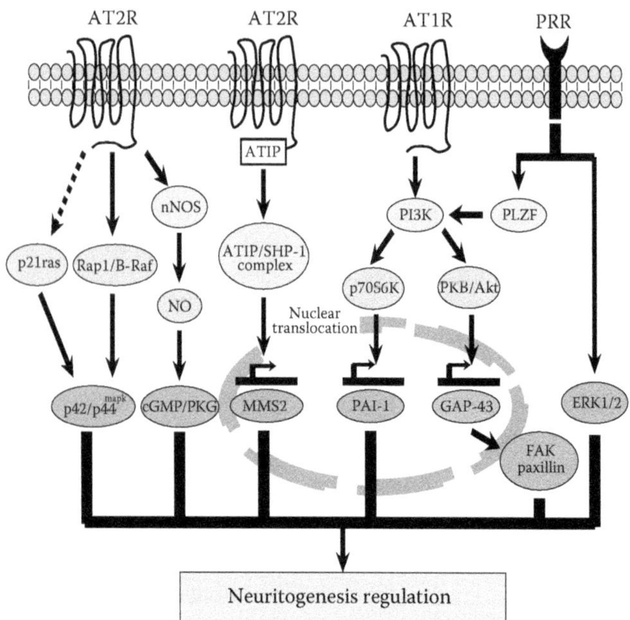

FIGURE 13.2 Schematic representation of RAS-mediated regulation of neuritogenesis.

activation produces contrary effects on ion currents. After stimulating AT1 receptor by ANG II, at least three intracellular signaling can be triggered: (1) activated PLC can stimulate calcium/calmodulin dependent protein kinase II (CaMKII) to phosphorylate Kv2.2 channel to decrease I_K in one way; in another way, it can stimulate PKC to phosphorylate Kv2.2, Kv1.4 and Ca^{2+} channel to inhibit I_K and I_A and increase I_{Ca}; (2) activated nicotinamide-adenine dinucleotide (phosphate)(NAD(P)H) oxidase/radical oxygen species (ROS) signaling to decrease I_K and increase I_{Ca}; (3) activated macrophage migration inhibitory factor can, in turn, stimulate thiol-protein oxidoreductase so as to increase I_K (Matsuura et al. 2006, 2007). Thus, in general, with respect to ionic regulation mediated by ANG II, AT1 and AT2 receptors have opposite roles in reciprocal balance of neuronal excitability (Figure 13.3).

Moreover, AT2 receptor activation induces neural migration and apoptosis (Côté et al. 1999; Yamada et al. 1996). Previous studies suggest that the apoptotic effect induced by AT2 receptor is an indirect inhibition of neurotropic effects by other factors such as epidermal growth factor (EGF) (Horiuchi et al. 1997; Shenoy et al. 1999). Emerging information indicate that ANG II also may trigger apoptosis mediated by NAD(P)H oxidase directly (Yamamoto et al. 2008). Additionally, Li et al. (2007) recently showed that AT2 activation could promote DNA repair processes through the mechanism by which MMS2 up-regulation could inhibit DNA binding 1 proteolysis via MMS2/Ubc-13 complex formation.

Together, accumulated data suggest that RAS plays explicit roles in brain development via integrating signals of AT1/AT2 receptor and PRR. Several neural behaviors during development, for instance, cellular migration, neuritogenesis, and cell survival, could be synergistically regulated by RAS. Notably, the mechanisms

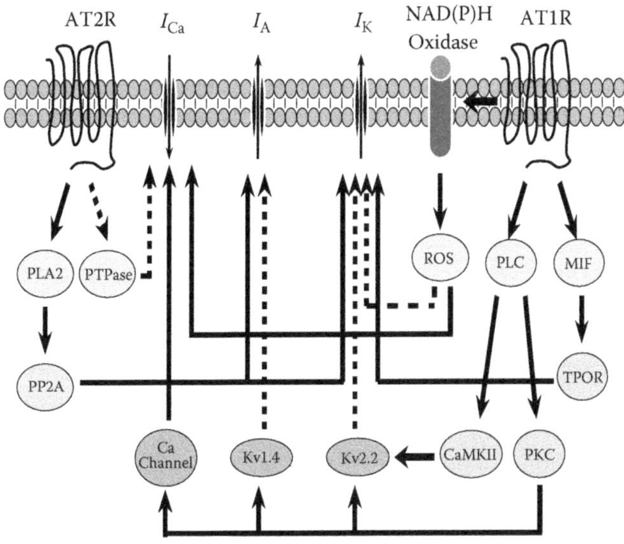

FIGURE 13.3 Schematic representation of RAS-mediated electrophysiological control in developing neurons.

discussed above are mainly from experiments of neurons originated in cell lines and cultured cells *in vitro*; whether they are consistent with the developing brain *in vivo* needs to be further investigated.

13.3.3 RAS and Development of Kidney

The organogenesis of mammalian kidney during embryonic development (i.e., nephrogenesis) can be divided into three probably overlapped stages: pronephros, mesonephros, and metanephros. The pronephros and mesonephros undergo regression in early embryonic days, and the metanephros further develops to form the permanent kidney. In the clinic, fetal exposure to ACE inhibitor or AT1 receptor blockers causes irreversible renal lesions such as renal tubular dysgenesis (RTD) (Barr and Cohen 1991; Bos-Thompson 2005; Saji et al. 2001). This finding also suggests that RAS may play important roles in the regulation of nephrogenesis, especially during metanephric kidney development.

13.3.3.1 Expression and Localization of RAS in Developing Kidney

Development of kidney in human and ovine fetuses share many similar mechanisms that will be described in this section; so we will explicitly mention other species, such as rodents, where applicable. ATG mRNA and protein are detected earlier in the proximal section of primitive tubules of mesonephros, and later in the proximal tubules of metanephros (Phat et al. 1981; Wintour et al. 1996). In rodents, ATG mRNA and protein are primarily expressed in the proximal tubules of the metanephric kidney, and transiently expressed in the ureteric bud and stroma during E15–17 in rats and E12–16 in mice (Niimura et al. 1997).

Renin mRNA and protein are also initially found in the mesonephric kidney, then in the vascular pool of glomeruli as well as arteries adjacent to glomeruli in the metanephric kidney, and finally shifted to the juxtaglomerular cells (Schütz et al. 1996; Wintour et al. 1996). In rodents, renin mRNA is detected first in the undifferentiated metanephric blastema on E12 before vascularization begins (about E14). After vessel formation begins, renin mRNA and protein are dynamically distributed along the walls of arteriolar vessels in the metanephric kidney, from renal artery, interlobar, arcuate to interlobular arteries, and in the juxtaglomerular cells (Jones et al. 1990; Sequeira Lopez et al. 2001). The renin expression pattern is parallel with the development of sympathetic innervation in the kidney, which further ascertains the centrifugal expression profile of renin during metanephric kidney development (Pupilli et al. 1991). Recent data indicate that expression of connexin 40 is a prerequisite for renin expression, whereas down-regulation of connexin 45 was associated with phenotype changes in renin-secreting cells (Kurt et al. 2011; Wagner et al. 2007). Thus, reciprocal expression of connexins involved in cell-to-cell communication could be closely associated with the centrifugal pattern of renin expression. In addition, intracellular cAMP increase and histone H4 acetylation at the cAMP-responsive element of the renin gene are critically important for acquisition and maintenance of embryonic cell expressing renin (Pentz et al. 2008). Moreover, expression of transforming growth factor-beta (TGF-β) type II receptor in stromal and vascular cells shifted in a similar manner as that of renin (Liu and Ballermann 1998). Whether there is a link between TGF-β receptor and changes in intracellular cAMP or epigenetic modification should be further studied, because transcriptional switch for on and off of gene expression might be an indispensable molecular mechanism for the centrifugal pattern of renin expression. Although transcriptional regulation of renin gene has been extensively studied (Pan and Gross 2005), more details need to be known about basal and induced promoter activity and their regulation during metanephric kidney development.

With immunohistochemical methods, the ACE protein is detected in the apical membrane of mesonephric tubule cells, and later in proximal convoluted tubules and collecting ducts of metanephros (Schütz et al. 1996; Wintour et al. 1996). In rodents, mRNA and protein expression of ACE is localized in proximal tubules, maturing gromeruli, and collecting ducts in metanephric kidney from E16 (Jung et al. 1993). Notably, renal ACE during early development is low and may not be capable of generating abundant intrarenal ANG II. Thus, ANG II formation in the developing kidney could be generated by pathways independent of renin and ACE via a novel serine protease, which might compensate for low renal ACE before birth (Yosypiv and El-Dahr 1996). However, whether it may be associated with the Ang (1–12) pathway should be properly investigated in the future. Furthermore, other various angiotensin formation enzymes such as tonin, cathepsins, aminopeptides, and carboxypeptidases could also generate ANG II independently from ACE. Nonetheless, none of those enzymes has been studied in renal development, which would be a fascinating topic for future studies.

ANG II AT1 receptor mRNA was first detected in presumptive mesangial cells of mesonephric glomeruli, and shown in maturing metanephros glomeruli, proximal tubules, and juxtaglomerular cells during metanephric kidney development (Butkus et al. 1997). Compared to that of AT1 receptor, AT2 receptor mRNA is expressed even earlier in mesonephros confined in undifferentiated mesenchyme surrounding preliminary tubules. In metanephrogenesis, AT2 receptor mRNA expression reached its peak in the first trimester and declined later on, as detected in undifferentiated mesenchyme surrounding tubules and immature glomeruli and epithelial sites in macula densa (Butkus et al. 1997). AT1 mRNA and protein is initially detected in ureteric bud and its surrounding mesenchyme on E12 in mice, and in maturing glomeruli, ureteric bud branches, and stromal mesenchyme in rats. From E16 in mice and E19 in rats, AT1 mRNA and protein are highly abundant in glomeruli, microvessles, proximal tubules, and collecting ducts, and gradually decline after birth (García-Villalba et al. 2003; Kakuchi et al. 1995). Similar to that in human and ovine fetuses, the AT2 receptor is expressed earlier than that of the AT1 receptor in rodents, and mainly distributed in mesenchyme cells surrounding ureteric buds. On E11.5–E13.5 in mice, AT2 receptor mRNA and protein are detected predominantly in ureteric buds and adjacent mesenchyme. In rats, mRNA and protein of the AT2 receptor are initially expressed on E14–E15 in ureteric branches and their derivatives. From then on, AT2 receptor expression is mainly confined to ureteric buds and stromal mesenchyme as well as medulla and inner cortical tubules in metanephrogenesis, and dramatically declines after birth (García-Villalba et al. 2003; Kakuchi et al. 1995). Very recently, Yu et al. (2010) showed that proteins of AT1 and AT2 receptors reciprocally changed after birth from neonate to adult in rats, with the former decreased and the latter increased, which seemed discrepant to the results discussed above. The data obtained by Yu et al. resulted from Western blotting in tissue of the whole kidney, whereas the other data derived from tissue sections using immunohistochemical methods. In addition, the whole kidney organ may contain other angiotensin components that could confuse total protein content measured by Western blotting. Indeed, all experiments performed to date in detecting developmental changes of RAS components showed their limitation in rigorous quantitative or tissue/cell specific. The single cell experimental technique should be more suitable for future studies about the explicit molecular development of RAS.

In adults, Ang (1–7) also plays important roles in regulating renal hemodynamics and functions of nephron via its specific Mas receptor. Furthermore, together with enzyme ACE2, Ang (1–7) could counteract pathogenesis of cardiovascular and renal diseases (Ferrario and Varagic 2010). Previous study showed that Ang (1–7) also could be detected in ovine fetuses (Moritz et al. 2001). However, the key components of Ang (1–7)/ACE2/Mas axis before birth are still largely unknown. Therefore, studies to gain further insight into the expression patterns of both ACE2 and Mas receptor in nephrogenesis should be an intriguing prospect, and may be important for understanding the potential roles of Ang (1–7)/ACE2/Mas axis in renal development.

13.3.3.2 Effects of RAS on Renal Organogenesis

As previously noted, treatments with ACE inhibitors or AT1 receptor blockers in pregnancy lead to fetal renal dysplasia. In animal models, similar pharmacological experiments further demonstrated that an intact RAS is critical to normal renal development (Friberg et al. 1994; Guron et al. 1997, 1999; Miyazaki et al. 1999). In rodents, nephrogenesis is not completed before birth, and it continues for about 2 weeks postnatally. Thus, neonatal rats and mice are extensively studied as a model for nephrogenesis. Administration of RAS blockers or inhibitors to neonatal rodents led to a delay of nephrogenesis and decrease in number and size of glomeruli. In addition, papillary atrophy and down-regulation of aquaporin-2 were found to be accompanied by persistent damage of urinary concentration (Friberg et al. 1994; Guron et al. 1999). Moreover, it could impair angiogenesis during nephrogenesis with reduced expression of important growth factors such as TGF-β and EGF (Friberg et al. 1994; Guron et al. 1997, 1999; Miyazaki et al. 1999).

Inactivation of ATG, renin, ACE, AT1, or AT2 receptor by genetic manipulations may result in abnormal nephrogenesis either prenatally or postnatally. Embryonic renal development was not apparently affected in $ATG^{-/-}$ mice, whereas hydronephrosis could be detected in neonates (Yoo et al. 1997). Certain mild anomalies in $ATG^{-/-}$ mice, such as dilated pelvis, renal papillae atrophy, coarse medulla, and hyperplasia of vascular smooth muscle cells (VSMC), were accompanied with up-regulation of platelet-derived growth factor B (PDGF-B) and TGF-β in the cortex and down-regulation of PDGF-A renal papillae (Nagata et al. 1996; Niimura et al. 1995). In $renin1c^{-/-}$ mice, although perinatal mortality was extremely high because of dehydration, morphology of renal development was only slightly affected except for thickening of renal vessels, hydronephrosis, and renal fibrosis scattered during the neonatal period (Takahashi et al. 2005). The intrarenal vasculature exhibited hyperplasia in the absence of ACE together with abnormal renin expression patterns in neonate mice. In $ACE^{-/-}$ mice, the morphological feature of nephrogenesis was rarely affected; however, abnormal renal tubules (obstruction, dilatation, and atrophy), hydronephrosis, and impaired medulla concentrating ability were detected (Esther et al. 1996; Hilgers et al. 1997). Some morphological anomalies such as ectopic ureters could have resulted from abnormal nephrogenesis of urinary tract in the absence of AT2 receptor (Oshima et al. 2001). Compared to the AT2 receptor, in AT1 receptor knockout mice, morphological anomalies was of little concern, whereas medullary and papillary development and intrarenal vasculature were extensively impaired. As a consequence, medullary hypoplasia and hydronephrosis were detected in AT1 $receptor^{-/-}$ mice (Gembardt et al. 2008; Oliverio et al. 1998; Tsuchida et al. 1998). In the clinical setting, an autosomal recessive RTD is one of the severe types of congenital anomalies of the kidney and urinary tract characterized by oligohydramnios *in utero* (Allanson et al. 1983). The predominant pathological features include incomplete development of the cortical proximal convoluted tubules accompanied by atrophic Henle's loops and collecting ducts, enlarged distal tubules adjacent to vascular pole of glomeruli, hyperplasia of intrarenal vasculature, and increased interstitial mesenchyme (Allanson et al. 1992; Kriegsmann et al. 2000). Notably, emerging evidence indicate that certain homozygous or compound heterozygous mutation of genes encoding ATG, renin, ACE, and AT1 receptor might be causally linked to the pathogenesis of RTD (Gribouval et al. 2005; Uematsu et al. 2006). However, as discussed above, phenotypes (e.g., morphological changes of nephrogenesis, renal functions before and after birth) in knockout mice models involved with RAS genes were mildly different from the human RTD, except for changes in intrarenal vessels. Thus, prenatal RAS could, at least in part, mediate development of renal vasculature in both human and rodents. Regarding the considerable differences in morphological changes between human diseases and rodent models, possible explanations could include species difference between human and rodents and timing of nephrogenesis: human nephrogenesis is almost accomplished after 38 gestational days, whereas it can last for 2 weeks after birth in rodents. Thus, it is possible that abnormal nephrogenesis before birth might be compensated for by changes in renal hemodynamics in neonate mice.

Researchers later found that ANG II could regulate ureteric bud branching *in vitro*, further indicating that RAS may play a role in regulating renal development (Iosipiv and Schroeder 2003). Before discussing RAS mechanisms in morphogenesis of ureteric bud, we would like to consider an important inducer for ureteric bud outgrowth and branching termed glial-derived neurotrophic factor (GDNF). GDNF is expressed and released from the mesenchyme surrounding ureteric buds and binds to and activate a protooncogene product called recombination during transfection (c-Ret). c-Ret localizes in the nephric duct and ureteric bud and belongs to the receptor tyrosine kinase (RTK) family (Costantini and Shakya 2006). To precisely regulate the GDNF–Ret pathway, both excitable and inhibitory signals are required. In one way, paired homeobox 2 (Pax2) in the intermediate mesoderm is a transcriptional factor for regulating expression of GDNF, which could be up-regulated by a protein wingless 11 (Wnt11) expressed in the ureteric bud tip. As a consequence, Pax2–GDNF–Ret–Wnt11 forms a local positive feedback loop. In another way, Sprouty 1 (Spryl) is an endogenous inhibitor for GDNF-Ret signaling that could be abrogated by bone marrow morphogenic protein 4 (BMP4) from stromal cells adjacent to nephric duct (Brophy et al. 2001). The ureteric bud branching is a complex process regulated by its intrinsic differentiation and inductive roles of surrounding mesenchyme (Al-Awqati and Goldberg 1998). In metanephric mesenchyme, ANG II could activate AT2 receptor to up-regulate Pax2 via Janus kinase-signal transducers and activators of transcription (JAK–STAT) signaling pathway (Porteous et al. 2000; Zhang et al. 2004). Recent study suggested that activation of AT2 receptor could increase expression of GDNF, c-Ret, and Wnt11, and decrease BMP4 (Song et al. 2010a). In ureteric buds, ANG II also could stimulate AT1 receptor to inhibit expression of Spryl during early metanephrogenesis in possible cooperation with AT2 receptor. This may explain why expression of Spryl was not changed in the AT2 knockout model, whereas it decreased in experiments using AT2 receptor blockers (Miura et al. 2010; Song et al. 2010b; Yosypiv et al. 2008). However, whether heterodimerization of AT1 and AT2 receptor could be a molecular mechanism by which Spryl is inhibited by ANG II remains to be elucidated.

Based on information discussed above, it is reasonable to consider that both positive regulators (GDNF, c-Ret, and Wnt11) and negative regulators (Spryl and BMP4) are involved in GDNF-Ret signaling. Outgrowth and branching of ureteric bud mediated by GDNF-Ret signaling could be augmented by ANG II via stimulation of both AT1 and AT2 receptors. In addition, ANG II also could induce phosphorylation of Tyr (1062) in Re, and, in turn, trigger PI3K-Akt and ERK1/2 pathways to enhance ureteric bud branching (Jain et al. 2010; Miura et al. 2010). Furthermore, ANG II is capable of triggering EGF RTK signaling, underlying possible mechanisms by which ANG II–stimulated AT1 receptor could transactivate EGF receptors (Yosypiv et al. 2006; Zhang et al. 2010). Notably, new evidence indicated that ANG II could activate Akt independent of GDNF-Ret signaling (Miura et al. 2010) (Figure 13.4). It is reasonable to consider that ANG II might either act as a partner of GDNF and EGF or independently regulate morphogenesis during metanephric kidney development. Some important questions are left for further investigation such as if these two pathways are overlapped or parallel and if quantitative regulation of GDNF-Ret signaling by ANG II is related to nuclear events.

A balance between cellular proliferation and apoptosis is also extremely important for nephrogenesis. However, there is limited information on proliferative and apoptotic events mediated by RAS in early renal development. Scattered data indicate that activation of AT1 and AT2 receptors affects proliferation and survival in metanephric kidney development (Song et al. 2010a; Watanabe et al. 1996). However, new research shows that growth retardation of developing kidney during late gestation induced by AT1 receptor antagonists is associated with an increased expression of AT2 receptor (Forhead et al. 2011), suggesting that growth of the fetal kidney was regulated via the AT1 receptor. Taken together, we assume that AT1 and AT2

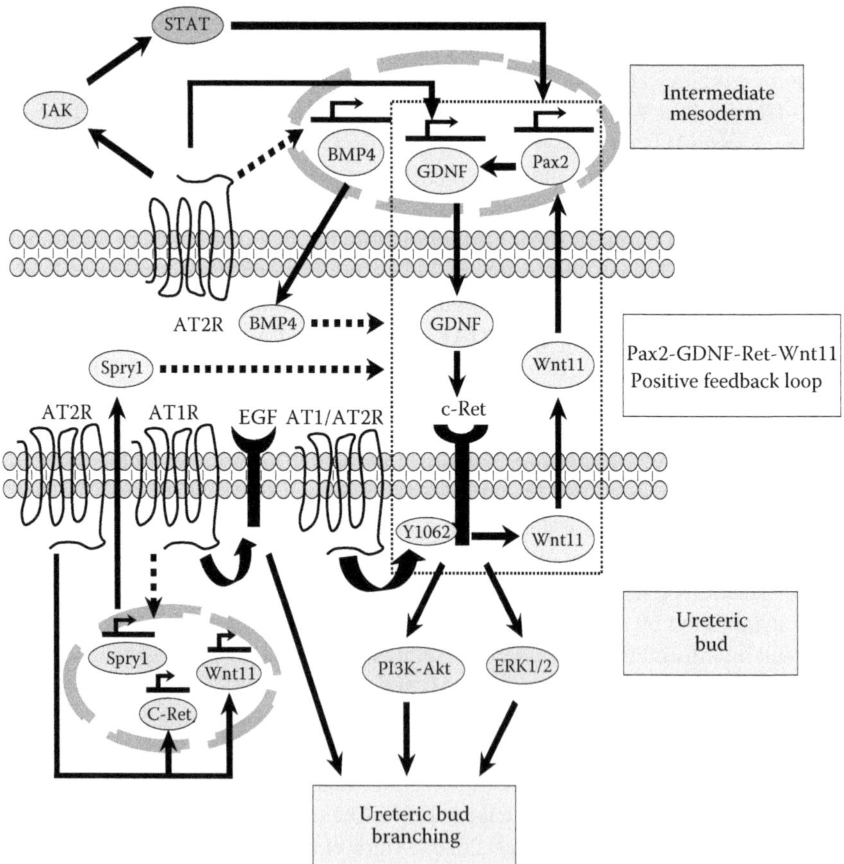

FIGURE 13.4 Schematic representation of the roles of RAS in control of ureteric bud branching in metanephrogenesis.

receptors could act cooperatively in synergistically promoting renal growth differentiation during early embryonic period; however, certain functions of AT2 receptor could be shifted to reciprocally regulate growth elicited by the AT1 receptor during fetal kidney development.

13.3.4 RAS and Development of Cardiovascular System

As demonstrated in the past two decades, nearly all key elements of RAS including renin, ACE, AT1, and AT2 receptors are critically important in control of cardiovascular systems in prenatal periods and probably in the pathogenesis of cardiovascular diseases after birth (Alwan et al. 2005; Ito et al. 1995; Gembardt et al. 2008). However, there has been limited information regarding whether RAS is linked to malformations of the heart and vasculature during development. Despite of this, the fact that several key cellular events such as proliferation and apoptosis could be regulated by RAS during development strongly suggests a role for RAS in cardiovascular remodeling.

13.3.4.1 Expression and Localization of RAS in Developing Cardiovascular System

13.3.4.1.1 Expression and Localization of RAS in Developing Heart

ANG II has been recognized as a growth factor, because of its own actions and interactions with other growth factors. ANG II has significant effects on both somatic and yolk sac/allantoic development. Those effects of ANG II could be blocked by PD123319, an AT2 blocker, but not by GR117289, an AT1 blocker, indicating a direct growth-promoting effect of ANG II via AT2 receptors during organogenesis in the rat embryo *in vitro* (Tebbs et al. 1999). Whether ATG is expressed in human fetal heart is unclear (Schütz et al. 1996). In ovine fetuses, mRNA of ATG is detected in both left and right ventricles (Reini et al. 2009). In rats, ATG mRNA is detected as early as E10.5 in the embryonic heart (Price et al. 1997). Renin mRNA is diffusely expressed in human embryonic heart and mainly located in the inner parts of the myocardium (Schütz et al. 1996). In other mammals, renin expression during embryonic periods has not been investigated. In humans, expression of ACE is detected in the inner layers of the embryonic heart at stage 14, later than that of renin (Schütz et al. 1996). In ovine fetuses, mRNA of both ACE and ACE2 was found in the left and right ventricles (Reini et al. 2009). Moreover, expression of ACE and ACE2 in the ovine fetal heart can be reciprocally changed during development. The ACE/ACE2 ratio at term is 15-fold more than that at 80 gestational days (Reini et al. 2009). In rats, ACE expression was detected on E19 in the cardiac vasculature and heart valves. ACE could also be expressed in the myocardium until the neonatal stage and increased during development (Hunt et al. 1995). Further evidence also indicated that ACE2 may be important in heart development in rodents, although its expression and localization during prenatal life has not been determined (Crackower et al. 2002). In humans, expression of AT1 receptor mRNA was detected in the myocardium around stage 15, whereas the AT2 receptor was preferentially expressed in the innermost layers of the myocardium and appeared earlier than that of the AT1 receptor (about stage 11–12) (Schütz et al. 1996). In ovine fetuses, mRNA expression of both AT1 and AT2 receptors were observed in the left and right ventricles (Reini et al. 2009). In rats, mRNA of both AT1 and AT2 receptors was seen in the embryonic heart and increased during development. Notably, the AT1 receptor could be present as early as E9.5 in the rat embryonic heart with a possible relationship in regulation of cardiac morphogenesis (Price et al. 1997; Hunt et al. 1995; Sechi et al. 1993). Recent evidence shows that Mas receptor could be a plausible factor regulating heart functions during development, despite the fact there have been very limited data on the expression and localization of the Mas receptor during prenatal periods (Santos et al. 2006).

13.3.4.1.2 Expression and Localization of RAS in Developing Vasculature

Almost all components of RAS in vasculature were determined in adults (Paul et al. 2006), and dynamical expression patterns of RAS during development of vasculature in embryonic human have not been properly investigated. Except for angiotensin receptors, other key elements of RAS in vasculature before birth, including ATG, renin, and ACE, have not been studied. In ovine fetuses, both AT1 and AT2 receptors could be detected in vasculatures around 120 gestational days. Notably, the AT1 receptor is predominantly expressed in VSMC of umbilical cord vessels, whereas the AT2 receptor seems to preferentially localize in VSMC of systemic blood vessels (Burrell et al. 2001; Cox and Rosenfeld 1999; Rosenfeld et al. 1993). In rats, expression of the AT1 receptor was found on E14 in the fetal aorta and was increasing until E16. In contrast with the AT1 receptor, expression of the AT2 receptor in the fetal aorta appeared later (about E16) and reached its peak between E19 and postnatal day 1; then its expression showed tendency of decrease (Hutchinson et al. 1999).

13.3.4.2 Developmental Effects of RAS on Cardiovascular System

13.3.4.2.1 Developmental Effects of RAS on the Heart

Embryonic development of the heart is initiated from induction of primary cardiac field formation at trilaminar embryonic disc stage. As mentioned above, current techniques can only detect the main components of RAS in C-shape embryo stage around E9.5 in rats. Thus, there is still a lack of information on whether RAS is involved in very early embryonic events such as cardiac induction and progenitor cell migration. However, there is evidence indicating that ANG II could increase cardiac loop inversion via the AT1 receptor in rats. In addition, the AT1 receptor was predominantly located in the convex area of the rat cardiac loop on E9.5, and might be involved in mechanical load-induced myocardium hyperplasia accompanied by myofibrillar differentiation. As a consequence, ANG II may mediate the processes of dextral looping patterns of cardiac embryogenesis via the AT1 receptor (Price et al. 1997). Recent research shows that deletion of the AT1 receptor in rats leads to a reduction of the heart/body weight ratio (Gembardt et al. 2008), indicating the effect of RAS on fetal heart growth. Furthermore, ATG or AT1 receptor deficiency resulted in oligo- or multiple-atrophy of myocardium characterized by enlarged interstitial space accompanied with pronounced dense reticular network (Van Esch et al. 2010), providing evidence of RAS-mediated cardiovascular remolding during development.

Further studies should consider functional testing in both prenatal and postnatal cardiovascular systems in order to provide novel information significant to long-term health influenced by RAS-mediated prenatal development.

It is well known that cellular events such as proliferation and apoptosis subserve fetal growth and development. Unlike that in adults, growth and maturation of fetal myocardium relies on mechanisms of cell division (proliferation) and cell enlargement (hypertrophy) (Oparil et al. 1984). Attributing to experimental models of ventricular cardiomyocytes from neonate rats, growing evidence shows that RAS, including its AT1/AT2 and Mas receptors, could synergistically affect hypertrophic and proliferative events in the heart in the pathogenesis of cardiovascular diseases (Tallant et al. 2005; Van Kesteren et al. 1997). Unfortunately, limitations of the available techniques limited our knowledge about the detailed cellular effects of ANG II on the embryonic an fetal myocardium. In spite of this, proliferative mode is a preferential pathway for heart maturation during early fetal periods. In ovine fetuses, ANG II stimulated, probably via the AT1 receptor, cellular proliferation by ERK and PI3K–Akt pathways without obvious hypertrophy in immature cardiomyocytes *in vitro* (O'Tierney et al. 2010; Sundgren et al. 2003). On the other hand, ANG II was capable of eliciting hypertrophy of myocardium in the pressure-overload fetal heart *in vivo*, particularly in the left ventricle (Segar et al. 1997, 2001). Those findings support the conception that ANG II influences cellular differentiation or proliferation resulting in ventricular hypertrophy in experimental models. However, several questions are still waiting to be addressed regarding the roles of RAS in dynamic heart development: (1) When to trigger and when to stop hypertrophic effects by ANG II *in vivo*, and what is the specific receptor and intracellular signaling? (2) What is the relationship between ANG II–mediated hypertrophic effects and functional maturation of the fetal heart? (3) Whether and when fibroblast and other cell types *in situ* could be involved in development of the fetal heart mediated by RAS. Furthermore, it has been accepted that activation of RAS, especially its AT2 receptor, played critical roles in apoptosis during development. Apoptotic effects induced by the AT2 receptor in cardiomyocytes under conditions of pathological changes have attracted attention. It is also worthwhile to carry out research on the effects and mechanisms of RAS-mediated apoptosis in heart development.

13.3.4.2.2 Developmental Effects of RAS on Vasculature

We mentioned in the preceding sections that some vasculature defects or malformations occur if the components of RAS are deleted. It seems that RAS serves as an indispensable morphogen for vascular development. Early evidence showed that RAS activation could be involved in the angiogenesis of the uteroplacental unit (Le Noble et al. 1991, 1993). Moreover, Nakajima et al. (1995) further demonstrated that overexpression of the AT2 receptor attenuated neointimal formation and negatively regulated DNA synthesis in the developing aorta, suggesting that RAS is involved with vasculogenesis concerning the development of both VSMC and endothelium.

In general, the AT2 receptor is considered to play roles in apoptosis and antiproliferation on VSMC for counteracting proliferative and hypertrophic influence via the AT1 receptor (Shi et al. 2010a; Xu et al. 2009a). However, during prenatal vasculogenesis, ANG II may stimulate the AT1 receptor, thus contributing to VSMC differentiation by up-regulating molecular markers such as smooth muscle α-actin, myosin heavy chain, and tropomyosin (Hautmann et al. 1997). In addition, AT2 activation mediates the up-regulation of calponin and high-molecular-weight caldesmon that was associated with VSMC differentiation and vasculogenesis (Yamada et al. 1999). Thus, unlike that in adults, AT1 and AT2 receptors might synergistically regulate the development of vasculature in prenatal periods. Furthermore, recent data show that the AT2 receptor may contribute to DNA synthesis and proliferation during vascular development in fetal mouse aorta (Perlegas et al. 2005). Notably, the effect of the AT2 receptor on growth was not accompanied by promoting differentiation in VSMC development (Perlegas et al. 2005). It remains to be elucidated in detail if AT1 and AT2 receptors are cooperators or competitors in the regulation of vasculogenesis. New research suggests that Ang (1–7) is capable of negatively regulating ERK1/2 pathway triggered by the AT1 receptor in ANG II–induced VSMC proliferation (Zhang et al. 2010). It is not known if Ang (1–7) could serve as an antiproliferative molecule in vascular development and fetal physiology. Based on the fact that Ang (1–7) is present in prenatal periods, and antiproliferative roles from AT1 and AT2 receptors seems limited in VSMC development, we may propose Ang (1–7) as a candidate to regulate ANG II–mediated growth of VSMC negatively via the AT1 and AT2 receptors.

ANG II could elicit antiapoptotic effects on bone marrow-derived endothelial progenitor cells *in vitro*, in association with regulation of angiogenesis and vascular regeneration in adults (Roks et al. 2011; Yin et al. 2008). However, the knowledge of its effects on vasculogenesis during embryogenesis is still limited. In the ovine fetal placental artery endothelial cells, ANG II stimulated dose-dependent proliferation via the AT1 receptor (Zheng et al. 1997, 1999), and the ERK1/2 pathway and oxidative stress have been suggested to be major intracellular signaling triggered by AT1 receptor activation (Zheng et al. 1997, 2006). In addition, increased growth factor secreted might in turn enhance the proliferation of endothelial cells (Zheng et al. 1997, 2006). However, underlying intracellular and nuclear events triggered by the AT1 receptor should be further illuminated.

13.3.5 RAS in Developmental Adrenal Gland

Activation of AT1a and AT1b receptors stimulates the release of aldosterone, which promotes sodium reabsorption in mineralocorticoid-responsive segments of the distal nephron (Aguilera 1992; Masilamani et al. 1999).

An intact local RAS in the adult adrenal gland has been demonstrated (Bader et al. 2001), but a similar local system in prenatal adrenal gland has not been fully established. Using immunohistochemical methods, renin was detected in a few cortical cells of the adrenal gland, and mainly localized in the Golgi region, in mouse fetuses on E13–E14. Later, on E16 and E18, renin expression increased and numerous tiny granules were found just below the cell membrane (Iwai et al. 1996). Renin expression was restricted in the cortical area, especially in the aortic side. Contrary to that on E13–E14, renin expression was decreased during development (Kon et al. 1990, 1991). Majority of ANG binding sites in primary cultures of human fetal and neocortex cells are of AT1 subtype (Rainey et al. 1992). However, autoradiographic studies show that ANG II receptors of human fetal adrenals are mainly of AT2 subtype, and so conditions such as culture environment could be considered causes of the difference (Breault et al. 1996). In contrast to observations in rats, the fetal adrenal medulla in humans did not show an expression of AT2 receptor mRNA (Shanmugam et al. 1994, 1995; Schütz et al. 1996). AT1 receptor mRNA could be detected in the neocortex of the human adrenal gland at 8 weeks and throughout gestation, whereas AT2 mRNA was expressed earlier (5–6 weeks) in a mass of condensed cells as initial tissue for adrenal zone in the developing kidney (Schütz et al. 1996; Wintour et al. 1996, 1999). In the ovine fetal adrenal gland, Wintour et al. found both mRNA and protein of AT1 receptor in the zona glomerulosa and fasciculata of the cortex, but not in the medulla by 60 days of gestation (Wintour et al. 1999). AT2 receptor mRNA was present in the same location (absent in the medulla) from 40 to 130 days and declined to extremely low levels after 140 days (term is 145–150 days), and application of ANG II caused a significant down-regulation of AT1 mRNA expression, whereas AT2 mRNA levels remained unchanged. Other studies showed a functional antagonistic effect of the ANG II receptors on aldosterone secretion by the midgestation in ovine fetuses (Moritz et al. 1999).

Growing evidence has shown a link between prenatal insults and differential modifications of AT1 and AT2 receptor genes in the adrenal gland and the resultant alteration of their expression patterns in adult life, which may ultimately lead to the development of hypertension (McMullen and Langley-Evans 2005; Singh et al. 2007). In maternal low protein diet rat models, expression of AT1b receptor gene in the adrenal gland is up-regulated by the first week of life, resulting in an increased expression of the receptor protein. Furthermore, the proximal promoter of the AT1b gene in the adrenal is significantly undermethylated, and that *in vitro*, AT1b gene expression is highly dependent on promoter methylation. These results indicate an interesting relationship between prenatal insults–induced epigenetic modification of genes and resultant alteration of gene expression in late life, leading ultimately to the development of diseases, including hypertension (Bogdarina et al. 2007).

13.4 FETAL RAS IN BODY FLUID HOMEOSTASIS

13.4.1 Fetal RAS and Swallowing

Fetal swallowing is an intrinsic behavior that develops *in utero* in human and ovine in preparation for neonatal fluid and food ingestion. Behavioral development such as fetal swallowing plays a role in fetal body fluid homeostasis, and stimulated fetal swallowing is a sign of the functional development in behavioral responses for the fetus in face of environmental challenges during pregnancy. Swallowing of amniotic fluid serves to regulate amniotic fluid volume and promotes fetal gastrointestinal growth and development (Mulvihill et al. 1985; Pritchard 1996).

The swallowed fluid is an admixture of amniotic fluid, lung liquid, and perhaps salivary secretions, and swallowing *in utero* may also contribute to fetal body fluid homeostasis (Gilbert and Brace 1989, 1990). In sheep and perhaps human fetuses, dipsogen-mediated swallowing responses develop during the last third of gestation (Ross and Nijland 1998).

Different approaches have provided opportunities to investigate fetal swallowing behavior and possible "dipsogenic" responses *in utero* in conscious animal models. Several studies have used electromagnetic flow probe in the near-term ovine fetal esophagus to examine swallowing (Bradley and Mistretta 1973), but also to demonstrate the swallowing activity associated with fetal neurobehavior. Other laboratories measured swallowing by the electromyogram activity of laryngeal adductor muscles associated with low-voltage, high-frequency electrocorticogram (ECoG) activity (Harding et al. 1984). In these studies, most swallowing bouts *in utero* were found during the last third of gestation (Harding et al. 1984). Concurrent electromyogram activity and esophageal fluid flow can be measured from wires inserted on the fetal thyrohyoid muscle and the nuchal and thoracic esophagus, with an ultrasonic flow probe sitting around the fetal thoracic esophagus. Volume swallowed is calculated by integral of the flow velocity waveform from the ultrasound flow meter. Analysis of the electromyogram waveforms detects individual muscle contractions and subsequently defines a fetal swallow as a timed sequence of waveform progression from the thyrohyoid to the thoracic esophagus (Ross and Nijland 1998).

Fetal swallowing occurs both spontaneously and induced by dipsogenic stimuli. Spontaneous ovine fetal swallowing activity and body fluid homeostasis are associated with fetal behavioral state (Ross et al. 1995). Multiple factors control fetal swallowing activity, including dipsogenic stimulation, amniotic fluid availability, and behavioral status (Brace et al. 1994; Harding et al. 1984; Kullama et al. 1994). Progress has been made in demonstrating that fetal swallowing activity is controlled by hyperosmotic, hypovolemic, and other chemical stimuli such as ANG II in the last third of gestation (Roberts et al. 2000; Ross et al. 1995; Xu et al. 2001b). Several studies have suggested that central ANG II–stimulated fetal swallowing may be mediated by nNOS and *N*-methyl-D-aspartate (NMDA) receptors. The fetal dipsogenic effect of ANG II is proposed to occur via releasing presynaptic glutamate, which binds to postsynaptic glutamate NMDA receptors with subsequent calcium influx, activation of nNOS, and nitric oxide release (El-Haddad et al. 1999, 2002).

Fetal studies have suggested that there is an elevated fetal threshold for osmotic stimulation compared with that of adults. *In utero* swallowing or dipsogenic responses are stimulated by intravenous or intracerebroventricular (icv) hypertonic saline, icv ANG II, and dehydration (Ross et al. 1995; Xu et al. 2001a). Hemorrhage, osmotic, and vasopressin stimulation also affect fetal renal fluid and electrolyte excretion (Desai et al. 2003; Gibson and Lumbers 1997; Gomez et al. 1984; Hoffman et al. 1977; Nijland et al. 1995). Notably, RAS mechanisms have been well demonstrated in hemorrhage or dehydration-induced water and salt intake in adults (Fitzsimons 1998). Although there has been limited experiments on RAS mechanisms in fetal body fluid regulation under conditions of *in utero* hemorrhage, the relationship between dehydration during pregnancy and the development of fetal RAS was investigated, and researchers noted that either fetal angiotensin peptides or its receptors could be affected by experimental dehydration (Zhu et al. 2009).

13.4.2 Fetal RAS in Brain in Control of Body Fluids

Excretion of water and electrolyte from the body are mainly via the kidney, whereas intake of water and salt can be controlled by the brain (Fitzsimons 1998). It is well known that the brain has its own local RAS and the brain RAS-mediated behavioral and endocrine responses that play a critical role in regulation of body fluid balance, including water and salt intake, renal excretion and reabsorption, vascular volumes related to water and salt balance (Fitzsimons 1998; Paul et al. 2006). Many central regions, including certain nuclei in the hypothalamus and medulla oblongata, are involved in dipsogenic regulations. In addition, main components of RAS have been detected in those regions of the brain in adults (Allen et al. 1987; Fitzsimons 1998). For example, tonin gene that can split ATG was found in the brain, as well as abundant ACE, to generate ANG II and ANG III (Araujo et al. 2002). The AT1 receptor is highly expressed in the hypothalamus and medulla oblongata in nuclei involved in the control of body fluid balance and blood pressure (Allen et al. 1987; Charron et al. 2002; Davisson et al. 1999; McKinley et al. 1986). In fetal studies, both AT1 and AT2 receptors were detected and measured in the brain with either immunohistological or Western blot analysis (Fitzsimons 1998; Mao et al. 2009a). From those data, intensive AT1 receptor expression in the fetal hypothalamus has been observed, especially in the PVN and supraoptic nuclei (SON) in the ovine fetus at near term and preterm (Mao et al. 2009a). Because both of those central nuclei in the hypothalamus are closely linked to neuroendocrine releases that are important for renal excretions and salt appetite, the data in fetal experiments strongly suggested that central RAS in the fetal brain plays an important role in body fluid balance.

In adults, accumulating evidence has demonstrated that the magnocellular neuroendocrine cells of the hypothalamus–neurohypophysial system are critical in the maintenance of body fluid homeostasis by releasing AVP and oxytocin (OT) in response to a variety of signals, including osmotic changes and central ANG II (Fitzsimons 1998). AVP and OT are two major neuropeptides synthesized and released by magnocellular neurosecretory neurons in the hypothalamus. Both of them are involved in vascular functions and hydromineral homeostasis.

Functional studies in the past two decades further confirmed that central application of ANG II or its precursor ANG I not only induced an increase in plasma AVP concentrations in the fetus, but also an enhancement of OT levels in the fetal circulation, providing evidence that central RAS has been well developed in the control of the neuroendocrine responses that are critical in body fluid regulation (Shi et al. 2010b). Recent studies have demonstrated that icv ANG II could elicit a significant increase in plasma AVP in both near-term (~135 gestational day) and pre-term (~100 gestational day) ovine fetuses without altered fetal plasma osmolality, sodium, and hematocrit levels (Ross and Nijland 1998; Shi et al. 2004a; Xu et al. 2005). The fetal plasma OT was also increased in response to icv ANG II during the last trimester of gestation. These results suggest that ANG II plays an important role in fetal neuroendocrine functions. Moreover, increased plasma AVP partly contributes to the central ANG II–induced pressor responses (Shi et al. 2004a).

Angiotensin receptor antagonist experiments also have demonstrated that losartan, but not PD123319, inhibited an increase of plasma AVP concentrations in response to icv ANG II (Shi et al. 2006). These data suggest that in the developing fetus, the brain AT1 mechanism is critical in the central ANG II–related neuroendocrine responses. Furthermore, icv ANG I significantly increased fetal plasma AVP, whereas application of captopril (a competitive inhibitor of ACE) significantly suppressed the increased plasma AVP. Double labeling showed colocalization of the AT1 receptor and c-*fos* expression in both SON and PVN after icv ANG I, indicating that central endogenous ACE has been functional at least at the last trimester of gestation and the endogenous brain RAS-mediated AVP release via AT1 receptor by acting at sites consistent with the dipsogenic network in the hypothalamus (Shi et al. 2010b).

Under the condition of administration of angiotensin peptides into the fetal brain in the conscious animal models, both AVP and OT neurons in the PVN and SON expressed c-*fos* as revealed by double immunostaining approaches (Shi et al. 2004a; Xu et al. 2003, 2004, 2005). In other words, central ANG II was able to activate AVP and OT cells in the fetal hypothalamus in association with the increased levels of those hormones in the fetal circulation.

Besides controlling neuroendocrine responses by the central RAS, other important contribution by central ANG II in body fluid regulation is via behavioral mechanisms. As described above, icv ANG II injected into the chronically catheterized sheep fetuses induces vigorous swallowing associated with low voltage ECoG high frequency at the late gestation (Xu et al. 2001a; El-Haddad et al. 2000, 2001, 2005). In addition, icv ANG II–induced fetal swallowing activity could be blocked by an angiotensin receptor antagonist (El-Haddad et al. 2001, 2005). These physiological experiments provide evidence that the fetal brain RAS is active at late gestation and appear to play an important role in the control of fetal dipsogenic responses during development.

Both ANG II and ANG I in the fetal brain are capable of increasing swallowing activity *in utero* (El-Haddad et al. 2001, 2005; Shi et al. 2010b). It is well known that water ingestion and salt intake are important in body fluid regulation, and there are no major technical problems in measuring either water or salt intake in experiments in adults (Fitzsimons 1998). However, current available technology does not permit to distinguish salt intake from water ingestion in measuring fetal behavior. This poses difficulty in the study of the functional development of salt appetite or salt ingestion–related body fluid regulation. In spite of this, researchers have demonstrated OT release into the circulation and OT neural activity in the fetal PVN and SON after administration of icv ANG II (Matthews 1999). Considering that OT has been shown to play an important role in regulating salt appetite in adults (Fitzsimons 1998), together with the evidence that central ANG II–induced neural activity marked with c-*fos* in the OT cells in the fetal hypothalamus, increased fetal plasma OT levels, and the stimulated fetal swallowing, the data indicate that the functional regulation for salt appetite could be developed *in utero* already.

Besides the SON and PVN mentioned above, other areas in the brain are also critical in the control of body fluids. For example, several nuclei located around the ventricular area, including the SFO, median preoptic nuclei, and organum vasculosum of the lamina terminals (OVLT), play an important role in water and salt intake in response to icv ANG II (Fitzsimons 1998; Mao et al. 2007). Those areas around the ventricular system are rich in ANG II receptors (Fitzsimons 1998). In fetal studies,

Shi et al. showed that the nuclei activated by icv ANG II in the brain of ovine fetuses not only included the SFO, median preoptic nucleus, OVLT, SON, and PVN in the forebrain, but also the NTS and lateral parabrachial nucleus in the hindbrain (Shi et al. 2004a, 2006; Xu et al. 2003, 2004). The results suggest that all those areas in the fetal brain could be activated by dipsogen ANG II or ANG I generated ANG II.

Recent studies have demonstrated that intravenous ANG II could induce cellular activity marked with intensive c-*fos* expression in the fetal SFO, OVLT, MPO, SON, and PVN (Shi et al. 2004a,b). The peripheral ANG II acts on the fetal hypothalamus indirectly via the circumventricular organs (CVOs) (such as SFO, OVLT in the forebrain), which lack the blood–brain barrier but are rich in ANG II receptors. These data suggest that pathways from the SFO, OVLT, and MPO to the fetal hypothalamus have been established, and are functional during the last trimester of gestation.

13.4.3 Fetal RAS in Renal Control of Body Fluid Homeostasis

Previous studies have demonstrated that fetal kidneys are relatively intact in histology, and relatively mature in physiological functions before birth. Thus, renal control of body fluid homeostasis actually appears *in utero* utilizing RAS as an important mechanism (Fitzsimons 1998). ANG II, as the main effector of RAS, maintains a high urine flow, ensures adequate volume of fetal fluids (amniotic and allantoic in sheep), increases blood pressure, and causes dieresis during midgestation (Lumbers et al. 1993; Xu et al. 2003). ANG II modulates renal blood flow and glomerular filtration rate through renal vasoconstriction acting directly on ANG II receptors in glomerular and proximal tubular cells (Lumbers 1995). An icv injection of ANG II can also elicit AVP release into the systemic circulation (Xu et al. 2005).

The reduction of availability of ANG II by converting enzyme inhibitors decreases glomerular filtration rate and renal vascular resistance while increasing fetal renal blood flow, resulting in low or no urine flow in the fetus at late gestation (Lumbers et al. 1993), whereas infusion of ANG II decreases proximal tubule sodium reabsorption. Previous studies have shown that adverse intrauterine environment, including those of drug exposure and poor nutrients, could affect the expression of angiotensin receptor subtypes AT1R and AT2R, as well as ATG, in the kidney of fetuses in late gestation (Dodic 2001a, 2002a). Recent studies have shown that hypoxia during pregnancy could influence fetal renal functions as revealed by alterations in fetal plasma creatinine, BUN/creatinine ratio, associated with reduced fetal kidney weight. In addition, mRNA and protein expression of AT1 and AT2 receptors in the fetal kidney was changed after exposure to hypoxia, indicating the altered renal functions by hypoxia may be mediated by changes in the local RAS in the fetal kidney (Mao et al. 2010).

13.5 FETAL RAS IN CARDIOVASCULAR CONTROL

13.5.1 Fetal ANG II Receptors and Cardiovascular Control

Compared with extensive studies on the relationship between RAS and cardiovascular systems in adults, investigations related to RAS-mediated cardiovascular regulation in the fetus are limited. However, development of RAS in normal and abnormal patterns before birth has attracted considerable attention. Increasing evidence has suggested that an overactivated RAS significantly contributes to fetal programming of hypertension in the adult (Langley-Evans and Jackson 1995; Sherman and Langley-Evans 2000). Dysfunction of the fetal tissue rich in ANG II receptors, such as the hypothalamus, could impact on fetal tissue maturation and hence have profound consequences in postnatal life (Chen et al. 2005).

AT1 receptor is expressed in all cells of the cardiovascular tissue, namely, endothelial cells, smooth muscle cells, fibroblasts, monocytes, macrophages, and cardiac myocytes. AT1 receptor mediates major ANG II effects (such as vasoconstriction, increased BP, cardiac contractility, and renal tubular absorption) (Mehta and Griendling 2007). In contrast, AT2 receptor is suggested to induce opposite effects, namely, vasodilatation, and possible inhibition of AT1 receptor (Levy 2004; Yamada et al. 1996). Expression of AT2 receptor is high in fetal tissues such as in the heart, aorta, kidney, lung, and liver, decreasing significantly after birth (Grady et al. 1991), suggesting that receptors might play an important part in fetal development (Shanmugam and Sandberg 1996; Shanmugam et al. 1995). However, a functional study of AT2 effects in fetal cardiovascular systems *in vivo* has been limited.

Changes in expression of ANG II receptor subtypes during pregnancy could affect both vascular and renal functions and thereby affect fetal blood pressure. ANG II activates not only AT1 receptor to produce vasoconstriction (Griendling et al. 1997; Raizada et al. 1995; Touyz et al. 1999, 2003), but also AT2 receptor to induce release of vasodilator substances (Gallinat et al. 2000; Gross et al. 2004; Kaschina and Unger 2003; Siragy 2004). The role of AT1 receptor in vascular contraction and hypertension is well characterized (Griendling et al. 1989; Reaves et al. 2003; Touyz and Berry 2002); in many instances, AT2 receptor could mediate responses that counteract the effects of AT1 receptor (Nakajima et al. 1995; Reaves et al. 2003). Studies have shown that AT2 receptor is expressed heavily in fetal and placental tissues compared with that in adult tissues (Amanda et al. 2009; Koukoulas et al. 2002). Recent work has demonstrated that AT2 receptor–mediated vascular relaxation was enhanced in vascular rings of pregnant rats, and the enhanced AT2 receptor–mediated vascular relaxation during pregnancy was associated with increased eNOS mRNA expression (Amanda et al. 2009). In brief, it is clear now that fetal AT1 receptor plays an important role in the regulation of vascular constriction, and increased tension of fetal vessels as well as blood pressure. The fetal data and information on AT2 receptor function for cardiovascular regulations is still very limited, in spite of the increasing evidence of their importance to the control of fetal cardiovascular function.

13.5.2 Central RAS and Fetal Blood Pressure

ANG II receptors are ubiquitous in the fetal brain and so found in the hypothalamus, thalamus, cerebellum, and cortex. AT1 receptors are present in the cardiovascular centers in the brain and seem important to control fetal cardiovascular responses (Gao et al. 2008; Mao et al. 2009a; Zucker et al. 2009). The timing of the appearance of these receptors supports the establishment of a functional RAS by late gestation

(Mungall et al. 1995). In general, both AT1 and AT2 receptors are widely located in the fetal brain, at least at late gestation. This differs from the local RAS in the adult brain predominated by AT1 receptors.

A series of studies have demonstrated that icv ANG I or ANG II in both near-term and pre-term ovine fetuses can produce an increase in systolic, diastolic, and mean arterial pressure, in association with a significant decrease in heart rate, indicating a central role for fetal RAS in the control of blood pressure. Additional studies show that V1 receptor blockers partially suppressed icv ANG II–increased fetal blood pressure, indicating that central RAS–induced fetal pressor effects depend at least partially on the hypothalamic release of AVP (Shi et al. 2004a; Xu et al. 2003, 2005). These studies have demonstrated that the local RAS in the fetal brain is functional at least in the last third of gestation, and its activation associated with central neuroendocrine pathways is critical for the control of fetal blood pressure.

ACE immunoreactivity has been demonstrated in the human fetal brain (Strittmatter et al. 1986). Both ACE mRNA and protein were detected in the choroid plexus and other brain regions of fetal rats and rabbits (Rogerson et al. 1995; Whiting et al. 1991). In contrast, ACE2's expression in the brain is absent until embryonic day 15, when a prompt increase in expression occurs (Crackower et al. 2002). ACE and ACE2 are coexpressed in many tissues, and alterations in their expressions and activities might participate in fetal programming of hypertension (Brosnihan et al. 2005; Shaltout et al. 2009). Although ACEs are expressed in the developing fetus before birth, the functional development of these enzymes are less understood. Recent studies have investigated the functional development of brain ACE related to the cardiovascular control *in utero* in the chronically prepared near-term ovine fetus via icv injection of ANG I (Shi et al. 2008, 2009). In addition, icv ANG I significantly increased the fetal blood pressure, and application of captopril (an inhibitor of ACE) significantly suppressed the increased fetal blood pressure (Shi et al. 2010a,b). These results show that the functional development of central endogenous ACE is established at least in the last trimester of gestation, and that endogenous brain RAS-stimulated pressor responses and AVP release are consistent with the hypothalamic cardiovascular network (Shi et al. 2010a,b).

13.6 FETAL RAS—PROGRAMMING TARGET OF HYPERTENSION

13.6.1 Low Birth Weight and Cardiovascular Diseases

Barker was the first to propose that influences from fetal environments could affect health and diseases in adult life (Barker et al. 1989). In the past 25 years of research, numerous epidemiological and experimental studies support a relationship between low birth weight (LBW) and hypertension (Huxley et al. 2000; Law et al. 2002; Lurbe et al. 2001; Yiu et al. 1999). On the basis of Barker's hypothesis, cumulating evidence has demonstrated that adverse conditions during pregnancy may lead to LBW-related hypertension and other cardiovascular diseases after birth. Notably, among studies of mechanisms for a relationship between prenatal environmental insults and cardiovascular diseases such as hypertension in the offspring, RAS increasingly attracted attention starting from 10 years ago. Several prenatal factors may affect fetal growth as well as increase risks for diseases after birth. Nutrition and oxygen supply limitations are major components of intrauterine environment that may restrict fetal growth and result in small for gestational age newborns. Fetal adaptations to oxygen and nutrient restriction as well as other prenatal stresses that occur during a critical period of fetal development may result in long-term influence and increased risk for development of cardiovascular diseases, including hypertension. This section reviews the influence of prenatal factors (hypoxia, malnutrition, glucocorticoid, dehydration, and nicotine) linked to an altered RAS on cardiovascular diseases in later life.

13.6.2 Hypoxia

Epidemiological and animal studies in the past two decades have shown a relatively clear association of adverse intrauterine environments with an increased risk of ischemic heart diseases and hypertension in adulthood (Barker and Osmond 1986; Bateson et al. 2004; Gluckman et al. 2008). Hypoxia is one of the most important and clinically relevant stresses that can adversely affect fetal development. There has been accumulating evidence of a link between hypoxia and fetal intrauterine growth restriction followed by an increased risk of cardiovascular diseases in the offspring (Heydeck et al. 1994; Jones et al. 2004; Peyronnet et al. 2002; Zhang 2005). Previous study demonstrated that chronic *in utero* hypoxia caused alternations in renal development as well as in the renal RAS of near-term ovine fetuses (Mao et al. 2010). Animal studies have demonstrated that hypoxia causes a significant increase in number and size of binucleated myocytes in the fetal heart (Bae et al. 2003), resulting in a higher susceptibility to acute ischemia and reperfusion injury in the adult male offspring in a sex-dependent manner (Xu et al. 2006; Xue and Zhang 2009).

ANG II plays a critical role in the regulation of cardiovascular responses, and has been implicated in programming of cardiovascular diseases induced by adverse *in utero* environments during fetal development (Bogdarina et al. 2007; Hadoke et al. 2006). Recent studies have demonstrated a link between fetal insults to differential epigenetic modifications of AT1 receptor genes in the adrenal gland and the resultant alteration of their expression patterns in adult life (Bogdarina et al. 2007), which may ultimately lead to an increased risk for hypertension (Bogdarina et al. 2007; McMullen and Langley-Evans 2005; Singh et al. 2007). The data show that fetal hypoxia causes the programming of increased AT2 receptor gene expression in the heart by down-regulating glucocorticoid receptors, which contribute to an increased ischemic vulnerability in the heart of adult rats (Xue and Zhang 2009). These lines of evidence provide interesting insight on an altered RAS by prenatal insults and its link to resultant cardiovascular consequences for the offspring.

13.6.3 Malnutrition

LBW, often as a consequence from prenatal nutrition, has been demonstrated to have an inverse association with blood pressure and risks in development of hypertension at various stages in later life. A number of epidemiological and experimental studies have shown that people who are small at birth tend to have higher blood pressure in later life. Recent reviews have emphasized that in both prepubertal children and adults, there is a consistent negative relationship between birth weight and systolic blood pressure (Huxley et al. 2000; Leon and Koupilova 1999). Babies with LBW often show accelerated rates of growth in infancy and early childhood, a phenomenon known as catch-up growth. Epidemiological studies that have linked LBW to raised blood pressure suggest that conceptions or strategies as well as approaches for early prevention of hypertension could be modified, at least for timing matter. In other words, early causes of essential hypertension developed probably in fetal origins could start during pregnancy and childhood, even before conception.

Notably, after the introduction of Barker's hypothesis about 25 years ago, most studies from different countries have used different subjects or models to test the theory of "adult diseases of fetal origins." As a result, accumulated evidence from epidemiological evidence in concert with experimental studies have consistently demonstrated that the offspring exposed to poor nutrition *in utero* are at increased risk for many adult-onset diseases, including renal diseases, heart diseases, hypertension, and diabetes[6,323]. For example, analysis of data from the Dutch hunger winter demonstrated that the period of gestation in which maternal nutrient restriction occurred had significant effects on birth weight and incidence of adult cardiovascular problems (Ravelli et al. 1998; Roseboom et al. 2000). The offspring exposed to the famine during gestation might have an increased risk of heart diseases (Roseboom et al. 2000).

In the past decade of research on fetal-origin diseases, researchers not only examined the prenatal causes for the programming of adult diseases, but also tried to answer why adult diseases could originate from fetal periods and could be attributable to prenatal insults. In other words, more attention has been paid to mechanisms behind the link between prenatal causes and postnatal health problems. Various models of nutritional stress such as global nutrient restriction, protein restriction, and overnutrition have been used to evaluate fetal adaptations and postnatal outcomes (Langley-Evans 2001; Langley-Evans and Jackson 1994; Nishina et al. 2003). Ozaki et al. (2000) showed hypertension in postpubescent offspring of nutrient-restricted sheep, a long gestation length animal with a long history of use in the investigation of fetal development; this report has indicated an increased blood pressure in 85-day-old offspring. Some possible mechanisms, including abnormalities of the vasculature or disruption of nephrogenesis, have been investigated (Brenner and Chertow 1994). For example, previous studies have shown that the developing kidney was susceptible to the effects of an altered growth trajectory induced by maternal nutrient restriction, and these effects have also been related to elevated blood pressure in the offspring in rats and sheep (Langley-Evans 2001; Ozaki et al. 2000; Woods et al. 2001). Among those mechanisms that may mediate programming cardiovascular diseases in fetal origins, RAS has attracted attention since 10 years ago.

Studies in rats have demonstrated elevated blood pressure, reduced nephron number, and altered gene expression of key components of the RAS in the offspring of dams subjected to gestational protein restriction (Langley-Evans 2001; Woods et al. 2001). Both systemic and local RAS have been implicated in the renal maldevelopment associated with the progression toward a hypertensive state. Perhaps the most notable effect of *in utero* perturbation of RAS was a reduced nephron endowment concomitant with elevated blood pressure after birth (Woods and Rasch 1998). Prior work has shown that effects initiated by AT1 receptor have a profound impact on renal development and subsequent morphology and physiology of the cardiovascular system (Woods and Rasch 1998). However, renal AT1 expression could remain unchanged in several forms of experimental hypertension (Harrison-Bernard et al. 1999; Licea et al. 2002), whereas the effects of the AT2 receptor expression during development appear interesting (Hein et al. 1995; Ichiki et al. 1995). In addition, the AT2 receptor showed increased expression in several models of adult renal injury (Ruiz-Ortega et al. 2003). Furthermore, expression of ACE, a key enzyme in the RAS, may have a profound impact on renal hemodynamics (Erdös 1976).

Malnutrition compromises development *in utero* and poses chronic health risks for the offspring. Protein restriction during prenatal periods increased rat offspring blood pressure by 20–30 mm Hg, associated with a reduction in nephron number and increased glomerular sensitivity to ANG II *in vivo*, indicating a possible mediation of RAS in the development of hypertension (Sahajpal and Ashton 2003; Woods et al. 2001). The elevated blood pressure was attributed to an increased AT1 receptor expression that may be a result from protein restriction (Sahajpal and Ashton 2003). Another study also added evidence for impaired nephrogenesis, decreased activity of the RAS, and the onset of hypertension in rats prenatally exposed to a maternal low protein diet (Langley and Jackson 1994). In the offspring rats exposed to maternal protein restriction during pregnancy, there was no change in renal renin activity, tissue ANG II, and plasma aldosterone concentrations, but AT1 receptor expression was increased (62%) and AT2 expression was decreased (35%) 1 month after birth, which is consistent with greater hemodynamic sensitivity to ANG II *in vivo* (Sahajpal and Ashton 2003). This may lead to an inappropriate reduction in glomerular filtration rate, salt and water retention, and finally an increase in blood pressure.

In addition to alterations in systemic and peripheral local RAS, those in central RAS may also influence hypertension of fetal origin. In the adult offspring with prenatal exposure to maternal low-protein (9%) diet, there was an increase in AT1 receptor expression in the hypothalamus, blood pressure was significantly higher, arterial baroreflex generated by intravenous infusion of phenylephrine was significantly shifted toward higher pressure, intravenous ACE inhibitor normalized blood pressure, and icv application of ACE inhibitor or AT1 antagonist also significantly reduced blood pressure (Pladys et al. 2004). These data demonstrate a major tonic role of the brain RAS in hypertension associated with prenatal nutrient deprivation.

Another concern in nutrition problems during pregnancy is consuming high salt diet. Contreras et al. (2000) showed that perinatal high salt could induce a lasting hypertension in Sprague–Dawley rats. They fed rats a high-salt diet before and during pregnancy, and during lactation. The results showed that hypertension is present in the offspring as early as 4 weeks after birth, and suggested that an increased sympathetic nervous activity secondary to enhanced activation of central AT1 receptor might be the mechanism (Swenson et al. 2004). Ding et al. (2010) recently demonstrated that after exposure to prenatal high salt, plasma ANG II decreased, and cardiac ANG II increased in the fetus. The S-phase for the cardiac cells was enhanced primarily via the AT1 receptor, whereas AT1 receptor protein and mRNA were increased. DNA methylation was found at the CpG sites that were related to

AT1b receptors in the fetal heart, suggesting a link between high salt diet during pregnancy and alterations of the cardiac cells and local RAS.

The RAS-related mechanisms of hypertension or other cardiovascular diseases due to *in utero* environmental insults such as malnutrition may be summarized as follows. (1) Embryonic or fetal RAS may mediate alternations in development of tissues in the kidney, brain, heart, vessels, and endocrine gland. As a result, those alternations could have long-term impact for the onset of hypertension as well as other cardiovascular diseases. (2) Prenatal environmental stresses such as poor nutrition may affect development of components in central and peripheral, systemic and local RAS. Functional changes of those RAS might influence regulation of vascular volumes and pressure in the offspring. (3) Epigenetic modifications such as DNA methylation may play an important role in alterations of expression of local RAS during early developmental periods, and result in molecular changes in fetal cardiac and vascular tissue that may have long-term impact.

13.6.4 Glucocorticoid

In addition to vasopressin release, ANG II in the brain can stimulate adrenocorticotropic hormone (ACTH) secretion (Ganong 1993), and circulating ANG II can act on the CVOs to increase corticotropin-releasing hormone secretion in adults (Ganong 2000), indicating a role for ANG II in hypothalamus–pituitary–adrenal (HPA) axis regulation. During fetal development, progressive maturation of the HPA axis is associated with an increase in plasma ACTH and cortisol (Currie and Brooks 1992). The fetal HPA axis responds to environmental stress during late gestational period (Ohkawa et al. 1991). Notably, both the brain RAS and HPA develop during late gestation and present certain functions that have been established in previous studies *in utero.*

Nearly 15 years ago, administration of large doses of carbenoloxone (a drug that inhibits endogenous glucocorticoid metabolism) to pregnant rats was shown to program adult hypertension (Lindsay et al. 1996), and treatment of pregnant rats with synthetic glucocorticoids was able to program hypertension in the offspring (Levitt et al. 1996). The subsequent research demonstrated that metyrapone (a chemical that inhibits glucocorticoid synthesis) treatment of the mother, with baseline corticosterone replacement, prevented hypertension expected to result from protein deprivation (Langley-Evans 1997). Other studies also showed that undernutrition, in both rats and guinea pigs, would increase maternal, and potentially fetal, glucocorticoid production (Dwyer and Stickland 1992; Lesage et al. 2001), indicating that glucocorticoid or its receptors may be involved in malnutrition-mediated cardiovascular diseases in late life. A question was raised during past decade when RAS mechanisms have attracted attention: is there any possible relationship between glucocorticoid pathway and RAS route in development of hypertension of fetal origin?

In addressing such questions, Dodic et al. (2001b) found that levels of mRNA for AT1 and AT2 receptors, as well as ATG, were increased in the kidney of dexamethasone-treated fetuses in late gestation in sheep, providing evidence that glucocorticoid was able to alter the renal RAS in the fetus. Their subsequent studies in sheep showed that dexamethasone administered between 26 and 28 days of gestation resulted in an increased expression of ATG in the fetal hypothalamus with higher expression of the

AT1 receptor in the medulla, and hypertension in the offspring (Dodic et al. 2002b). Moderately elevated maternal cortisol levels late in gestation (115 to 130 days gestation) causes enlargement of fetal sheep heart. Expression of mineralocorticoid receptor mRNA abundance in fetal hearts was also found (Reini et al. 2006). Although there was no significant changes in glucocorticoid receptor, 11beta-hydroxysteroid dehydrogenases 1 (11beta-HSD1) and IGF-1R expression were decreased in the high cortisol group and 11beta-HSD2 expression negatively correlated to left ventricular wall thickness. There was also a significant increase in AT2 receptors and decrease in AT1 receptors in the ventricle of the high cortisol group, suggesting that glucocorticoid could induce enlargement of the fetal heart and that RAS may play a role (Reini et al. 2006).

These data not only demonstrate that even brief prenatal exposure of pregnant mothers to dexamethasone may lead to hypertension in the adult offspring, but also suggest central and peripheral RAS mechanisms associated with glucocorticoids leading to fetal programming of hypertension.

13.6.5 Nicotine

Prenatal exposure to nicotine is closely associated with a variety of adverse outcomes, including premature delivery, LBW, and sudden infant death syndrome (Coleman et al. 2004; Salihu et al. 2005). Exposure to nicotine during pregnancy induces fetal intrauterine hypoxia (Suzuki et al. 1971) and affects fetal heart rate, umbilical blood flow (Kirschbaum et al. 1970; Lindblad et al. 1988), and brain development (Guan et al. 2009a; Lv et al. 2008; Mao et al. 2008). Although their basal levels of blood pressure remains unchanged, the offspring rats with prenatal history of exposure to nicotine during pregnancy show a significantly higher blood pressure to ANG II simulation in association with alterations of the ANG II receptor levels in the heart (Xiao et al. 2008). Furthermore, maternal exposure to nicotine reduces renal AT2 mRNA and protein in the neonate and adult offspring, and the affected gene of the ANG II receptors may play a critical role in "programming" renal and cardiovascular alterations of fetal origins (Mao et al. 2009b). Together, these data have demonstrated that prenatal exposure to nicotine or smoking may induce, associated with alterations in the RAS, increased cardiovascular risks in later life. Although these studies and findings are important and interesting, the question was also raised as to how prenatal exposure to nicotine could affect blood pressure via RAS mechanisms. At least two pathways have been considered. First, chemical stressors such as exposure to nicotine or smoking could induce *in utero* hypoxia that may affect the development of major components of RAS during embryonic or fetal periods; second, there may exist interactions between nicotine stimulation and growth factors, including angiotensin peptides or their receptors, which have an impact on molecular or functional development in the cardiovascular system.

13.6.6 Dehydration

Dehydration during pregnancy is commonly attributable to a number of factors, including vomiting, diarrhea, bleeding, and certain maternal diseases. Water deprivation can cause imbalance in body fluids, and intracellular and extracellular dehydration (McKinley et al. 1983; Salas et al. 2004). A number of studies suggest that water deprivation can reduce blood volume and increase plasma sodium and osmolality, leading to a series of responses in the RAS, including increased plasma renin activity and ANG II concentrations (Chatelain et al. 2003; De Luca et al. 2002; Deloof et al. 1999; Gottlieb et al. 2006; Di Nicolantonio and Mendelsohn 1986). In the study of the influence of dehydration on fetal development, previous experiments showed that hemorrhage can affect fetal cardiovascular responses. For example, ANG II infusion resulted in an increase in plasma ANG II concentrations that were similar to those observed after hemorrhage in fetal lambs. Fetal blood pressure and heart rate were significantly increased. Fetal cardiac output was measured during infusion of ANG II by the radionuclide-labeled microsphere technique. ANG II constricted the umbilical–placental circulation, and increased blood flow to the myocardium and the pulmonary circulation, indicating that ANG II, at the plasma levels achieved with moderate hemorrhage, has marked influences on the circulation, and the RAS may be important in the fetal response to hemorrhage-related dehydration (Iwamoto and Rudolph 1981). Maternal water deprivation in sheep increases fetal blood osmolality and sodium, plasma ANG II, aldosterone, and AVP levels. Moreover, intensive Fos and Fos-B expression occurs in the MPO and PVN of the fetal brain after exposure to maternal water deprivation. These data suggest that RAS-mediated regulatory mechanisms play a role in the control of fetal body fluid homeostasis, in response to maternal dehydration (Zhu et al. 2009).

Whether prenatal stresses such as dehydration may impact on cardiovascular functions at adult stages is an important question. In addressing this question, studies showed that maternal water deprivation could affect the fetal RAS, as evidenced by changes in fetal plasma ANG II and the ANG II receptors in the heart, and importantly, prenatal water deprivation also increased expression of both ATG and ANG II receptors in the brain and dipsogenic responses in male offspring. The altered RAS in the fetus was also associated with an increased risk of hypertension in later life, as ANG II–stimulated increase in blood pressure was enhanced and baroreflex activity was decreased in the young offspring (Guan et al. 2009b; Zhang et al. 2011). Maternal water deprivation not only significantly increased fetal plasma sodium, osmolality, and hematocrit, but also increased the ANG II receptors in the fetal and offspring brain, resulting in a significant increase in icv ANG II–induced water intake in the offspring (Zhang et al. 2011). The accumulated evidence demonstrates an RAS-mediated link between body fluid homeostasis during pregnancy and cardiovascular alterations in adults. Such a link may involve persistent alterations in central dipsogenic areas and in the kidney, which are rich in components of local RAS. Prenatal dehydration may affect the central dipsogenic areas rich in components of RAS, and influence fluid intake, as well as impact on renal RAS in the control of fluid balance.

13.7 CONCLUSIONS AND FUTURE DIRECTIONS

Evidence generated during the past two decades show that the local RAS in the fetus is as important as the systemic RAS for the control of blood pressure and body fluid homeostasis. The peptides of RAS, especially ANG II, can act as growth factors to control cell proliferation, differentiation, and apoptosis in the cardiovascular as well as in the central nervous system. An important concept emerges from those studies, that the development of RAS can be altered at its molecular and functional levels during embryonic and fetal periods. Most importantly, such RAS alterations affect not only fetal mechanisms *in utero*, but may also impact critically on behavioral and renal control of body fluid homeostasis and cardiovascular functions of the offspring. Growing evidence strongly supports the hypothesis that the RAS plays an important role in the "programming" of cardiovascular diseases *in utero*. In spite of the progresses in the study of fetal RAS, many questions remain for future investigation. Among them, how environmental insults or stressors alter RAS and its functions during early developmental stages, and how such alterations link to epigenetic mechanisms to produce a long-term impact on the health of the offspring.

ACKNOWLEDGMENTS

This study was supported in part by grants (2013BAI04B05, NSFC 81030006, 30973211, 81070540; BK2009122) and Jiangsu Province's Key Discipline/Laboratory of Medicine (XK201119) and Chuan Xin Tainui grants and Chuan Xin Tainui grants.

REFERENCES

Abramić, M., D. Schleuder, L. Dolovcak et al. 2000. Human and rat dipeptidyl peptidase III: Biochemical and mass spectrometric arguments for similarities and differences. *Biol Chem* 381:1233–43.

Aguilera, G. 1992. Role of angiotensin II receptor subtypes on the regulation of aldosterone secretion in the adrenal glomerulosa zone in the rat. *Mol Cell Endocrinol* 90:53–60.

Ahmad, S., J. Varagic, B. M. Westwood, M. C. Chappell, and C. M. Ferrario. 2011. Uptake and metabolism of the novel peptide angiotensin-(1–12) by neonatal cardiac myocytes. *PLoS One* 6:e15759.

Al-Awqati, Q., and M. R. Goldberg. 1998. Architectural patterns in branching morphogenesis in the kidney. *Kidney Int* 54:1832–42.

Albiston, A. L., S. G. McDowall, D. Matsacos et al. 2001. Evidence that the angiotensin IV (AT(4)) receptor is the enzyme insulin-regulated aminopeptidase. *J Biol Chem* 276:48623–6.

Alenina, N., P. Xu, B. Rentzsch, E. L. Patkin, and M. Bader. 2008. Genetically altered animal models for Mas and angiotensin-(1–7). *Exp Physiol* 93:528–37.

Allanson, J. E., A. G. Hunter, G. S. Mettler, and C. Jimenez. 1992. Renal tubular dysgenesis: A not uncommon autosomal recessive syndrome: A review. *Am J Med Genet* 43:811–4.

Allanson, J. E., J. T. Pantzar, and P. M. MacLeod. 1983. Possible new autosomal recessive syndrome with unusual renal histopathological changes. *Am J Med Genet* 16:57–60.

Allen, A. M., S. Y. Chai, P. M. Sexton et al. 1987. Angiotensin II receptors and angiotensin converting enzyme in the medulla oblongata. *Hypertension* 9:198–205.

Alwan, S., J. E. Polifka, and J. M. Friedman. 2005. Angiotensin II receptor antagonist treatment during pregnancy. *Birth Defects Res A Clin Mol Teratol* 73:123–30.

Amanda, K. S., Q. Xiaoying, E. F. Anthony, V. K. Vera, and A. K. Raouf. 2009. Increased vascular angiotensin type 2 receptor expression and NOS-mediated mechanisms of vascular relaxation in pregnant rats. *Am J Physiol Heart Circ Physiol* 296:H745–55.

Amsterdam, A., R. M. Nissen, Z. Sun, E. C. Swindell, S. Farrington, and N. Hopkins. 2004. Identification of 315 genes essential for early zebrafish development. *Proc Natl Acad Sci USA* 101:12792–7.

Araujo, R. C., M. P. Lima, E. S. Lomez et al. 2002. Tonin expression in the rat brain and tonin-mediated central production of angiotensin II. *Physiol Behav* 76:327–33.

Arnold, A. C., K. Isa, H. A. Shaltout et al. 2010. Angiotensin-(1–12) requires angiotensin converting enzyme and AT1 receptors for cardiovascular actions within the solitary tract nucleus. *Am J Physiol Heart Circ Physiol* 299:H763–71.

Azaryan, A., N. Barkhudaryan, A. Galoyan, and A. Lajtha. 1985. Action of brain cathepsin B, cathepsin D, and high-molecular-weight aspartic proteinase on angiotensins I and II. *Neurochem Res* 10:1525–32.

Bacani, C., and W. H. Frishman. 2006. Chymase: A new pharmacologic target in cardiovascular disease. *Cardiol Rev* 14:187–93.

Bader, M., J. Peters, O. Baltatu, D. N. Müller, F. C. Luft, and D. Ganten. 2001. Tissue renin-angiotensin systems: New insights from experimental animal models in hypertension research. *J Mol Med (Berl)* 79:76–102.

Bae, S., Y. Xiao, G. Li, C. A. Casiano, and L. Zhang. 2003. Effect of maternal chronic hypoxic exposure during gestation on apoptosis in fetal rat heart. *Am J Physiol Heart Circ Physiol* 285:H983–90.

Barker, D. J. 2002. Fetal programming of coronary heart disease. *Trends Endocrinol Metab* 13:364–8.

Barker, D. J., and C. Osmond. 1986. Infant mortality, childhood nutrition, and ischaemic heart disease in England and Wales. *Lancet* 1: 1077–81.

Barker, D. J., P. D. Winter, C. Osmond, B. Margetts, and S. J. Simmonds. 1989. Weight in infancy and death from ischaemic heart disease. *Lancet* 2:577–80.

Barr Jr., M., and M. M. Cohen Jr. 1991. ACE inhibitor fetopathy and hypocalvaria: The kidney-skull connection. *Teratology* 44:485–95.

Bateson, P., D. Barker, T. Clutton-Brock et al. 2004. Developmental plasticity and human health. *Nature* 430:419–21.

Bichu, P., R. Nistala, A. Khan, J. R. Sowers, and A. Whaley-Connell. 2009. Angiotensin receptor blockers for the reduction of proteinuria in diabetic patients with overt nephropathy: Results from the AMADEO study. *Vasc Health Risk Manag* 5:129–40.

Bogdarina, I., S. Welham, P. J. King, S. P. Burns, and A. J. Clark. 2007. Epigenetic modification of the renin–angiotensin system in the fetal programming of hypertension. *Circ Res* 100:520–6.

Bos-Thompson, M. A., D. Hillaire-Buys, F. Muller et al. 2005. Fetal toxic effects of angiotensin II receptor antagonists: Case report and follow-up after birth. *Ann Pharmacother* 39:157–61.

Brace, R. A., M. E. Wlodek, M. L. Cock, and R. Harding. 1994. Swallowing of lung liquid and amniotic fluid by the ovine fetus under normoxic and hypoxic conditions. *Am J Obstet Gynecol* 171:764–70.

Bradley, R. M., and C. Mistretta. 1973. Swallowing in fetal sheep. *Science* 179:1016–7.

Breault, L., J. G. Lehoux, and N. Gallo-Payet. 1996. The angiotensin AT2 receptor is present in the human fetal adrenal gland throughout the second trimester of gestation. *J Clin Endocrinol Metab* 81:3914–22.

Brenner, B. M., and G. M. Chertow. 1994. Congenital oligonephropathy and the etiology of adult hypertension and progressive renal injury. *Am J Kidney Dis* 23:171–5.

Brophy, P. D., L. Ostrom, K. M. Lang, and G. R. Dressler. 2001. Regulation of ureteric bud outgrowth by Pax2-dependent activation of the glial derived neurotrophic factor gene. *Development* 128:4747–56.

Brosnihan, K. B., L. A. Neves, and M. C. Chappell. 2005. Does the angiotensin-converting enzyme (ACE)/ACE2 balance contribute to the fate of angiotensin peptides in programmed hypertension? *Hypertension* 46:1097–9.

Brunet, A., A. Bonni, M. J. Zigmond et al. 1999. Akt promotes cell survival by phosphorylating and inhibiting a Forkhead transcription factor. *Cell* 96:857–68.

Buisson, B., S. P. Bottari, M. de Gasparo, N. Gallo-Payet, and M. D. Payet. 1992. The angiotensin AT2 receptor modulates T-type calcium current in non-differentiated NG108-15 cells. *FEBS Lett* 309:161–4.

Buisson, B., L. Laflamme, S. P. Bottari, M. de Gasparo, N. Gallo-Payet, and M. D. Payet. 1995. A G protein is involved in the angiotensin AT2 receptor inhibition of the T-type calcium current in non-differentiated NG108-15 cells. *J Biol Chem* 270:1670–4.

Bujak-Gizycka, B., R. Olszanecki, M. Suski, J. Madek, A. Stachowicz, and R. Korbut. 2010. Angiotensinogen metabolism in rat aorta: Robust formation of proangiotensin-12. *J Physiol Pharmacol* 61:679–82.

Burcklé, C., and M. Bader. 2006. Prorenin and its ancient receptor. *Hypertension* 48:549–51.

Burrell, J. H., B. D. Hegarty, J. R. McMullen, and E. R. Lumbers. 2001. Effects of gestation on ovine fetal and maternal angiotensin receptor subtypes in the heart and major blood vessels. *Exp Physiol* 86:71–82.

Butkus, A., A. Albiston, D. Alcorn et al. 1997. Ontogeny of angiotensin II receptors, types 1 and 2, in ovine mesonephros and metanephros. *Kidney Int* 52:628–36.

Cardoso, C. C., N. Alenina, A. J. Ferreira et al. 2010. Increased blood pressure and water intake in transgenic mice expressing rat tonin in the brain. *Biol Chem* 391:435–41.

Carey, R. M., and H. M. Siragy. 2003. Newly recognized components of the renin–angiotensin system: Potential roles in cardiovascular and renal regulation. *Endocr Rev* 24:261–71.

Carey, R. M., Z. Q. Wang, and H. M. Siragy. 2000. Role of the angiotensin type 2 receptor in the regulation of blood pressure and renal function. *Hypertension* 35:155–63.

Casamassima, A., and E. Rozengurt. 1998. Insulin-like growth factor I stimulates tyrosine phosphorylation of p130 (Cas), focal adhesion kinase, and paxillin. Role of phosphatidylinositol 3-kinase and formation of a p130 (Cas) crk complex. *J Biol Chem* 273: 26149–56.

Charron, G., S. Laforest, C. Gagnon, G. Drolet, G., and D. Mouginot. 2002. Acute sodium deficit triggers plasticity of the brain angiotensin type 1 receptors. *FASEB J* 16:610–2.

Chatelain, D., V. Montel, A. Dickes-Coopman, A. Chatelain, and S. Deloof. 2003. Trophic and steroidogenic effects of water deprivation on the adrenal gland of the adult female rat. *Regul Pept* 110:249–55.

Chen, K., L. C. Carey, J. Liu, J., N. K. Valego, S. B. Tatter, and J. C. Rose. 2005. The effect of hypothalamo–pituitary disconnection on the renin–angiotensin system in the late-gestation fetal sheep. *Am J Physiol Regul Integr Comp Physiol* 288:R1279–87.

Clauser, E., K. M. Curnow, E. Davies et al. 1996. Angiotensin II receptors: Protein and gene structures, expression and potential pathological involvements. *Eur J Endocrinol* 134:403–11.

Coleman, T., J. Britton, and J. Thornton. 2004. Nicotine replacement therapy in pregnancy. *BMJ* 328:965–6.

Contrepas, A., J. Walker, A. Koulakoff et al. 2009. A role of the (pro)renin receptor in neuronal cell differentiation. *Am J Physiol Regul Integr Comp Physiol* 297:R250–7.

Contreras, R. J., D. L. Wong, R. Henderson, K. S. Curtis, and J. C. Smith. 2000. High dietary NaCl early in development enhances mean arterial pressure of adult rats. *Physiol Behav* 71:173–81.

Costantini, F., and R. Shakya. 2006. GDNF/Ret signaling and the development of the kidney. *Bioessays* 28:117–27.

Côté, F., T. H. Do, L. Laflamme, J. M. Gallo, and N. Gallo-Payet. 1999. Activation of the AT(2) receptor of angiotensin II induces neurite outgrowth and cell migration in microexplant cultures of the cerebellum. *J Biol Chem* 274:31686–92.

Cox, B. E., and C. R. Rosenfeld. 1999. Ontogeny of vascular angiotensin II receptor subtype expression in ovine development. *Pediatr Res* 45:414–24.

Crackower, M. A., R. Sarao, G. Y. Oudit et al. 2002. Angiotensin-converting enzyme 2 is an essential regulator of heart function. *Nature* 417:822–8.

Cruciat, C. M., B. Ohkawara, S. P. Acebron et al. 2010. Requirement of prorenin receptor and vacuolar H^+-ATPase-mediated acidification for Wnt signaling. *Science* 327:459–63.

Cui, T., H. Nakagami, M. Iwai et al. 2001. Pivotal role of tyrosine phosphatase SHP-1 in AT2 receptor-mediated apoptosis in rat fetal vascular smooth muscle cell. *Cardiovasc Res* 49:863–71.

Currie, I. S., and A. N. Brooks. 1992. Corticotrophin-releasing factors in the hypothalamus of the developing fetal sheep. *J Dev Physiol* 17:241–6.

Davis, M. H., and J. Pieringer. 1987. Regulation of dipeptidyl aminopeptidase I and angiotensin converting enzyme activities in cultured murine brain cells by cortisol and thyroid hormone. *J Neurochem* 48:447–54.

Davisson, R. L., Y. Ding, D. E. Stec, J. F. Catterall, and C. D. Sigmund. 1999. Novel mechanism of hypertension revealed by cell-specific targeting of human angiotensinogen in transgenic mice. *Physiol Genomics* 1:3–9.

De Luca Jr. L. A., Z. Xu, Z., G. H. M. Schoorlemmer et al. 2002. Water deprivation-induced sodium appetite: Humoral and cardiovascular mediators and immediate early genes. *Am J Physiol Regul Integr Comp Physiol* 282:R552–59.

Deloof, S., C. De Seze, V. Montel, and A. Chatelain, A. 1999. Effects of water deprivation on atrial natriuretic peptide secretion and density of binding sites in adrenal glands and kidneys of maternal and fetal rats in late gestation. *Eur J Endocrinol* 141:160–8.

Desai, M., Z. Xu, C. Guerra, C., N. Kallichanda, and M. G. Ross. 2003. Maternal DDAVP-induced hyponatremia preserves fetal urine flow during acute fetal hemorrhage. *Am J Physiol Regul Integr Comp Physiol* 285:R373–9.

Di Nicolantonio, R., and F. A. O. Mendelsohn. 1986. Plasma renin and angiotensin in dehydrated and rehydrated rats. *Am J Physiol Regul Integr Comp Physiol* 250:R898–901.

Ding, Y., J. Lv, C. Mao et al. 2010. High-salt diet during pregnancy and angiotensin-related cardiac changes. *J Hypertens* 28:1290–7.

Dodic, M., R. Baird, V. Hantzis et al. 2001a. Organs/systems potentially involved in one model of programmed hypertension in sheep. *Clin Exp Pharmacol Physiol* 28:952–6.

Dodic, M., C. Samuel, K. Moritz et al. 2001b. Impaired cardiac functional reserve and left ventricular hypertrophy in adult sheep after prenatal dexamethasone exposure. *Circ Res* 89:623–9.

Dodic, M., T. Abouantan, A. O'Connor, E. M. Wintour, and K. M. Moritz. 2002a. Programming effects of short prenatal exposure to dexamethasone in sheep. *Hypertension* 40:729–34.

Dodic, M., V. Hantzis, J. Duncan et al. 2002b. Programming effects of short prenatal exposure to cortisol. *FASEB J* 16:1017–26.

Dwyer, C. M., and N. C. Stickland. 1992. The effects of maternal undernutrition on maternal and fetal serum insulin-like growth factors, thyroid hormones and cortisol in the guinea pig. *J Dev Physiol* 18:303–13.

El-Haddad, M. A., C. R. Chao, S. X. Ma, and M. G. Ross. 1999. Nitric oxide modulates spontaneous swallowing behavior in near-term ovine fetus. *Am J Physiol Regul Integr Comp Physiol* 277:R981–6.

El-Haddad, M. A., C. R. Chao, S. Ma, and M. G. Ross. 2000. Nitric oxide modulates angiotensin II-induced drinking behavior in the near-term ovine fetus. *Am J Obstet Gynecol* 182:713–9.

El-Haddad, M. A., C. R. Chao, A. A. Sayed, H. El-Haddad, and M. G. Ross, M.G. 2001. Effects of central angiotensin II receptor antagonism on fetal swallowing and cardiovascular activity. *Am J Obstet Gynecol* 185:828–33.

El-Haddad, M. A., C. R. Chao, S. X. Ma, and M. G. Ross. 2002. Neuronal NO modulates spontaneous and ANG II-stimulated fetal swallowing behavior in the near-term ovine fetus. *Am J Physiol Regul Integr Comp Physiol* 282:R1521–7.

El-Haddad, M. A., Y. Ismail, D. Gayle, and M. G. Ross. 2005. Central angiotensin II AT1 receptors mediate fetal swallowing and pressor responses in the near-term ovine fetus. *Am J Physiol Regul Integr Comp Physiol* 288:R1014–20.

Erdös, E. G. 1976. Conversion of angiotensin I to angiotensin II. *Am J Med* 60:749–59.

Esther Jr. C. R., T. E. Howard, E. M. Marino, J. M. Goddard, M. R. Capecchi, and K. E. Bernstein. 1996. Mice lacking angiotensin-converting enzyme have low blood pressure, renal pathology, and reduced male fertility. *Lab Invest* 74:953–65.

Feldt, S., W. W. Batenburg, I. Mazak et al. 2008. Prorenin and renin-induced extracellular signal-regulated kinase 1/2 activation in monocytes is not blocked by aliskiren or the handle-region peptide. *Hypertension* 51:682–8.

Ferrario, C. M. 2011. ACE2: More of Ang-(1–7) or less ANG II? *Curr Opin Nephrol Hypertens* 20:1–6.

Ferrario, C. M., and J. Varagic. 2010. The ANG-(1–7)/ACE2/mas axis in the regulation of nephron function. *Am J Physiol Renal Physiol* 298:F1297–305.

Ferrario, C. M., J. Varagic, J. Habibi et al. 2009. Differential regulation of angiotensin-(1–12) in plasma and cardiac tissue in response to bilateral nephrectomy. *Am J Physiol Heart Circ Physiol* 296:H1184–92.

Fitzsimons, J.T. 1998. Angiotensin, thirst, and sodium appetite. *Physiol Rev* 78:583–686.

Forhead, A. J., J. K. Jellyman, K. Gillham, J. W. Ward, D. Blache, and A. L. Fowden. 2011. Renal growth retardation following angiotensin II type 1 (AT_1) receptor antagonism is associated with increased AT_2 receptor protein in fetal sheep. *J Endocrinol* 208:137–45.

Forte, G., M. Minieri, P. Cossa et al. 2006. Hepatocyte growth factor effects on mesenchymal stem cells: Proliferation, migration, and differentiation. *Stem Cells* 24:23–33.

Foulon, T., S. Cadel, and P. Cohen. 1999. Aminopeptidase B (EC 3.4.11.6). *Int J Biochem Cell Biol* 31:747–50.

Friberg, P., B. Sundelin, B., S. O. Bohman et al. 1994. Renin–angiotensin system in neonatal rats: Induction of a renal abnormality in response to ACE inhibition or angiotensin II antagonism. *Kidney Int* 45:485–92.

Gallinat, S., S. Busche, M. K. Raizada, and C. Sumners. 2000. The angiotensin II type 2 receptor: An enigma with multiple variations. *Am J Physiol Endocrinol Metab* 278:E357–74.

Ganong, W. F. 1993. Blood, pituitary, and brain renin–angiotensin systems and regulation of secretion of anterior pituitary gland. *Front Neuroendocrinol* 14:233–49.

Ganong, W. F. 2000. Circumventricular organs: Definition and role in the regulation of endocrine and autonomic function. *Clin Exp Pharmacol Physiol* 27:422–7.

Gao, L., W. Z. Wang, W. Wang, and I. H. Zucker. 2008. Imbalance of angiotensin type 1 receptor and angiotensin II type 2 receptor in the rostral ventrolateral medulla: Potential mechanism for sympathetic overactivity in heart failure. *Hypertension* 52:708–14.

García-Villalba, P., N. D. Denkers, C. T. Wittwer, C. Hoff, R. D. Nelson, and T. J. Mauch. 2003. Real-time PCR quantification of AT1 and AT2 angiotensin receptor mRNA expression in the developing rat kidney. *Nephron Exp Nephrol* 94:e154–9.

Gembardt, F., S. Heringer-Walther, J. H. van Esch et al. 2008. Cardiovascular phenotype of mice lacking all three subtypes of angiotensin II receptors. *FASEB J* 22:3068–77.

Gendron, L., L. Laflamme, N. Rivard, C. Asselin, M. D. Payet, and N. Gallo-Payet. 1999. Signals from the AT2 (angiotensin type 2) receptor of angiotensin II inhibit p21ras and activate MAPK (mitogen-activated protein kinase) to induce morphological neuronal differentiation in NG108-15 cells. *Mol Endocrinol* 13:1615–26.

Gendron, L., F. Côté, M. D. Payet, and N. Gallo-Payet. 2002. Nitric oxide and cyclic GMP are involved in angiotensin II AT(2) receptor effects on neurite outgrowth in NG108-15 cells. *Neuroendocrinology* 75:70–81.

Gendron, L., J. F. Oligny, M. D. Payet, and N. Gallo-Payet. 2003. Cyclic AMP-independent involvement of Rap1/B-Raf in the angiotensin II AT2 receptor signaling pathway in NG108-15 cells. *J Biol Chem* 278:3606–14.

Gibson, K. J., and E. R. Lumbers. 1997. Ovine fetal cardiovascular, renal, and fluid balance responses to 3 days of high arginine vasopressin levels. *Am J Physiol Regul Integr Comp Physiol* 272:R1069–76.

Gilbert, W. M., and R. A. Brace. 1989. The missing link in amniotic fluid volume regulation: Intramembranous flow. *Obstet Gynecol* 74:748–54.

Gilbert, W. M., and R. A. Brace. 1990. Novel determination of filtration coefficient of ovine placenta and intramembranous flow pathway. *Am J Physiol* 259:R1281–8.

Gluckman, P. D., M. A. Hanson, C. Cooper, and K. L. Thornburg. 2008. Effect of in utero and early-life conditions on adult health and disease. *N Engl J Med* 359:61–73.

Gomez, R. A., J. G. Meernik, W. D. Kuehl, and J. E. Robillard. 1984. Developmental aspects of the renal response to hemorrhage during fetal life. *Pediatr Res* 18:40–46.

Gorelik, G., L. A. Carbini, and A. G. Scicli. 1998. Angiotensin 1–7 induces bradykinin-mediated relaxation in porcine coronary artery. *J Pharmacol Exp Ther* 286:403–10.

Gottlieb, H. B., L. L. Ji, H. Jones, M. L. Penny, T. Fleming, and J. T. Cunningham. 2006. Differential effects of water and saline intake on water deprivation-induced c-Fos staining in the rat. *Am J Physiol Regul Integr Comp Physiol* 290:R1251–61.

Grady, E. F., L. A. Sechi, C. A. Griffin, M. Schambelan, and J. E. Kalinyak. 1991. Expression of AT2 receptors in the developing rat fetus. *J Clin Invest* 88:921–33.

Gribouval, O., M. Gonzales, T. Neuhaus et al. 2005. Mutations in genes in the renin–angiotensin system are associated with autosomal recessive renal tubular dysgenesis. *Nat Genet* 37: 964–8.

Griendling, K. K., T. Tsuda, B. C. Berk, and R. W. Alexander. 1989. Angiotensin II stimulation of vascular smooth muscle cells. Secondary signalling mechanisms. *Am J Hypertens* 2:659–65.

Griendling, K. K., M. Ushio-Fukai, B. Lassegue, and R. W. Alexander. 1997. Angiotensin II signaling in vascular smooth muscle. New concepts. *Hypertension* 29:366–73.

Grobe, J. L., D. Xu, and C. D. Sigmund. 2008. An intracellular renin–angiotensin system in neurons: Fact, hypothesis, or fantasy. *Physiology* 23:187–93.

Gross, V., M. Obst, and F. C. Luft. 2004. Insights into angiotensin II receptor function through AT2 receptor knockout mice. *Acta Physiol Scand* 181:487–94.

Guan, J., C. Mao, F. Xu et al. 2009a. Low doses of nicotine induced fetal cardiovascular responses, hypoxia, and brain cellular activation in ovine fetuses. *Neurotoxicology* 30:290–7.

Guan, J., C. Mao, F. Xu et al. 2009b. Prenatal dehydration affected renin–angiotensin system associated with angiotensin-increased blood pressure in young offspring. *Hypertens Res* 32:1104–11.

Guo, D. F., and T. Inagami. 1994. The genomic organization of the rat angiotensin II receptor AT1B. *Biochim Biophys Acta* 1218:91–4.

Guron, G., M. A. Adams, B. Sundelin, and P. Friberg. 1997. Neonatal angiotensin-converting enzyme inhibition in the rat induces persistent abnormalities in renal function and histology. *Hypertension* 29:91–7.

Guron, G., A. Nilsson, N. Nitescu et al. 1999. Mechanisms of impaired urinary concentrating ability in adult rats treated neonatally with enalapril. *Acta Physiol Scand* 165:103–12.

Hackenthal, E., R. Hackenthal, and U. Hilgenfeldt. 1978. Isorenin, pseudorenin, cathepsin D and renin. A comparative enzymatic study of angiotensin-forming enzymes. *Biochim Biophys Acta* 522:574–88.

Hadoke, P. W., R. S. Lindsay, J. R. Seckl, B. R. Walker, and C. J. Kenyon. 2006. Altered vascular contractility in adult female rats with hypertension programmed by prenatal glucocorticoid exposure. *J Endocrinol* 188:435–42.

Han, H. J., J. Y. Han, J. S. Heo, S. H. Lee, M. Y. Lee, and Y. H. Kim. 2007. ANG II-stimulated DNA synthesis is mediated by ANG II receptor-dependent Ca(2+)/PKC as well as EGF receptor-dependent PI3K/Akt/mTOR/p70S6K1 signal pathways in mouse embryonic stem cells. *J Cell Physiol* 211:618–29.

Handa, R. K., C. M. Ferrario, and J. W. Strandhoy. 1996. Renal actions of angiotensin-(1–7): In vivo and in vitro studies. *Am J Physiol Renal Physiol* 270:F141–47.

Harding, J. W., V. I. Cook, A. V. Miller-Wing et al. 1992. Identification of an AII(3–8) [AIV] binding site in guinea pig hippocampus. *Brain Res* 583:340–3.

Harding, R., J. N. Sigger, Poore, and P. Johnson. 1984. Ingestion in fetal sheep and its relation to sleep states and breathing movements. *Q J Exp Physiol* 69:477–86.

Harrison-Bernard, L. M., S. S. El-Dahr, D. F. O'Leary, and L. G. Navar. 1999. Regulation of angiotensin II type 1 receptor mRNA and proteinin angiotensin II-induced hypertension. *Hypertension* 33:340–6.

Hautmann, M. B., M. M. Thompson, E. A. Swartz, E. N. Olson, and G. K. Owens. 1997. Angiotensin II-induced stimulation of smooth muscle alpha-actin expression by serum response factor and the homeodomain transcription factor MHox. *Circ Res* 81:600–10.

Hein, L., G. S. Barsh, R. E. Pratt, V. J. Dzau, and B. K. Kobilka. 1995. Behavioural and cardiovascular effects of disrupting the angiotensin II type-2 receptor in mice. *Nature* 377: 744–7.

Heydeck, D., J. Roigas, C. Roigas, B. Papies, and A. Lun. 1994. The catecholamine sensitivity of adult rats is enhanced after prenatal hypoxia. *Biol Neonate* 66:106–11.

Hilgers, K. F., V. Reddi, J. H. Krege, O. Smithies, and R. A. Gomez. 1997. Aberrant renal vascular morphology and renin expression in mutant mice lacking angiotensin-converting enzyme. *Hypertension* 29:216–21.

Hiraiwa, M. 1999. Cathepsin A/protective protein: An unusual lysosomal multifunctional protein. *Cell Mol Life Sci* 56:894–907.

Hoffman, W. E., M. I. Philips, P. G. Schmid, J. Falcon, and J. F. Weet. 1977. Antidiuretic hormone release and the pressor response to central angiotensin II and cholinergic stimulation. *Neuropharmacology* 16:463–72.

Horiuchi, M., W. Hayashida, T. Kambe, T. Yamada, and V. J. Dzau. 1997. Angiotensin type 2 receptor dephosphorylates Bcl-2 by activating mitogen-activated protein kinase phosphatase-1 and induces apoptosis. *J Biol Chem* 272:19022–6.

Hu, F., P. Morrissey, J. Yao, and Z. Xu. 2004. Development of AT1 and AT2 receptors in the ovine fetal brain. *Brain Res Dev Brain Res* 150:51–61.

Huang, Y., S. Wongamorntham, J. Kasting et al. 2006. Renin increases mesangial cell transforming growth factor-beta1 and matrix proteins through receptor-mediated, angiotensin II-independent mechanisms. *Kidney Int* 69:105–13.

Hunt, R. A., G. M. Ciuffo, J. M. Saavedra, and D. C. Tucker. 1995. Quantification and localisation of angiotensin II receptors and angiotensin converting enzyme in the developing rat heart. *Cardiovasc Res* 29:834–40.

Hutchinson, H. G., L. Hein, M. Fujinaga, and R. E. Pratt. 1999. Modulation of vascular development and injury by angiotensin II. *Cardiovasc Res* 41:689–700.

Huxley, R. R., A. W. Shiell, and C. M. Law. 2000. The role of size at birth and postnatal catch-up growth in determining systolic blood pressure: A systematic review of the literature. *J Hypertens* 18:815–31.

Ichiki, T., C. L. Herold, Y. Kambayashi, S. Bardhan, and T. Inagami. 1994. Cloning of the cDNA and the genomic DNA of the mouse angiotensin II type 2 receptor. *Biochim Biophys Acta* 1189:247–50.

Ichiki, T., P. A. Labosky, C. Shiota et al. 1995. Effects on blood pressure and exploratory behaviour of mice lacking angiotensin II type-2 receptor. *Nature* 377:748–50.

Iosipiv, I. V., and M. Schroeder. 2003. A role for angiotensin II AT1 receptors in ureteric bud cell branching. *Am J Physiol Renal Physiol* 285:F199–207.

Ito, M., M. I. Oliverio, P. J. Mannon et al. 1995. Regulation of blood pressure by the type 1A angiotensin II receptor gene. *Proc Natl Acad Sci* 92:3521–5.

Iwai, N., T. Inagami, N. Ohmichi, and M. Kinoshita. 1996. Renin is expressed in rat macrophage/monocyte cells. *Hypertension* 27:399–403.

Iwamoto, H. S., and A. M. Rudolph. 1981. Effects of angiotensin II on the blood flow and its distribution in fetal lambs. *Circ Res* 48:183–9.

Jain, S., A. Knoten, M. Hoshi et al. 2010. Organotypic specificity of key RET adaptor-docking sites in the pathogenesis of neurocristopathies and renal malformations in mice. *J Clin Invest* 120:778–90.

Jones, C. A., C. D. Sigmund, R. A. McGowan, C. M. Kane-Haas, and K. W. Gross. 1990. Expression of murine renin genes during fetal development. *Mol Endocrinol* 4:375–83.

Jones, R. D., A. H. Morice, and C. J. Emery. 2004. Effects of perinatal exposure to hypoxia upon the pulmonary circulation of the adult rat. *Physiol Res* 53:11–7.

Jung, F. F., B. Bouyounes, R. Barrio, S. S. Tang, D. Diamant, and J. R. Ingelfinger. 1993. Angiotensin converting enzyme in renal ontogeny: Hypothesis for multiple roles. *Pediatr Nephrol* 7:834–40.

Kageyama, T., M. Ichinose, and S. Yonezawa. 1995. Processing of the precursors to neurotensin and other bioactive peptides by cathepsin E. *J Biol Chem* 270:19135–40.

Kakuchi, J., T. Ichiki, S. Kiyama et al. 1995. Developmental expression of renal angiotensin II receptor genes in the mouse. *Kidney Int* 47:140–7.

Karamyan, V. T., and R. C. Speth. 2007. Enzymatic pathways of the brain renin–angiotensin system: Unsolved problems and continuing challenges. *Regul Pept* 143:15–27.

Kaschina, E., and T. Unger. 2003. Angiotensin AT1/AT2 receptors: Regulation, signaling and function. *Blood Press* 12:70–88.

Kim, Y. H., J. M. Ryu, Y. J. Lee, and H. J. Han. 2010. Fibronectin synthesis by high glucose level mediated proliferation of mouse embryonic stem cells: Involvement of ANG II and TGF-beta1. *J Cell Physiol* 223:397–407.

Kirschbaum, T. H., P. V. Dilts, and C. R. Brinkman. 1970. Some acute effects of smoking in sheep and their fetuses. *Obstet Gynecol* 35:527–36.

Kobori, H., M. Nangaku, L. G. Navar, and A. Nishiyama. 2007. The intrarenal renin–angiotensin system: From physiology to the pathobiology of hypertension and kidney disease. *Pharmacol Rev* 59:251–87.

Kon, Y., Y. Hashimoto, H. Kitagawa, and N. Kudo. 1989. An immunohistochemical study on the embryonic development of renin-containing cells in the mouse and pig. *Anat Histol Embryol* 18:14–26.

Kon, Y., Y. Hashimoto, H. Kitagawa, M. Sugimura, and K. Murakami. 1990. Renin immunohistochemistry in the adrenal gland of the mouse fetus and neonate. *Anat Rec* 227:124–31.

Kon, Y., Y. Hashimoto, H. Kitagawa, M. Sugimura, and K. Murakami. 1991. Intracellular production of adrenal renin in the fetal mouse. An immuno-electron microscopical study. *J Anat* 176:23–33.

Kops, G. J., N. D. de Ruiter, A. M. De Vries-Smits, D. R. Powell, J. L. Bos, and B. M. Burgering. 1999. Direct control of the Forkhead transcription factor AFX by protein kinase B. *Nature* 398:630–34.

Koukoulas, I., T. Mustafa, R. Douglas-Denton, and E. M. Wintour. 2002. AT1R and AT2R expression peaks when placental growth is maximal in sheep. *Am J Physiol Regul Integr Comp Physiol* 283:R972–82.

Kriegsmann, J., W. Coerdt, F. Kommoss, R. Beetz, C. Hallermann, and H. Müntefering. 2000. Renal tubular dysgenesis (RTD)—an important cause of the oligohydramnion-sequence. Report of 3 cases and review of the literature. *Pathol Res Pract* 196:861–5.

Kullama, L. K., C. L. Agnew, L. Day, M. G. Ervin, and M. G. Ross. 1994. Ovine fetal swallowing and renal responses to oligohydramnios. *Am J Physiol Regul Integr Comp Physiol* 266:R972–8.

Kumar, R., and M. A. Boim. 2009. Diversity of pathways for intracellular angiotensin II synthesis. *Curr Opin Nephrol Hypertens* 18:33–9.

Kumar, R., V. P. Singh, and K. M. Baker. 2007. The intracellular renin–angiotensin system: A new paradigm. *Trends Endocrinol Metab* 18:208–14.

Kurt, B., L. Kurtz, M. L. Sequeira-Lopez et al. 2011. Reciprocal expression of connexin 40 and 45 during phenotypical changes in renin-secreting cells. *Am J Physiol Renal Physiol* 300:F743–8.

Laflamme, L., M. Gasparo, J. M. Gallo, M. D. Payet, and N. Gallo-Payet. 1996. Angiotensin II induction of neurite outgrowth by AT2 receptors in NG108-15 cells. Effect counteracted by the AT1 receptors. *J Biol Chem* 271:22729–35.

Langley, S. C., and A. A. Jackson. 1994. Increased systolic blood pressure in adult rats induced by fetal exposure to maternal low protein diets. *Clin Sci (Lond)* 86:217–22.

Langley-Evans, S. C. 1997. Hypertension induced by foetal exposure to a maternal low-protein diet, in the rat, is prevented by pharmacological blockade of maternal glucocorticoid synthesis. *J Hypertension* 15:537–44.

Langley-Evans, S. C. 2001. Fetal programming of cardiovascular function through exposure to maternal undernutrition. *Proc Nutr Soc* 60:505–13.

Langley-Evans, S. C., and A. A. Jackson. 1995. Captopril normalises systolic blood pressure in rats with hypertension induced by fetal exposure to maternal low protein diets. *Comp Biochem Physiol A Physiol* 110:223–8.

Law, C. M., A. W. Shiell, C. A. Newsome et al. 2002. Fetal, infant, and childhood growth and adult blood pressure: A longitudinal study from birth to 22 years of age. *Circulation* 105:1088–92.

Lazartigues, E. 2009. A map and new directions for the (pro)renin receptor in the brain: Focus on "A role of the (pro)renin receptor in neuronal cell differentiation". *Am J Physiol Regul Integr Comp Physiol* 297:R248–9.

Le Noble, F. A., J. W. Hekking, H. W. Van Straaten, D. W. Slaaf, and H. A. Struyker Boudier. 1991. Angiotensin II stimulates angiogenesis in the chorio-allantoic membrane of the chick embryo. *Eur J Pharmacol* 195:305–6.

Le Noble, F. A., N. H. Schreurs, H. W. van Straaten et al. 1993. Evidence for a novel angiotensin II receptor involved in angiogenesis in chick embryo chorioallantoic membrane. *Am J Physiol Regul Integr Comp Physiol* 264:R460–5.

Leon, D. A., and I. Koupilova. 1999. Birth weight, blood pressure and hypertension: Epidemiological studies. In: *Fetal Origins of Cardiovascular and Llung Disease*, ed. D. J. P. Barker. Bethesda, MD: National Institutes of Health.

Lesage, J., B. Blondeau, M. Grino, B. Bréant, and J. P. Dupouy. 2001. Maternal undernutrition during late gestation induces fetal overexposure to glucocorticoids and intrauterine growth retardation, and disturbs the hypothalamo–pituitary adrenal axis in the newborn rat. *Endocrinology* 142:1692–702.

Levitt, N. S., R. S. Lindsay, M. C. Holmes, and J. R. Seckl. 1996. Dexamethasone in the last week of pregnancy attenuates hippocampal glucocorticoid receptor gene expression and elevates blood pressure in the adult offspring in the rat. *Neuroendocrinology* 64:412–8.

Levy, B. I. 2004. Can angiotensin II type 2 receptors have deleterious effects in cardiovascular disease? Implications for therapeutic blockade of the renin–angiotensin system. *Circulation* 109:8–13.

Lew, R. A. 2004. The zinc metallopeptidase family: New faces, new functions. *Protein Pept Lett* 11:407–14.

Lew, R. A., T. Mustafa, S. Ye, S. G. McDowall, S.Y. Chai, and A. L. Albiston. 2003. Angiotensin AT4 ligands are potent, competitive inhibitors of insulin regulated aminopeptidase (IRAP). *J Neurochem* 86:344–50.

Li, J. M., M. Mogi, K. Tsukuda et al. 2007. Angiotensin II-induced neural differentiation via angiotensin II type 2 (AT2) receptor-MMS2 cascade involving interaction between AT2 receptor-interacting protein and Src homology 2 domain-containing protein-tyrosine phosphatase 1. *Mol Endocrinol* 21:499–511.

Licea, H., M. R. Walters, and L. G. Navar. 2002. Renal nuclear angiotensin II receptors in normal and hypertensive rats. *Acta Physiol Hung* 89:427–38.

Lindblad, A., K. Marsál, and K. E. Andersson. 1988. Effect of nicotine on human fetal blood flow. *Obstet Gynecol* 72:371–82.

Lindsay, R. S., R. M. Lindsay, C. R. Edwards, and J. R. Seckl. 1996. Inhibition of 11-beta-hydroxysteroid dehydrogenase in pregnant rats and the programming of blood pressure in the offspring. *Hypertension* 27:1200–4.

Liu, A., and B. J. Ballermann. 1998. TGF-beta type II receptor in rat renal vascular development: Localization to juxtaglomerular cells. *Kidney Int* 53:716–25.

Lumbers, E. R. 1995. Functions of the renin–angiotensin system during development. *Clin Exp Pharmacol Physiol* 22:499–505.

Lumbers, E. R., J. H. Burrell, R. I. Menzies, and A. D. Stevens. 1993. The effects of a converting enzyme inhibitor (captopril) and angiotensin II on fetal renal function. *Br J Pharmacol* 110:821–7.

Lurbe, E., I. Torro, C. Rodriguez, V. Alvarez, and J. Redon. 2001. Birth weight influences blood pressure values and variability in children and adolescents. *Hypertension* 38:389–93.

Lv, J., C. Mao, L. Zhu, L. et al. 2008. The effect of prenatal nicotine on expression of nicotine receptor subunits in the fetal brain. *Neurotoxicology* 29:722–6.

Mao, C., J. Hou, J. Ge et al. 2010. Changes of renal AT1/AT2 receptors and structures in ovine fetuses following exposure to long-term hypoxia. *Am J Nephrol* 31:141–50.

Mao, C., J. Lv, H. Zhu et al. 2007. Fetal functional capabilities in response to maternal hypertonicity associated with altered central and peripheral angiotensinogen mRNA in rats. *Peptides* 28:1178–84.

Mao, C., L. Shi, F. Xu, L. Zhang, and Z. Xu. 2009a. Development of fetal brain renin–angiotensin system and hypertension programmed in fetal origins. *Prog Neurobiol* 87:252–63.

Mao, C., J. Wu, D. Xiao et al. 2009b. Effect of fetal and neonatal nicotine exposure on renal development of AT1 and AT2 receptors. *Reprod Toxicol* 27:149–54.

Mao, C., H. Zhang, D. Xia et al. 2008. Perinatal nicotine exposure alters AT1 and AT2 receptor expression pattern in the brain of fetal and offspring rats. *Brain Res* 1243:47–52.

Masilamani, S., G. H. Kim, C. Mitchell, J. B. Wade, and M. A. Knepper. 1999. Aldosterone-mediated regulation of ENaC alpha, beta, and gamma subunit proteins in rat kidney. *J Clin Invest* 104:R19–23.

Matsumoto, H., T. Rogi, K. Yamashiro et al. 2000. Characterization of a recombinant soluble form of human placental leucine aminopeptidase/oxytocinase expressed in Chinese hamster ovary cells. *Eur J Biochem* 267:46–52.

Matsuura, T., R. A. Harrison, A. D. Westwell, H. Nakamura, A. F. Martynyuk, and C. Sumners. 2007. Basal and angiotensin II-inhibited neuronal delayed-rectifier K^+ current are regulated by thioredoxin. *Am J Physiol Cell Physiol* 293:C211–7.

Matsuura, T., C. Sun, L. Leng et al. 2006. Macrophage migration inhibitory factor increases neuronal delayed rectifier K^+ current. *J Neurophysiol* 95:1042–48.

Matthews, S. G. 1999. Hypothalamic oxytocin in the developing ovine fetus: Interaction with pituitary–adrenocortical function. *Brain Res* 820:92–100.

McKinley, M. J., A. Allen, J. Clevers, D. A. Denton, and F. A. Mendelsohn. 1986. Autoradiographic localization of angiotensin receptors in the sheep brain. *Brain Res* 375:373–6.

McKinley, M. J., D. A. Denton, J. F. Nelson, and R. S. Weisinger. 1983. Dehydration induces sodium depletion in rats, rabbits, and sheep. *Am J Physiol Regul Integr Comp Physiol* 45:R287–92.

McMullen, S., and S. C. Langley-Evans. 2005. Maternal low-protein diet in rat pregnancy programs blood pressure through sex-specific mechanisms. *Am J Physiol Regul Integr Comp Physiol* 288:R85–R90.

Meffert, S., M. Stoll, U. M. Steckelings, S. P. Bottari, and T. Unger. 1996. The angiotensin II AT2 receptor inhibits proliferation and promotes differentiation in PC12W cells. *Mol Cell Endocrinol* 122:59–67.

Mehta, P. K., and K. K. Griendling. 2007. Angiotensin II cell signaling: Physiological and pathological effects in the cardiovascular system. *Am J Physiol Cell Physiol* 292:C82–97.

Miura, S., Y. Matsuo, Y. Kiya, S. S. Karnik, and K. Saku. 2010. Molecular mechanisms of the antagonistic action between AT1 and AT2 receptors. *Biochem Biophys Res Commun* 391:85–90.

Miyazaki, Y., S. Tsuchida, A. Fogo, and I. Ichikawa. 1999. The renal lesions that develop in neonatal mice during angiotensin inhibition mimic obstructive nephropathy. *Kidney Int* 55:1683–95.

Moritz, K. M., W. C. Boon, and E. M. Wintour. 1999. Aldosterone secretion by the mid-gestation ovine fetus: Role of the AT2 receptor. *Mol Cell Endocrinol* 157:153–60.

Moritz, K. M., D. J. Campbell, and E. M. Wintour. 2001. Angiotensin-(1–7) in the ovine fetus. *Am J Physiol Regul Integr Comp Physiol* 280:R404–9.

Mukoyama, M., M. Nakajima, M. Horiuchi, H. Sasamura, R. E. Pratt, and V. J. Dzau. 1993. Expression cloning of type 2 angiotensin II receptor reveals a unique class of seven-transmembrane receptors. *J Biol Chem* 268:24539–42.

Mulvihill, S. J., M. D. Stone, H. T. Debas, and E. W. Fonkal. 1985. The role of amniotic fluid in fetal nutrition. *J Pediat Surg* 20:668–72.

Mungall, B. A., T. A. Shinkel, and C. Sernia. 1995. Immunocytochemical localization of angiotensinogen in the fetal and neonatal rat brain. *Neuroscience* 67:505–24.

Nagata, M., K. Tanimoto, A. Fukamizu et al. 1996. Nephrogenesis and renovascular development in angiotensinogen-deficient mice. *Lab Invest* 75:745–53.

Nagata, S., J. Kato, K. Sasaki, N. Minamino, T. Eto, and K. Kitamura. 2006. Isolation and identification of proangiotensin-12, a possible component of the renin–angiotensin system. *Biochem Biophys Res Commun* 350:1026–31.

Nakajima, M., H. G. Hutchinson, M. Fujinaga et al. 1995. The angiotensin II type 2 (AT2) receptor antagonizes the growth effects of the AT1 receptor: Gain-of-function study using gene transfer. *Proc Natl Acad Sci USA* 92:10663–67.

Niimura, F., P. A. Labosky, J. Kakuchi et al. 1995. Gene targeting in mice reveals a requirement for angiotensin in the development and maintenance of kidney morphology and growth factor regulation. *J Clin Invest* 96:2947–54.

Niimura, F., S. Okubo, A. Fogo, and I. Ichikawa. 1997. Temporal and spatial expression pattern of the angiotensinogen gene in mice and rats. *Am J Physiol Regul Integr Comp Physiol* 272: R142–7.

Nijland, M. J., M. G. Ross, L. K. Kullama, K. Bradley, and M. G. Ervin. 1995. DDAVP-induced maternal hyposmolality increases ovine fetal urine flow. *Am J Physiol Regul Integr Comp Physiol* 268:R358–65.

Nilsson, E., G. Stålberg, P. Lichtenstein, S. Cnattingius, P. O. Olausson, and C. M. Hultman. 2005. Fetal growth restriction and schizophrenia: A Swedish twin study. *Twin Res Hum Genet* 8:402–8.

Nishina, H., L. R. Green, H. H. McGarrigle, D. E. Noakes, L. Poston, and M. A. Hanson. 2003. Effect of nutritional restriction in early pregnancy on isolated femoral artery function in mid-gestation fetal sheep. *J Physiol* 553:637–47.

Nouet, S., and C. Nahmias. 2000. Signal transduction from the angiotensin II AT2 receptor. *Trends Endocrinol Metab* 11:1–6.

Nuyt, A. M., Z. Lenkei, P. Corvol, M. Palkovits, and C. Llorens-Cortes. 2001. Ontogeny of angiotensin II type 1 receptor mRNAs in fetal and neonatal rat brain. *J Comp Neurol* 440:192–203.

Nuyt, A. M., Z. Lenkei, M. Palkovits, P. Corvol, and C. Llorens-Cortes. 1999. Ontogeny of angiotensin II type 2 receptor mRNA expression in fetal and neonatal rat brain. *J Comp Neurol* 407:193–206.

Ohkawa, T., W. Rohde, S. Takeshita, G. Dörner, K. Arai, and S. Okinaga. 1991. Effect of an acute maternal stress on the fetal hypothalamo–pituitary–adrenal system in late gestational life of the rat. *Exp Clin Endocrinol* 98:123–9.

Oliverio, M. I., H. S. Kim, M. Ito et al. 1998. Reduced growth, abnormal kidney structure, and type 2 (AT2) angiotensin receptor-mediated blood pressure regulation in mice lacking both AT1A and AT1B receptors for angiotensin II. *Proc Natl Acad Sci USA* 95:15496–501.

Oparil, S., S. P. Bishop, and F. J. Clubb. 1984. Myocardial cell hypertrophy or hyperplasia. *Hypertension* 6:III38–43.

Oshima, K., Y. Miyazaki, J. W. Brock, M. C. Adams, I. Ichikawa, and J. C. Pope. 2001. Angiotensin type II receptor expression and ureteral budding. *J Urol* 166:1848–52.

O'Tierney, P. F., N. N. Chattergoon, S. Louey, G. D. Giraud, and K. L. Thornburg. 2010. Atrial natriuretic peptide inhibits angiotensin II-stimulated proliferation in fetal cardiomyocytes. *J Physiol* 588:2879–89.

Ozaki, T., P. Hawkins, H. Nishina, C. Steyn, L. Poston, and M. A. Hanson. 2000. Effects of undernutrition in early pregnancy on systemic small artery function in late-gestation fetal sheep. *Am J Obstet Gynecol* 183:1301–7.

Padia, S. H., N. L. Howell, B. A. Kemp, M. C. Fournie-Zaluski, B. P. Roques, and R. M. Carey. 2010. Intrarenal aminopeptidase N inhibition restores defective angiotensin II type 2-mediated natriuresis in spontaneously hypertensive rats. *Hypertension* 55:474–80.

Padia, S. H., B. A. Kemp, N. L. Howell, M. C. Fournie-Zaluski, B. P. Roques, and R. M. Carey. 2008. Conversion of renal angiotensin II to angiotensin III is critical for AT2 receptor-mediated natriuresis in rats. *Hypertension* 51:460–5.

Pan, L., and K. W. Gross. 2005. Transcriptional regulation of renin: An update. *Hypertension* 45:3–8.

Paul, M., A. Poyan Mehr, and R. Kreutz. 2006. Physiology of local renin–angiotensin systems. *Physiol Rev* 86:747–803.

Paula, R. D., C. V. Lima, R. R. Britto, M. J. Campagnole-Santos, M. C. Khosla, and R. A. Santos. 1999. Potentiation of the hypotensive effect of bradykinin by angiotensin-(1-7)-related peptides. *Peptides* 20:493–500.

Peach, M. J. 1977. Renin–angiotensin system: Biochemistry and mechanisms of action. *Physiol Rev* 57:313–70.

Pentz, E. S., M. L. Lopez, M. Cordaillat, and R. A. Gomez. 2008. Identity of the renin cell is mediated by cAMP and chromatin remodeling: An in vitro model for studying cell recruitment and plasticity. *Am J Physiol Heart Circ Physiol* 294:H699–707.

Perlegas, D., H. Xie, S. Sinha, A. V. Somlyo, and G. K. Owens. 2005. ANG II type 2 receptor regulates smooth muscle growth and force generation in late fetal mouse development. *Am J Physiol Heart Circ Physiol* 288:H96–102.

Peyronnet, J., Y. Dalmaz, M. Ehrstrom et al. 2002. Long-lasting adverse effects of prenatal hypoxia on developing autonomic nervous system and cardiovascular parameters in rats. *Pflugers Arch* 443:858–65.

Phat, V. N., J. P. Camilleri, J. Bariety et al. 1981. Immunohistochemical characterization of renin-containing cells in the human juxtaglomerular apparatus during embryonal and fetal development. *Lab Invest* 45:387–90.

Pladys, P., I. Lahaie, G. Cambonie et al. 2004. Role of brain and peripheral angiotensin II in hypertension and altered arterial baroreflex programmed during fetal life in rat. *Pediatr Res* 55:1042–9.

Porteous, S., E. Torban, N. P. Cho et al. 2000. Primary renal hypoplasia in humans and mice with PAX2 mutations: Evidence of increased apoptosis in fetal kidneys of Pax2(1Neu) +/– mutant mice. *Hum Mol Genet* 9:1–11.

Price, R. L., W. Carver, D. G. Simpson et al. 1997. The effects of angiotensin II and specific angiotensin receptor blockers on embryonic cardiac development and looping patterns. *Dev Biol* 192:572–84.

Pritchard, J. A. 1966. Fetal swallowing and amniotic fluid volume. *Obstet Gynecol* 28: 606–16.

Prosser, H. C., M. E. Forster, A. M. Richards, and C. J. Pemberton. 2009. Cardiac chymase converts rat proAngiotensin-12 (PA12) to angiotensin II: Effects of PA12 upon cardiac haemodynamics. *Cardiovasc Res* 82:40–50.

Pupilli, C., R. A. Gomez, J. B. Tuttle, M. J. Peach, and R. M. Carey. 1991. Spatial association of renin-containing cells and nerve fibers in developing rat kidney. *Pediatr Nephrol* 5:690–5.

Rainey, W. E., I. M. Bird, J. I. Mason, and B. R. Carr. 1992. Angiotensin II receptors on human fetal adrenal cells. *Am J Obstet Gynecol* 167:1679–85.

Raizada, M. K., D. Lu, and C. Sumners, C. 1995. AT1 receptors and angiotensin actions in the brain and neuronal cultures of normotensive and hypertensive rats. *Adv Exp Med Biol* 377:331–48.

Ramaha, A., and P. A. Patston. 2002. Release and degradation of angiotensin I and angiotensin II from angiotensinogen by neutrophil serine proteinases. *Arch Biochem Biophys* 397:77–83.

Ravelli, A. C., J. H. van der Meulen, R. P. Michels et al. 1998. Glucose tolerance in adults after prenatal exposure to famine. *Lancet* 351:173–7.

Re, R. N. 2004a. Mechanisms of disease: Local renin–angiotensin–aldosterone systems and the pathogenesis and treatment of cardiovascular disease. *Nat Clin Pract Cardiovasc Med* 1:42–7.

Re, R. N. 2004b. Tissue renin angiotensin systems. *Med Clin North Am* 88:19–38.

Réaux, A., N. de Mota, S. Zini et al. 1999. PC18, a specific aminopeptidase N inhibitor, induces vasopressin release by increasing the half-life of brain angiotensin III. *Neuroendocrinology* 69:370–6.

Reaux, A., M. C. Fournie-Zaluski, and C. Llorens-Cortes. 2001. Angiotensin III: A central regulator of vasopressin release and blood pressure. *Trends Endocrinol Metab* 12:157–62.

Reaux, A., X. Iturrioz, G. Vazeux et al. 2000. Aminopeptidase A, which generates one of the main effector peptides of the brain renin–angiotensin system, angiotensin III, has a key role in central control of arterial blood pressure. *Biochem Soc Trans* 28:435–40.

Reaves, P. Y., C. R. Beck, H. W. Wang, M. K. Raizada, and M. J. Katovich. 2003. Endothelial-independent prevention of high blood pressure in L-NAME-treated rats by angiotensin II type I receptor antisense gene therapy. *Exp Physiol* 88:467–73.

Reini, S.A., C. E. Wood, E. Jensen, and M. Keller-Wood. 2006. Increased maternal cortisol in late-gestation ewes decreases fetal cardiac expression of 11beta-HSD2 mRNA and the ratio of AT1 to AT2 receptor mRNA. *Am J Physiol Regul Integr Comp Physiol* 291:R1708–16.

Reini, S. A., C. E. Wood, and M. Keller-Wood. 2009. The ontogeny of genes related to ovine fetal cardiac growth. *Gene Expr Patterns* 9:122–8.

Rice, G. I., D. A. Thomas, P. J. Grant, A. J. Turner, and N. M. Hooper. 2004. Evaluation of angiotensin-converting enzyme (ACE), its homologue ACE2 and neprilysin in angiotensin peptide metabolism. *Biochem J* 383:45–51.

Roberts, T. J., A. Caston-Balderrama, M. J. Nijland, and M. G. Ross. 2000. Central neuropeptide Y stimulates ingestive behavior and increase urine output in the ovine fetus. *Am J Physiol Endocrinol Metab* 279:E494–500.

Rogerson, F. M., I. Schlawe, G. Paxinos, S. Y. Chai, M. J. McKinley, and F. A. Mendelsohn. 1995. Localization of angiotensin converting enzyme by in vitro autoradiography in the rabbit brain. *J Chem Neuroanat* 8:227–43.

Roks, A. J., K. Rodgers, and T. Walther. 2011. Effects of the renin angiotensin system on vasculogenesis-related progenitor cells. *Curr Opin Pharmacol* 11:162–74.

Roseboom, T., J. H. P. van der Meulen, C. Osmond et al. 2000. Coronary heart disease after prenatal exposure to the Dutch famine, 1944-45. *Heart* 84:595–8.

Rosenfeld, C. R., B. E. Cox, R. R. Magness, and P. W. Shaul. 1993. Ontogeny of angiotensin II vascular smooth muscle receptors in ovine fetal aorta and placental and uterine arteries. *Am J Obstet Gynecol* 168:1562–9.

Ross, M. G., L. K. Kullama, O. A. Ogundipe, K. Chan, and M. G. Ervin. 1995. Ovine fetal swallowing response to icv hypertonic saline. *J Appl Physiol* 78:2267–71.

Ross, M. G., and M. J. Nijland. 1998. Development of ingestive behavior. *Am J Physiol Regul Integr Comp Physiol* 274:R879–93.

Ruiz-Ortega, M., V. Esteban, Y. Suzuki et al. 2003. Renal expression of angiotensin type 2 (AT2) receptors during kidney damage. *Kidney Int Suppl* (86):S21–6.

Sahajpal, V., and N. Ashton. 2003. Renal function and angiotensin AT1 receptor expression in young rats following intrauterine exposure to a maternal low-protein diet. *Clin Sci (Lond)* 104:607–14.

Saji, H., M. Yamanaka, A. Hagiwara, and R. Ijiri. 2001. Losartan and fetal toxic effects. *Lancet* 357:363.

Sakoda, M., A. Ichihara, Y. Kaneshiro et al. 2007. (Pro)renin receptor-mediated activation of mitogen-activated protein kinases in human vascular smooth muscle cells. *Hypertens Res* 30:1139–46.

Salas, S. P., A. Giacaman, and C. P. Vio. 2004. Renal and hormonal effects of water deprivation in late term pregnant rats. *Hypertension* 44:334–9.

Salihu, H. M., M. H. Aliyu, and R. S. Kirby. 2005. In utero nicotine exposure and fetal growth inhibition among twins. *Am J Perinatol* 22:421–7.

Sampaio, W. O., R. A. Souza dos Santos, R. Faria-Silva, L. T. da Mata Machado, E. L. Schiffrin, and R. M. Touyz. 2007. Angiotensin-(1–7) through receptor Mas mediates endothelial nitric oxide synthase activation via Akt-dependent pathways. *Hypertension* 49:185–92.

Santos, R. A., C. H. Castro, F. Gava, E. et al. 2006. Impairment of in vitro and in vivo heart function in angiotensin-(1–7) receptor MAS knockout mice. *Hypertension* 47:996–1002.

Santos, R. A., A. C. Simoes e Silva, C. Maric et al. 2003. Angiotensin-(1–7) is an endogenous ligand for the G protein-coupled receptor Mas. *Proc Natl Acad Sci USA* 100:8258–63.

Saris, J. J., P. A. Hoen, I. M. Garrelds et al. 2006. Prorenin induces intracellular signaling in cardiomyocytes independently of angiotensin II. *Hypertension* 48:564–71.

Savoia, C., M. D'Agostino, F. Lauri, and M. Volpe. 2011. Angiotensin type 2 receptor in hypertensive cardiovascular disease. *Curr Opin Nephrol Hypertens* 20:125–32.

Schefe, J. H., M. Menk, J. Reinemund et al. 2006. A novel signal transduction cascade involving direct physical interaction of the renin/prorenin receptor with the transcription factor promyelocytic zinc finger protein. *Circ Res* 99:1355–66.

Schütz, S., J. M. Le Moullec, P. Corvol P, and J. M. Gasc. 1996. Early expression of all the components of the renin–angiotensin-system in human development. *Am J Pathol* 149:2067–79.

Sechi, L. A., G. Sechi, S. De Carli, C. A. Griffin, M. Schambelan, and E. Bartoli. 1993. Angiotensin receptors in the rat myocardium during pre- and postnatal development. *Cardiologia* 38:471–6.

Seckl, J. R., and M. C. Holmes. 2007. Mechanisms of disease: Glucocorticoids, their placental metabolism and fetal 'programming' of adult pathophysiology. *Nat Clin Pract Endocrinol Metab* 3:479–88.

Segar, J. L., G. B. Dalshaug, K. A. Bedell, O. M. Smith, and T. D. Scholz. 2001. Angiotensin II in cardiac pressure-overload hypertrophy in fetal sheep. *Am J Physiol Regul Integr Comp Physiol* 281:R2037–47.

Segar, J. L., T. D. Scholz, K. A. Bedell, O. M. Smith, D. J. Huss, and E. N. Guillery. 1997. Angiotensin AT1 receptor blockade fails to attenuate pressure-overload cardiac hypertrophy in fetal sheep. *Am J Physiol Regul Integr Comp Physiol* 273:R1501–8.

Sequeira Lopez, M. L., E. S. Pentz, B. Robert, D. R. Abrahamson, and R. A. Gomez. 2001. Embryonic origin and lineage of juxtaglomerular cells. *Am J Physiol Renal Physiol* 281:F345–56.

Serizawa, A., P. M. Dando, and A. J. Barrett. 1995. Characterization of a mitochondrial metallopeptidase reveals neurolysin as a homologue of thimet oligopeptidase. *J Biol Chem* 270:2092–8.

Sernia, C., T. Zeng, D. Kerr, and B. Wyse. 1997. Novel perspectives on pituitary and brain angiotensinogen. *Front Neuroendocrinol* 18:174–208.

Shaltout, H. A., J. P. Figueroa, J. C. Rose, D. I. Diz, and M. C. Chappell. 2009. Alterations in circulatory and renal angiotensin-converting enzyme and angiotensin-converting enzyme 2 in fetal programmed hypertension. *Hypertension* 53:404–8.

Shanmugam, S., P. Corvol, and J. M. Gasc. 1994. Ontogeny of the two angiotensin II type 1 receptor subtypes in rats. *Am J Physiol Endocrinol Metab* 267:E828–36.

Shanmugam, S., C. Llorens-Cortes, E. Clauser, P. Corvol, and J. M. Gasc. 1995. Expression of angiotensin II AT2 receptor mRNA during development of rat kidney and adrenal gland. *Am J Physiol Renal Physiol* 268:F922–30.

Shanmugam, S., and K. Sandberg. 1996. Ontogeny of angiotensin II receptors. *Cell Biol Int* 20:169–76.

Shenoy, U. V., E. M. Richards, X. C. Huang, and C. Sumners. 1999. Angiotensin II type 2 receptor-mediated apoptosis of cultured neurons from newborn rat brain. *Endocrinology* 140:500–9.

Sherman, R. C., and S. C. Langley-Evans. 2000. Antihypertensive treatment in early postnatal life modulates prenatal dietary influences upon blood pressure in the rat. *Clin Sci (Lond)* 98:269–75.

Shi, L., C. Guerra, J. Yao, and Z. Xu. 2004a. Vasopressin mechanism-mediated pressor responses caused by central angiotensin II in the ovine fetus. *Pediatr Res* 56:756–62.

Shi, L., J. Yao, L. Stewart, and Z. Xu. 2004b. Brain C-FOS expression and pressor responses after I.V. or I.C.V. angiotensin in the near-term ovine fetus. *Neuroscience* 126:979–87.

Shi, L., C. Mao, J. Wu, P. Morrissey, J. Li, and Z. Xu. 2006. Effects of i.c.v. losartan on the angiotensin II-mediated vasopressin release and hypothalamic fos expression in near-term ovine fetuses. *Peptides* 27:2230–8.

Shi, L., C. Mao, F. Zeng, Y. Zhang, and Z. Xu. 2008. Central cholinergic signal-mediated neuroendocrine regulation of vasopressin and oxytocin in ovine fetuses. *BMC Dev Biol* 8:95.

Shi, L., C. Mao, F. Zeng, L. Zhu, and Z. Xu. 2009. Central cholinergic mechanisms mediate swallowing, renal excretion, and c-fos expression in the ovine fetus near term. *Am J Physiol Regul Integr Comp Physiol* 296:R318–25.

Shi, L., C. Mao, Z. Xu, and L. Zhang. 2010a. Angiotensin-converting enzymes and drug discovery in cardiovascular diseases. *Drug Discov Today* 15:332–41.

Shi, L., C. Mao, F. Zeng, J. Hou, H. Zhang, and Z. Xu. 2010b. Central angiotensin I increases fetal AVP neuron activity and pressor responses. *Am J Physiol Endocrinol Metab* 298:E1274–82.

Singh, R. R., L. A. Cullen-McEwen, M. M. Kett et al. 2007. Prenatal corticosterone exposure results in altered AT1/AT2, nephron deficit and hypertension in the rat offspring. *J Physiol* 579:503–13.

Siragy, H. M. 2004. AT1 and AT2 receptor in the kidney: Role in health and disease. *Semin Nephrol* 24:93–100.

Song, R., M. Spera, C. Garrett, S. S. El-Dahr, and I. V. Yosypiv. 2010a. Angiotensin II AT2 receptor regulates ureteric bud morphogenesis. *Am J Physiol Renal Physiol* 298:F807–17.

Song, R., M. Spera, C. Garrett, and I. V. Yosypiv. 2010b. Angiotensin II-induced activation of c-Ret signaling is critical in ureteric bud branching morphogenesis. *Mech Dev* 127: 21–7.

Sood, P. P., M. Panigel, and R. Wegmann. 1987. The existence of renin–angiotensinogen system in the rat fetal brain: I. Immunocytochemical localization of renin-like activity at the 19th day of gestation. *Cell Mol Biol* 33:675–80.

Sood, P. P., M. Panigel, and R. Wegmann. 1989. Co-existence of renin-like immunoreactivity in the rat maternal and fetal neocortex. *Neurochem Res* 14:499–502.

Sood, P. P., J. P. Richoux, M. Panigel, J. Bouhnik, and R. Wegmann. 1990. Angiotensinogen in the developing rat fetal hindbrain and spinal cord from 18th to 20th day of gestation: An immunocytochemical study. *Neuroscience* 37:517–22.

Steckelings, U. M., E. Kaschina, and T. Unger. 2005. The AT2 receptor—a matter of love and hate. *Peptides* 26:1401–9.

Strittmatter, S. M., D. R. Lynch, and S. H. Snyder. 1986. Differential ontogeny of rat brain peptidases: Prenatal expression of enkephalin converytase and postnatal development of angiotensin-converting enzyme. *Brain Res* 394:207–15.

Sundgren, N. C., G. D. Giraud, P. J. Stork, J. G. Maylie, and K. L. Thornburg. 2003. Angiotensin II stimulates hyperplasia but not hypertrophy in immature ovine cardiomyocytes. *J Physiol* 548:881–91.

Suzuki, K., T. Horiguchi, A. C. Comas-Urrutia, E. Mueller-Heubach, H. O. Morishima, and K. Adamsons. 1971. Pharmacologic effects of nicotine upon the fetus and mother in the rhesus monkey. *Am J Obstet Gynecol* 111:1092–101.

Swanson, G. N., J. M. Hanesworth, M. F. Sardinia et al. 1992. Discovery of a distinct binding site for angiotensin II (3–8), a putative angiotensin IV receptor. *Regul Pept* 40:409–19.

Swenson, S. J., R. C. Speth, and J. P. Porter. 2004. Effect of a perinatal high-salt diet on blood pressure control mechanisms in young Sprague–Dawley rats. *Am J Physiol Regul Integr Comp Physiol* 286:R764–70.

Takahashi, N., M. L. Lopez, J. E. Cowhig Jr. et al. 2005. Ren1c homozygous null mice are hypotensive and polyuric, but heterozygotes are indistinguishable from wild-type. *J Am Soc Nephrol* 16:125–32.

Tallant, E. A., C. M. Ferrario, and P. E. Gallagher. 2005. Angiotensin-(1–7) inhibits growth of cardiac myocytes through activation of the mas receptor. *Am J Physiol Heart Circ Physiol* 289:H1560–6.

Tan, F., P. W. Morris, R. A. Skidgel, and E. G. Erdös. 1993. Sequencing and cloning of human prolylcarboxypeptidase (angiotensinase C). Similarity to both serine carboxypeptidase and prolylendopeptidase families. *J Biol Chem* 268:16631–8.

Tebbs, C., M. K. Pratten, and F. Broughton Pipkin. 1999. Angiotensin II is a growth factor in the peri-implantation rat embryo. *J Anat* 195:75–86.

Tigerstedt, R., and P. G. Bergman. 1898. Niere und Kreislauf. *Skand Arch Physiol* 8: 223–71.

Touyz, R. M., F. Tabet, and E. L. Schiffrin. 2003. Redox-dependent signalling by angiotensin II and vascular remodelling in hypertension. *Clin Exp Pharmacol Physiol* 30:860–6.

Touyz, R. M., and C. Berry. 2002. Recent advances in angiotensin II signaling. *Braz J Med Biol Res* 35:1001–15.

Touyz, R. M., G. He, L. Y. Deng, and E. L. Schiffrin. 1999. Role of extracellular signal-regulated kinases in angiotensin II-stimulated contraction of smooth muscle cells from human resistance arteries. *Circulation*. 26; 99:392–9.

Trask, A. J., D. B. Averill, D. Ganten, M. C. Chappell, and C. M. Ferrario. 2007. Primary role of angiotensin-converting enzyme-2 in cardiac production of angiotensin-(1–7) in transgenic Ren-2 hypertensive rats. *Am J Physiol Heart Circ Physiol* 292:H3019–24.

Trask, A. J., J. A. Jessup, M. C. Chappell, and C. M. Ferrario. 2008. Angiotensin-(1–12) is an alternate substrate for angiotensin peptide production in the heart. *Am J Physiol Heart Circ Physiol* 294:H2242–47.

Tsuchida, S., T. Matsusaka, X. Chen et al. 1998. Murine double nullizygotes of the angiotensin type 1A and 1B receptor genes duplicate severe abnormal phenotypes of angiotensinogen nullizygotes. *J Clin Invest* 101:755–60.

Tsujimoto, M., and A. Hattori. 2005. The oxytocinase subfamily of M1 aminopeptidases. *Biochim Biophys Acta* 1751:9–18.

Tsutsumi, K., A. Seltzer, and J. M. Saavedra. 1993. Angiotensin II receptor subtypes and angiotensin-converting enzyme in the fetal rat brain. *Brain Res* 631:212–20.

Tsutsumi, K., M. Viswanathan, C. Stromberg, and J. M. Saavedra. 1991. Type-1 and type-2 angiotensin II receptors in fetal rat brain. *Eur J Pharmacol* 198:89–92.

Uematsu, M., O. Sakamoto, T. Nishio et al. 2006. A case surviving for over a year of renal tubular dysgenesis with compound heterozygous angiotensinogen gene mutations. *Am J Med Genet A* 140:2355–60.

Van Esch, J. H., F. Gembardt, A. Sterner-Kock et al. 2010. Cardiac phenotype and angiotensin II levels in AT1a, AT1b, and AT2 receptor single, double, and triple knockouts. *Cardiovasc Res* 86:401–9.

Van Kesteren, C. A., H. A. van Heugten, J. M. Lamers, P. R. Saxena, M. A. Schalekamp, and A. H. Danser. 1997. Angiotensin II-mediated growth and antigrowth effects in cultured neonatal rat cardiac myocytes and fibroblasts. *J Mol Cell Cardiol* 29:2147–57.

Vickers, C., P. Hales, V. Kaushik et al. 2002. Hydrolysis of biological peptides by human angiotensin-converting enzyme-related carboxypeptidase. *J Biol Chem* 277:14838–43.

Von Bohlen und Halbach, O., T. Walther, M. Bader, and D. Albrecht. 2001. Genetic deletion of angiotensin AT2 receptor leads to increased cell numbers in different brain structures of mice. *Regul Pept* 99:209–16.

Wagner, C., C. de Wit, L. Kurtz, C. Grünberger, A. Kurtz, and F. Schweda. 2007. Connexin40 is essential for the pressure control of renin synthesis and secretion. *Circ Res* 100:556–63.

Walther, T., D. Balschun, J. P. Voigt et al. 1998. Sustained long term potentiation and anxiety in mice lacking the Mas protooncogene. *J Biol Chem* 273:11867–73.

Walther, T., N. Wessel, N. Kang et al. 2000. Altered heart rate and blood pressure variability in mice lacking the Mas protooncogene. *Braz J Med Biol Res* 33:1–9.

Watanabe, G., R. J. Lee, C. Albanese, W. E. Rainey, D. Batlle, and R. G. Pestell. 1996. Angiotensin II activation of cyclin D1-dependent kinase activity. *J Biol Chem* 271: 22570–7.

Welches, W. R., R. A. Santos, M. C. Chappell, K. B, Brosnihan, L. J. Greene, and C. M. Ferrario. 1991. Evidence that prolyl endopeptidase participates in the processing of brain angiotensin. *J Hypertens* 9:631–8.

Whiting, P., S. Nava, L. Mozley, H. Eastham, and J. Poat. 1991. Expression of angiotensin converting enzyme mRNA in rat brain. *Brain Res Mol Brain Res* 11:93–6.

Wintour, E. M., D. Alcorn, A. Butkus et al. 1996. Ontogeny of hormonal and excretory function of the meso- and metanephros in the ovine fetus. *Kidney Int* 50:1624–33.

Wintour, E. M., K. Moritz, A. Butkus, R. Baird, A. Albiston, and N. Tenis. 1999. Ontogeny and regulation of the AT1 and AT2 receptors in the ovine fetal adrenal gland. *Mol Cell Endocrinol* 157:161–70.

Wintour, E. M., S. Schutz, J. M. Le Moullec, P. Corvol, and J. M. Gasc. 1996. Early expression of all the components of the renin–angiotensin-system in human development. *Am J Pathol* 149:2067–79.

Woods, L. L., J. R. Ingelfinger, J. R. Nyengaard, and R. Rasch. 2001. Maternal protein restriction suppresses the newborn renin–angiotensin system and programs adult hypertension in rats. *Pediatr Res* 49:460–67.

Woods, L.L., and R. Rasch. 1998. Perinatal ANG II programs adult blood pressure, glomerular number, and renal function in rats. *Am J Physiol* 275:R1593–9.

Wright, J. W., and J. W. Harding. 2004. The brain angiotensin system and extracellular matrix molecules in neural plasticity, learning, and memory. *Prog Neurobiol* 72:263–93.

Wright, J. W., E. Tamura-Myers, W. L. Wilson et al. 2003. Conversion of brain angiotensin II to angiotensin III is critical for pressor response in rats. *Am J Physiol Regul Integr Comp Physiol* 284:R725–33.

Wright, J. W., B. J. Yamamoto, and J. W. Harding. 2008. Angiotensin receptor subtype mediated physiologies and behaviors: New discoveries and clinical targets. *Prog Neurobiol* 84:157–81.

Xiao, D., Z. Xu, X. Huang, L. D. Longo, S. Yang, and L. Zhang. 2008. Prenatal nicotine exposure causes a gender-related increase of blood pressure response to angiotensin II in adult offspring. *Hypertensin* 51:1239–47.

Xu, F., C. Mao, C. Rui, Z. Xu, and L. Zhang. 2009a. Cardiovascular effects of losartan and its relevant clinical application. *Curr Med Chem* 16:3841–57.

Xu, F., C. Mao, Y. Liu, L. Wu, Z. Xu, and L. Zhang. 2009b. Losartan chemistry and its effects via AT1 mechanisms in the kidney. *Curr Med Chem* 16:3701–15.

Xu, P., A. C. Costa-Goncalves, M. Todiras et al. 2008. Endothelial dysfunction and elevated blood pressure in MAS gene-deleted mice. *Hypertension* 51:574–80.

Xu, Y., S. J. Williams, D. O'Brien, and S. T. Davidge. 2006. Hypoxia or nutrient restriction during pregnancy in rats leads to progressive cardiac remodeling and impairs postischemic recovery in adult male offspring. *FASEB J* 20:1251–3.

Xu, Z., C. Glenda, L. Day, J. Yao, and M. G. Ross. 2001a. Central angiotensin induction of fetal brain c-fos expression and swallowing activity. *Am J Physiol Regul Integr Comp Physiol* 280:R1837–43.

Xu, Z., M. J. Nijland, and M. Ross. 2001b. Plasma osmolality dipsogenic thresholds and c-fos expression in the near-term ovine fetus. *Pediatr Res* 49:678–85.

Xu, Z., F. Hu, L. Shi, W. Sun, J. Wu, P. Morrussey, and J. Yao. 2005. Central angiotensin-mediated vasopressin release and activation of hypothalamic neurons in younger fetus at pre-term. *Peptides* 26:307–14.

Xu, Z., L. Shi, F. Hu, R. White, L. Stewart, and J. Yao. 2003. In utero development of central ANG-stimulated pressor response and hypothalamic fos expression. *Brain Res* 145:169–76.

Xu, Z., L. Shi, and J. Yao. 2004. Central angiotensin II induced pressor responses and neural activity in utero and hypothalamic angiotensin receptors in preterm ovine fetus. *Am J Physiol Heart Circ Physiol* 286:H1507–14.

Xue, Q., and L. Zhang. 2009. Prenatal hypoxia causes a sex-dependent increase in heart susceptibility to ischemia and reperfusion injury in adult male offspring: Role of protein kinase C epsilon. *J Pharmacol Exp Ther* 330:624–32.

Yamada, H., M. Akishita, M. Ito et al. 1999. AT2 receptor and vascular smooth muscle cell differentiation in vascular development. *Hypertension* 33:1414–9.

Yamada, T., M. Horiuchi, and V. J. Dzau. 1996. Angiotensin II type 2 receptor mediates programmed cell death. *Proc Natl Acad Sci USA* 93:156–60.

Yamamoto, B. J., P. D. Elias, J. A. Masino et al. 2010. The angiotensin IV analog Nle-Tyr-Leu-psi-(CH_2-NH_2)3-4-His-Pro-Phe (norleual) can act as a hepatocyte growth factor/c-Met inhibitor. *J Pharmacol Exp Ther* 333:161–73.

Yamamoto, E., N. Tamamaki, T. Nakamura et al. 2008. Excess salt causes cerebral neuronal apoptosis and inflammation in stroke-prone hypertensive rats through angiotensin II-induced NADPH oxidase activation. *Stroke* 39:3049–56.

Yamazaki, M., K. Chiba, and T. Mohri. 2005. Fundamental role of nitric oxide in neuritogenesis of PC12h cells. *Br J Pharmacol* 146:662–9.

Yang, H., G. Shaw, and M. K. Raizada. 2002. ANG II stimulation of neuritogenesis involves protein kinase B in brain neurons. *Am J Physiol Regul Integr Comp Physiol* 283:R107–14.

Yang, H., W. Y. Wang, and M. K. Raizada. 2001. Characterization of signal transduction pathway in neurotrophic action of angiotensin II in brain neurons. *Endocrinology* 142:3502–11.

Yin, T., X. Ma, L. Zhao, K. Cheng, and H. Wang. 2008. Angiotensin II promotes NO production, inhibits apoptosis and enhances adhesion potential of bone marrow-derived endothelial progenitor cells. *Cell Res* 18:792–9.

Yiu, V., S. Buka, D. Zurakowski, M. McCormick, B. Brenner, and K. Jabs. 1999. Relationship between birthweight and blood pressure in childhood. *Am J Kidney Dis* 33:253–60.

Yoo, K. H., J. T. Wolstenholme, and R. L. Chevalier. 1997. Angiotensin-converting enzyme inhibition decreases growth factor expression in the neonatal rat kidney. *Pediatr Res* 42: 588–92.

Yosypiv, I. V., and S. S. El-Dahr. 1996. Activation of angiotensin-generating systems in the developing rat kidney. *Hypertension* 27:281–6.

Yosypiv, I. V., M. K. Boh, M. A. Spera, S. S. El-Dahr. 2008. Downregulation of Spry-1, an inhibitor of GDNF/Ret, causes angiotensin II-induced ureteric bud branching. *Kidney Int* 74:1287–93.

Yosypiv, I. V., M. Schroeder, S. S. El-Dahr. 2006. Angiotensin II type 1 receptor-EGF receptor cross-talk regulates ureteric bud branching morphogenesis. *J Am Soc Nephrol* 17:1005–14.

Young, D., K. O'Neill, T. Jessell, and M. Wigler. 1988. Characterization of the rat mas oncogene and its high-level expression in the hippocampus and cerebral cortex of rat brain. *Proc Natl Acad Sci USA* 85:5339–42.

Young, D., G. Waitches, C. Birchmeier, O. Fasano, and M. Wigler. 1986. Isolation and characterization of a new cellular oncogene encoding a protein with multiple potential transmembrane domains. *Cell* 45:711–9.

Yu, L., M. Zheng, W. Wang, G. J. Rozanski, I. H. Zucker, and L. Gao. 2010. Developmental changes in AT1 and AT2 receptor-protein expression in rats. *J Renin Angiotensin Aldosterone Syst* 11:214–21.

Zelezna, B., B. Rydzewski, D. Lu et al. 1992. Angiotensin-II induction of plasminogen activator inhibitor-1 gene expression in astroglial cells of normotensive and spontaneously hypertensive rat brain. *Mol Endocrinol.* 6:2009–17.

Zhang, F., Y. Hu, Q. Xu, and S. Ye. 2010. Different effects of angiotensin II and angiotensin-(1–7) on vascular smooth muscle cell proliferation and migration. *PLoS One* 5:e12323.

Zhang, H., Y. Fan, F. Xia et al. 2011. Prenatal water deprivation alters brain angiotensin system and dipsogenic changes in the offspring. *Brain Res* 13:128–36.

Zhang, J., and R. E. Pratt. 1996. The AT2 receptor selectively associates with Gialpha2 and Gialpha3 in the rat fetus. *J Biol Chem* 271:15026–33.

Zhang, J. H., J. M. Hanesworth, M. F. Sardinia, J. A. Alt, J. W. Wright, and J. W. Harding. 1999. Structural analysis of angiotensin IV receptor (AT4) from selected bovine tissues. *J Pharmacol Exp Ther* 289:1075–83.

Zhang, L. 2005. Prenatal hypoxia and cardiac programming. *J Soc Gynecol Investig* 12:2–13.

Zhang, S. L., B. Moini, and J. R. Ingelfinger. 2004. Angiotensin II increases Pax-2 expression in fetal kidney cells via the AT2 receptor. *J Am Soc Nephrol* 15:1452–65.

Zhang, Z., E. Pascuet, P. A. Hueber et al. 2010. Targeted inactivation of EGF receptor inhibits renal collecting duct development and function. *J Am Soc Nephrol* 21:573–8.

Zheng, J., I. M. Bird, D. B. Chen, and R. R. Magness. 2005. Angiotensin II regulation of ovine fetoplacental artery endothelial functions: Interactions with nitric oxide. *J Physiol* 565:59–69.

Zhe, R.R. 1999. Activation of the mitogen-activated protein kinase cascade is necessary but not sufficient for basic fibroblast growth factor- and epidermal growth factor-stimulated expression of endothelial nitric oxide synthase in ovine fetoplacental artery endothelial cells. *Endocrinology* 140:1399–407.

Zheng, J., K. E. Vagnoni, I. M. Bird, and R. R. Magness. 1997. Expression of basic fibroblast growth factor, endothelial mitogenic activity, and angiotensin II type-1 receptors in the ovine placenta during the third trimester of pregnancy. *Biol Reprod* 56:1189–97.

Zheng, J., Y. Wen, J. L. Austin, and D. B. Chen. 2006. Exogenous nitric oxide stimulates cell proliferation via activation of a mitogen-activated protein kinase pathway in ovine fetoplacental artery endothelial cells. *Biol Reprod* 74:375–82.

Zhu, L., C. Mao, J. Wu et al. 2009. Ovine fetal hormonal and hypothalamic neuroendocrine responses to maternal water deprivation at late gestation. *Int J Dev Neurosci* 27:385–91.

Zhu, M., C. H. Gelband, J. M. Moore, P. Posner, and C. Sumners. 1998. Angiotensin II type 2 receptor stimulation of neuronal delayed-rectifier potassium current involves phospholipase A2 and arachidonic acid. *J Neurosci* 18:679–86.

Zhuo, J. L., and X. C. Li. 2011. New insights and perspectives on intrarenal renin–angiotensin system: Focus on intracrine/intracellular angiotensin II. *Peptides* 32:1551–65.

Zucker, I. H., H. D. Schultz, K. P. Patel, W. Wang, and L. Gao. 2009. Regulation of central angiotensin type 1 receptors and sympathetic outflow in heart failure. *Am J Physiol Heart Circ Physiol* 297:H1557–66.

14 Sodium Appetite Sensitization

Seth W. Hurley, Robert L. Thunhorst, and Alan Kim Johnson

CONTENTS

14.1 INTRODUCTION AND BACKGROUND

Sodium is necessary for the survival of most, if not all, animals. It is the most prevalent of extracellular solutes, and because it is largely excluded from the intracellular fluid compartment by the cell plasma membrane, it in a sense comprises the "true grit" of the extracellular fluid. The maintenance of near-ideal concentrations of intracellular constituents is necessary for optimum cellular fitness and function. The concentration of impermeable solutes such as sodium in the extracellular fluid compartment is the major determinant of water movement across the plasma membrane. A high concentration of sodium in the extracellular space causes water to move out of cells and produces dehydration of the intracellular fluid compartment, whereas a low concentration of extracellular solute causes overhydration of intracellular space. In addition to its importance in establishing the distribution of fluid across semipermeable membranes, the relative concentration of extra- and intracellular sodium can also influence cellular functions. For example, the resting membrane potential of cells is in large part dependent on the relative concentrations of intracellular versus extracellular sodium, and the shape of action potentials generated by neurons is affected by the ratio of these two sodium concentrations.

Animals cannot synthesize sodium, so that when the ion is lost from the body it must be acquired from external sources and ingested in order to restore homeostasis.

Many mammalian species, especially omnivores and herbivores, display motivated behaviors of seeking and ingesting sodium-containing substances when they are sodium deficient. Collectively, the set of behaviors and the motivational state (i.e., drive) that mobilizes and directs animals to finding and consuming substances containing sodium is most often referred to as *sodium* or *salt appetite*. A common *operational definition* of sodium appetite is simply a significant increase in the intake of unpalatable saline solutions (usually 1.5% to 3% NaCl w/v) over a specified period in response to a deficit. Such hypertonic saline solutions are normally consumed in minimal quantities, or not consumed at all, by sodium-replete animals.

14.2 PARAMETERS OF SODIUM APPETITE AND EXPERIMENTAL APPROACHES

In the case of most strains of laboratory rat, substances containing high concentrations of sodium (e.g., hypertonic saline solutions) appear to be unpleasant and are rejected when they are tasted in the absence of a salt appetite. However, such eschewed salty substances become acceptable to salt-hungry animals. In other words, in the transition from being salt sated to a state of sodium appetite, many animals, including man, exhibit a hedonic shift where acceptability and preference for hypertonic saline solutions increase (Bare 1949; Smith and Stricker 1969; Takamata et al. 1994). Using a method that characterizes species-specific oral and forelimb motor behaviors associated with either appetitive or aversive responses to palatable or unpalatable solutions, Berridge and colleagues demonstrated that hypertonic saline delivered intraorally evokes a constellation of aversive behaviors in sodium replete rats. Such negative responses change to a set of pleasure-associated actions when animals are in a condition of sodium deficiency (Berridge et al. 1984). Sodium appetite is also associated with enhanced taste sensitivity (for review, see Fregly and Rowland 1985) and changes in activity in taste-related neurons in regions such as the nucleus of the solitary tract (Contreras 1977; Jacobs et al. 1988).

Curt Richter, the father of sodium appetite research, induced the motivated state and increased salt intake by reducing dietary sodium and by performing adrenalectomies to remove the source of sodium retaining hormones such as aldosterone (Richter 1936; see Table 14.1). In classic studies, he also induced a salt appetite with pharmacological doses of the mineralocorticoid agonist, deoxycorticosterone acetate (DOCA) (Rice and Richter 1943). It is interesting to note that with the methods used by Richter, as well as with several other techniques developed later, it requires a few to several days for a salt appetite to be reliably expressed. Fitzsimons (1961) used hyperoncotic colloids for intraperitoneal dialysis to sequester extracellular fluid in the abdominal cavity in order to create a relative sodium deficit in the rest of the body. With this method, it was possible to reduce the latency for the onset of sodium appetite to between 5 and 10 h. Denying rats dietary sodium elicits a sodium appetite on a time course of 8 days in male rats (Stricker et al. 1991). The latency to induce sodium appetite can be reduced and the magnitude of sodium appetite increased by

TABLE 14.1
Selected Treatments that Induce Salt Appetite and Latency to Salt Appetite Onset

Type I. Sodium Deficiency

Sodium Deficiency	Approximate Latency for Salt Appetite Onset
Sodium deprivation	3–4 days
Adrenalectomy	2–3 days
Furosemide + sodium-free diet	24 h
Hypovolemia – PEG	6–8 h
Intraperitoneal dialysis	<5–12 h
Furosemide + captopril (acute)	1 h
Water deprivation – partial repletion	12–24 h

Type II. Stimulated – Hormone or Drug Induced

Drug or Hormone	Approximate Latency for Salt Appetite Onset
ACTH	2–3 days
DOCA	2–3 days
Aldosterone	2–3 days
Corticosterone	2–3 days
Fludrocortisone	2–3 days
Chronic IV Ang II infusions	2–3 days
Oral captopril	1–2 days
Renin (ICV)	1–2 h
Yohimbine (SC)	<5 h

Note: Type I treatments evoke a need-induced salt appetite by causing a sodium deficiency. Type II treatments evoke a salt appetite through administration of drugs or hormones. PEG, polyethylene glycol; IV, intravenous; Ang II, angiotensin II; ICV, intracerebroventricular; SC, subcutaneous.

restricting rats' sodium access for a few days before extracellular depletion (Jalowiec 1974; Jalowiec and Stricker 1973; Stricker 1981). Subsequent studies demonstrated that the latency for the onset of sodium appetite could be decreased in animals with extracellular dehydration to as little as 1 h by concurrently lowering blood pressure (Fitts and Masson 1989; Masson and Fitts 1989; Thunhorst and Johnson 1994; Thunhorst et al. 1994).

There is a reasonable body of evidence to support the idea that both circulating aldosterone and angiotensin II along with central angiotensin II (Yang and Epstein 1991) are used as signals for central information processing in the generation of sodium appetite (Epstein 1982; Fluharty and Epstein 1983; Sakai et al. 1986; Schoorlemmer et al. 2001; Thunhorst and Fitts 1994). Renin is released from the kidneys in states of hypovolemia, hypotension, or hyponatremia, and acts on constitutively present plasma angiotensinogen (renin substrate) to form angiotensin I. Angiotensin I is converted by the angiotensin converting enzyme, primarily located in the lungs, into angiotensin II, which—in addition to many other actions—stimulates the adrenals to release aldosterone. Elevated angiotensin II and aldosterone levels can act synergistically in many cases to evoke physiological regulatory responses, including the induction of salt appetite (Epstein 1991). The integrity of the renin–angiotensin–aldosterone system (RAAS) is necessary to elicit a sodium deficit-induced salt appetite (Moe et al. 1984; Sakai et al. 1986; Fitts and Thunhorst 1996).

Sodium intake occurring under the conditions where animals are in a physiological state of absolute sodium deficiency or where a sodium deficit is simulated by an experimental treatment (e.g., administration of angiotensin and aldosterone; Table 14.1) is referred to as *need-induced sodium appetite.* A second type of increased hypertonic saline intake can be produced under basal conditions, where animals have *ad libitum* access to adequate dietary salt and are sodium replete. This type of enhanced salt intake is defined as *need-free sodium appetite.* Enhanced need-free salt appetite can be generated by first inducing a need-induced sodium appetite followed by recovery of normal sodium homeostasis where animals are maintained with *ad libitum* access to water, normal laboratory chow, and saline. In this case, need-free sodium appetite is expressed as a chronic increase in daily salt intake (Sakai et al. 1989).

Although many protocols have been used to induce sodium appetite (Table 14.1), this review will focus primarily on experiments using three models to generate enhanced salt intake: the diuretic/natriuretic, furosemide model; furosemide plus the hypotensive drug, captopril (furo/cap) model; and the mineralocorticoid agonist, DOCA model. In the furosemide model, ambient sodium is removed and animals are typically given two injections of the diuretic furosemide separated by 1 or 2 h (Frankmann et al. 1986; Sakai et al. 1987). Twenty-four hours after the first furosemide injection, access to saline is restored and sodium appetite is assessed. In this protocol, animals have access to water between the furosemide treatments and the time when access to sodium is provided. Thus, when tested, rats exhibit only sodium appetite.

In the furo/cap model, animals are given furosemide and a low dose of captopril within a few minutes of each other, and both salt and water are returned an hour later (Fitts and Masson 1989; Masson and Fitts 1989). The concomitant negative fluid–mineral balance and captopril-induced decrease in blood pressure and increase in renin secretion lead to a rapid onset, robust sodium appetite accompanied by thirst (Thunhorst and Johnson 1994).

DOCA is a precursor for the mineralocorticoid aldosterone. In this model, successive daily DOCA injections elevate aldosterone levels and cause both sodium retention and sodium appetite (Rice and Richter 1943). Sodium appetite in the DOCA protocol appears to be paradoxical as animals retain excess sodium yet exhibit a hunger for salt.

14.3 SODIUM APPETITE, RESPONSE SENSITIZATION, AND NEUROPLASTICITY

Falk made the original observation in rats that, when given two successive bouts of intraperitoneal dialysis with sucrose, the second dialysis produced significantly more saline intake than the first treatment (Falk 1965). Importantly, the enhanced sodium intake appeared to be independent of associative learning as rats still elevated their salt intake even when sodium was restored by means other than ingestion (Falk 1966). This enhancement of sodium intake due to repeated episodes of depletion is a form of behavioral sensitization (also known as reverse tolerance). That is, there is an increase in the magnitude of a behavior in response to identical treatments when administered at two different times. Sodium appetite sensitization progressively develops over three or four successive depletions until intake plateaus and is seemingly lifelong (Sakai et al. 1987, 1989).

The water deprivation with partial repletion (WDPR) protocol is another model of hypovolemia that can induce sodium appetite sensitization. In this model, rats are water deprived, which causes natriuresis to maintain the appropriate water and sodium ratio in the face of dehydration (De Luca et al. 2002). Dehydration combined with sodium depletion ultimately results in animals entering a state of hypovolemia. Rats are then allowed to drink water to satiation, and then provided access to sodium. Water and sodium intake can be dissociated because fluids are offered separately. Repeated WDPR treatments result in enhanced need-free and need-induced sodium intake (De Luca et al. 2010). Furthermore, rats with a history of furo/cap treatments exhibit enhanced sodium and water intake in the WDPR protocol (Pereira et al. 2010). Therefore, WDPR is a reliable model for sodium appetite sensitization.

The nervous system exhibits plasticity, or the capacity to adapt by changes in structure and function. Plasticity allows for learning—a change in behavior owing to past experience. Behavioral and neural plasticity have traditionally been studied primarily in the context of learning, development, and memory. However, in recent decades, neural plasticity has been recognized to be involved in diverse phenomena such as pain, addiction, drug tolerance, and exercise (Cotman and Berchtold 2002; Kauer and Malenka 2007; Koob and Le Moal 2005; Trujillo and Akil 1991; Woolf and Salter 2000). It is likely that central nervous system (CNS) plasticity mediates sodium appetite sensitization. This hypothesis is supported by evidence demonstrating functional similarities and overlap of CNS neural and neurochemical systems related to both psychomotor and sodium appetite sensitization.

14.3.1 Motivation, Reward, Sodium Appetite, and Sodium Appetite Sensitization

Addictive drugs, primarily psychomotor stimulants and opioids, when administered to experimental animals promote locomotor behavior. Upon repeated administration, drug-induced locomotor behavior is enhanced, a process known as behavioral sensitization. Central neuronal plasticity mediates behavioral sensitization (for review, see Kauer and Malenka 2007; Wolf et al. 2004). Dr. Mitchell Roitman, working in Dr. Ilene Bernstein's laboratory at the University of Washington, initially demonstrated that multiple sodium depletions sensitized the locomotor response of rats to the psychomotor stimulant amphetamine (Roitman et al. 2002). Furthermore, repeated amphetamine injections enhanced subsequent furosemide-induced sodium appetite. Therefore, sodium depletions and psychomotor stimulants can substitute for each other to produce sensitization, a phenomenon known as *reciprocal cross-sensitization* (Clark and Bernstein 2004). Cross-sensitization treatments involving enhanced sodium appetite and locomotor behavior have been extended to other addictive drugs, morphine and cocaine, and another method of inducing salt appetite, daily DOCA administration (Acerbo and Johnson 2011; Na et al. 2012) (Figure 14.1). These studies indicate that drug-induced locomotor and sodium appetite sensitization are likely to share common neural substrates and mechanisms. Fortunately, there is a substantial body of research on the mechanisms underlying behavioral sensitization, which can provide insight into mechanisms mediating sodium appetite sensitization.

Brain areas typically involved in motivation and reward, such as the nucleus accumbens, ventral tegmental area (VTA), and amygdala, are critical for behavioral sensitization. Many of these same nuclei have also been implicated in sodium appetite. For example, repeated sodium depletion enhances dendritic length and the number of synaptic terminals in nucleus accumbens shell neurons (Roitman et al. 2002). Compared to a single sodium depletion, repeated sodium depletions increase expression of the immediate early gene c-*fos* in limbic circuitry, including the bed nucleus of the stria terminalis, nucleus accumbens shell, and basolateral amygdala (Figure 14.2) (Na et al. 2007). One hypothesis that emerges from these data is that sodium appetite sensitization may be mediated by plasticity in the same brain areas that are important for psychomotor stimulant sensitization. Consistent with this idea is that lesions of the central nucleus of the amygdala that attenuate sodium appetite (Galaverna et al. 1992; Zardetto-Smith et al. 1994) also block psychomotor stimulant sensitization (Wolf et al. 1995).

Much of the work on sodium appetite has centered on the hormonal and neuromodulatory roles of angiotensin II and aldosterone. However, neuromodulators that are critical for motivation, reward, and addiction should be considered as candidates for sodium appetite sensitization. Evidence indicates that sodium appetite sensitization is not due to the progressive increase of levels in circulating angiotensin II or aldosterone with each successive sodium depletion (Sakai et al. 1989). Therefore, it is likely that there are alterations in CNS processes that are beyond the body-to-brain signaling role played by angiotensin II and aldosterone. Central opioids and the mesolimbic dopamine system are critical for all motivated behaviors including sodium appetite.

Administration of opioid agonists enhances sodium appetite under need-induced conditions (Figure 14.3a) (Kuta et al. 1984; Hubbell and McCutcheon 1993; Na et al. 2012). Conversely, administration of opioid receptor antagonists decreases intake of saline solutions (Figure 14.3b) (Cooper and Gilbert 1984; Gosnell and Majchrzak 1990; Hubbell and McCutcheon 1993; Kuta et al. 1984; Na et al. 2012). Both μ- and δ-opioid receptors have a role in sodium appetite. Central infusions of the μ-opioid receptor agonist DAMGO increase need-free hypotonic and hypertonic saline consumption (Gosnell and Majchrzak 1990). Rats with a sodium hunger induced by DOCA or furosemide treatment display elevated enkephalin mRNA in the nucleus accumbens shell, but only if given sodium access (Lucas et al. 2000, 2003). Recently, we found that in addition to reducing hypertonic saline intake in a need-free state, the opioid receptor antagonist naltrexone reduces positive hedonic behaviors and increases the expression of negative hedonic behaviors associated with intraorally infused hypertonic saline in sodium-depleted rats (Na et al. 2012). This suggests

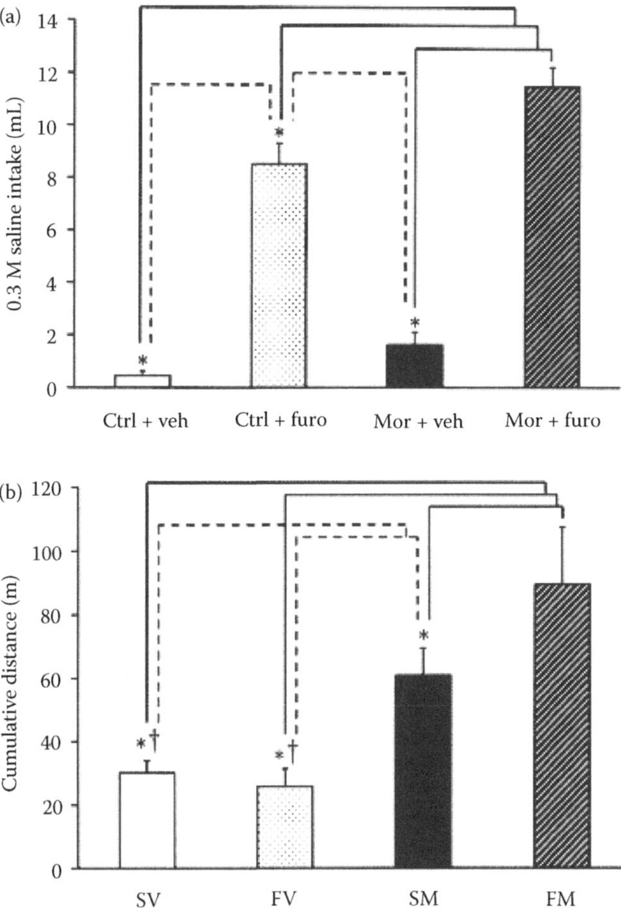

FIGURE 14.1 Reciprocal cross-sensitization between morphine and sodium depletion. (a) morphine pretreatment sensitizes sodium appetite. Rats were treated for 5 days with either morphine (Mor) or vehicle (veh) and then received a sham (veh) or furosemide (furo) depletion. Furosemide depleted rats with a history of morphine treatment consumed significantly more sodium than furosemide depleted rats with a history of vehicle injections (* $p < 0.05$ vs. respective comparisons). (b) sodium depletions sensitize morphine induced locomotor behavior. Rats had a history of 3 sham or sodium depletions and afterward received morphine or vehicle. Locomotor behavior was then assessed in an open field. Rats with history of sodium depletions (FM) displayed elevated locomotor behavior compared to sham-depleted animals that received morphine (SM), animals with history of sodium depletions that received vehicle (FV), and sham-depleted animals that received vehicle (SV) (* $p < 0.05$ vs. respective comparisons, † $p < 0.05$ vs. respective comparisons). (Reprinted from *Pharmacol Biochem Behav*, 93, Na, E. S., M. J. Morris, and A. K. Johnson, Behavioral cross-sensitization between morphine-induced locomotion and sodium depletion-induced salt appetite, 368–74, Copyright 2009, with permission from Elsevier.)

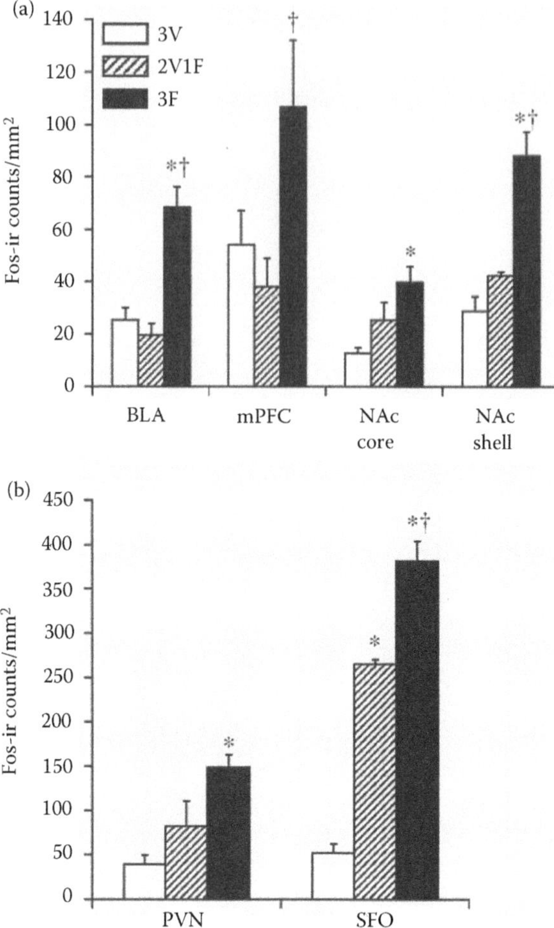

FIGURE 14.2 c-*fos* expression in brain regions after three successive sham sodium depletions (3V), two sham depletions followed by a single furosemide depletion (2V1F), or three sodium depletions (3F). (a) *fos* expression in limbic brain nuclei. *fos* is significantly elevated in basolateral nucleus of the amygdala (BLA) and nucleus accumbens shell (NAc shell) in 3F-treated animals compared to 3V and 2V1F. Group 3F expresses higher amounts of *fos* in nucleus accumbens core (NAc core) compared to group 3V. Group 3F expressed significantly more *fos* in medial prefrontal cortex (mPFC) compared to group 2V1F. (b) *fos* expression in brain nuclei associated with body fluid homeostasis. *fos* is significantly elevated in subfornical organ (SFO) in rats with history of three sodium depletions compared to a single sodium depletion. *fos* expression in the paraventricular nucleus of the hypothalamus (PVN) is significantly elevated in animals with history of sodium depletions compared to sham-treated animals (†$p < 0.05$ vs. 2V1F, *$p < 0.05$ vs. 3V). (Reprinted from *Brain Research*, 1171, Na, E. S., M. J. Morris, R. F. Johnson, T. G. Beltz, and A. K. Johnson, The neural substrates of enhanced salt appetite after repeated sodium depletions, 104–10, Copyright 2007, with permission from Elsevier.)

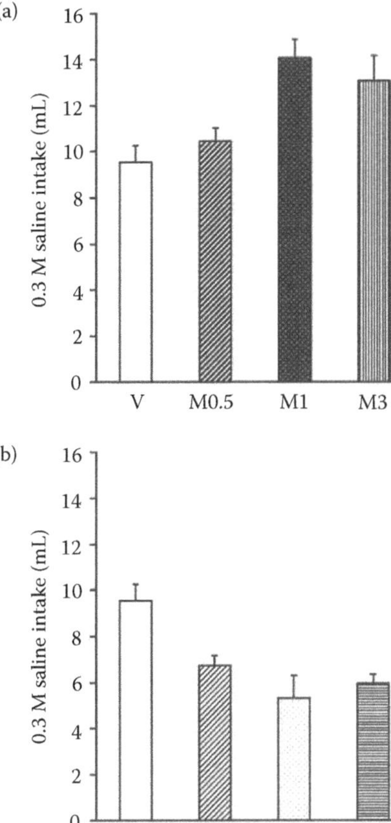

FIGURE 14.3 Effect of morphine and naltrexone on furosemide-induced sodium intake. (a) Morphine significantly enhances sodium intake at 1 and 3 mg/kg doses (V, vehicle; M.5, 0.5 mg/kg morphine; M1, 1 mg/kg morphine; M3, 3 mg/kg morphine). (b) Naltrexone significantly attenuates need-induced sodium appetite at all doses tested (V, vehicle; N1, 1 mg/kg naltrexone; N3, 3 mg/kg naltrexone; N10, 10 mg/kg naltrexone). (Reprinted from *Physiol Behavior*, 106, Na, E. S., M. J. Morris, and A. K. Johnson, Opioid mechanisms that mediate the palatability of and appetite for salt in sodium replete and deficient states, 164–70, Copyright 2012, with permission from Elsevier.)

that opioids mediate the hedonic shift that accompanies sodium deficiency, and this hedonic shift is necessary for the expression of sodium appetite.

Systemic administration of a δ-opioid receptor antagonist mildly attenuates furosemide-induced sodium appetite (Lucas et al. 2003), but robustly attenuates sodium appetite when infused directly into the VTA (Lucas et al. 2007). The reason for the modest attenuation observed with systemic δ-opioid antagonist may be attributable to site-specific actions of δ-opioids. Agonist infused into the nucleus accumbens shell reduces need-free sodium intake; conversely, need-free intake is potentiated when the same agonist is infused into the VTA (Lucas et al. 2007).

Although most studies find that manipulations that increase need-induced sodium intake also increase need-free sodium intake (Sakai et al. 1989; Galaverna et al. 1992; Pereira et al. 2010; Zardetto-Smith et al. 1994), the physiological underpinnings appear to be dissociable. For example, injection of the δ-opioid receptor agonist DSLET into the VTA has no effect on furosemide-induced sodium intake but enhances need-free sodium intakes (Lucas et al. 2007).

Dopamine has a critical role in sodium appetite. In mice, both D_1 and D_2 antagonists decrease need-induced sodium intake (Liedtke et al. 2011). D_2 receptor blockade decreases sodium intake under sham-feeding conditions (Roitman et al. 1997). In addition, sodium depleted rats with access to saline also show greater dopamine transporter occupancy (Lucas et al. 2003). Of particular relevance is the fact that D_2 receptor antagonists block the expression of cross-sensitization between sodium appetite and amphetamine (Clark and Bernstein 2006a). There are several reports of dopamine antagonists failing to suppress need-induced sodium intake (Roitman et al. 1997; Lucas et al. 2003). This may be attributable to the difficulty of attenuating sodium appetite after the appetitive state is well established. Manipulations may be successful in blocking sodium appetite in models in which the intake of salt is rapid, such as the furo/cap model. Acute drug manipulations are more likely to affect early events between the detection of the signals of disrupted fluid homeostasis (i.e., elevated angiotensin II and aldosterone) than when full-blown salt appetite has been established.

In relation to affective processes, Berridge (1996) developed the concepts of "wanting" and "liking" in response to taste stimuli. *Wanting* is associated with the desire to seek and consume a goal object, such as food or addictive drugs. *Liking* is associated with affective sense of pleasure during goal object consumption. In animal models developed by Grill and Norgren (1978), positive and negative appraisal of intraorally infused solutions is measured by changes in facial, oral, and forelimb behavioral responses. Different types of species-specific motor patterns can be categorized as pleasurable or aversive and have been validated by their association with changes in the palatability of various solutions (e.g., pleasurable responses to sucrose solutions; aversive responses to quinine solutions). Evidence indicates that opioids potentiate the reward, or liking, associated with goal object consumption, and dopamine mediates goal object wanting (for review, see Berridge et al. 2009). Sensitized animals do not display indications of incremental increases in liking salt as a result of repeated sodium depletions, but do increase bar pressing to obtain sodium in an operant progressive ratio task considered to be indicative of increased wanting (Clark and Bernstein 2006b). A provocative finding is that the genes expressed during unsatiated sodium appetite overlap with those showing increased expression induced by addictive drugs (Liedtke et al. 2011). It may be that observed alterations in gene expression are involved in the craving of goal objects.

14.3.2 Role of RAAS in Sensitization of Sodium Appetite

Several studies have demonstrated the importance of angiotensin II and aldosterone in the genesis of sodium appetite sensitization. Initially it was found that repeated injections of renin, isoproteranol, or intracerebroventricular (ICV) angiotensin

II caused sensitized sodium and water intake (Bryant et al. 1980). These authors also found that chronic administration of angiotensin II elicited robust increases in need-free sodium and water intake that continued after cessation of angiotensin II administration. Sakai and colleagues (1987) reported that ICV angiotensin II and peripheral aldosterone, in doses capable of eliciting a sodium appetite, enhance subsequent need-induced sodium appetite. Furthermore, these investigators found that coadministering the angiotensin converting enzyme inhibitor captopril and the mineralocorticoid receptor (MR) antagonist RU-28318, in doses that block need-induced sodium appetite, abolishes sensitization produced by furosemide-induced sodium and water depletion. Pereira et al. (2010) demonstrated that the AT_1 receptor antagonist losartan blocks both furo/cap-induced sodium appetite sensitization and sensitization to water drinking. These studies suggest that angiotensin II and aldosterone actions are critical for sodium appetite sensitization. The emergence of need-induced sodium appetite has been hypothesized to be the result of synergistic action between central angiotensin II and peripheral aldosterone (Epstein 1982; Sakai et al. 1986).

Aldosterone may elicit sodium appetite sensitization through MR action. Repeated DOCA administration causes a progressive increase in sodium appetite over the course of treatments (Acerbo and Johnson 2011; Rice and Richter 1943). Support for the role of aldosterone in the sensitization of salt appetite comes from Thornton and Nicolaidis' (1994) laboratory, where the investigators studied electrophysiological changes induced by DOCA pretreatment. Three days of systemic DOCA administration increased the firing rate and spontaneous activity of neurons in the median preoptic area and septum (Thornton and Nicolaidis 1994). Furthermore, DOCA pretreatment prolonged the excitation of angiotensin II–sensitive neurons. A very interesting finding from this study was that local infusion of aldosterone caused rapid neuronal excitation and enabled neurons to respond to angiotensin II on a timescale of minutes, suggesting a nongenomic action of aldosterone (Karst et al. 2005; Sakai et al. 2000). Pretreatment with DOCA facilitates the activity of sodium taste selective neurons in the anteromedial septum and median preoptic area (Mousseau et al. 1996). Mineralocorticoid-induced changes in neuronal physiology may be related to neural plasticity underlying salt appetite sensitization. Behavioral data support the notion that DOCA-induced sodium appetite sensitization shares common mechanisms with sodium depletion sensitization. Lesions of the central nucleus of the amygdala, which attenuate the enhanced need-free sodium appetite that was produced by a previous series of sensitizing sodium depletions, also block DOCA-induced sodium appetite (Galaverna et al. 1992). Furthermore, sensitized DOCA-induced salt appetite and cocaine-sensitized locomotor behavior cross-sensitize (Acerbo and Johnson 2011).

A possible mechanism of MR-mediated sensitization is angiotensin type 1 receptor (AT_1) up-regulation. Brain AT_1 receptors are up-regulated after DOCA or angiotensin II treatment in areas involved in body fluid homeostasis such as the area postrema, septum, and tissues surrounding the anteroventral third ventricle (King et al. 1988; Wilson et al. 1986). Adrenalectomy reduces AT_1 receptors in the subfornical organ, paraventricular nucleus, and area postrema (Shelat et al. 1998). AT_1 receptor levels can be restored by administration of aldosterone or corticosterone (Aguilera et al. 1995; Castrén and Saavedra 1989).

Both mineralocorticoids and glucocorticoids bind to the MR. Thus, the critical factor in adrenal steroid-induced AT_1 receptor regulation is occupation of MRs, independent of the occupying ligand. Corticosterone also appears to enhance sodium appetite, supporting the idea that MR activation is sufficient to enhance sodium appetite. Behaviorally, glucocorticoids potentiate sodium appetite induced by DOCA or angiotensin II injections (Ganesan and Sumners 1989; Ma et al. 1993; Shelat et al. 2000). However, glucocorticoid administration also increases water and sodium excretion, evoking a negative fluid balance that may be the direct cause for glucocorticoid-enhanced sodium appetite (Thunhorst et al. 2007). Administration of the potent MR and glucocorticoid receptor agonist dexamethasone promotes AT_1 receptor expression in the CNS (Aguilera et al. 1995). Furthermore, glucocorticoids increase brain angiotensinogen levels (Bunnemann et al. 1993). There is a cooperative action of mineralocorticoid and glucocorticoid agonists on sodium appetite. Beyond DOCA treatment alone, coadministration of dexamethasone and DOCA enhances AT_1 receptor expression in areas involved in sodium appetite and thirst, including the paraventricular nucleus, subfornical organ, and area postrema (Shelat et al. 2000).

14.3.3 Stress, Stressors, and Sodium Appetite Sensitization

Stress is a process associated with an actual or anticipated disruption of homeostasis (Ulrich-Lai and Herman 2009), and stressors are the stimuli or conditions that trigger this process. By their very nature, alterations in body fluid homeostasis constitute stress. Stressors can be putatively classified as either physiological or psychological. Physiological stressors involve homeostatic disruptions which, if sufficiently severe or sustained, will lead to failure of vital functions (e.g., circulatory shock). Psychological stressors (also known as processive stressors; see Herman et al. 2003) involve the processing of cognitive information that is correctly or incorrectly interpreted as being a threat to an animal's physical integrity or well-being. Recent studies have developed the idea that, depending on modality, stressors may use different neural circuitry to elicit a stress response (Dayas et al. 2001; Li et al. 1996). Sodium depletion is a physiological stressor arising from the bodily loss of the ion or its sequestration at some site in the body [e.g., experimental localized edema (Stricker 1981) or intraperitoneal dialysis (Falk 1965, 1966)]; however, the persistent craving for sodium that accompanies an unsatiated sodium appetite may be a psychological stressor (Morris et al. 2006, 2008, 2010). In this context, it is important to note that repeated stressors sensitize cocaine-induced locomotor behavior (Antelman and Chiodo 1983; Antelman et al. 1980) and that biochemical changes induced by stress exposure overlap with those from psychomotor stimulant administration (Ortiz et al. 1996). Therefore, it is possible that it is the repeated exposure to the psychological stressor of a sustained "wanting" or craving of salt that leads to sodium appetite sensitization, rather than an insult that is specific to sodium or a disruption in fluid homeostasis *per se*.

It is argued by many that both stress and sodium intake have a role in the etiology of hypertension (Dahl 1961, 1972; Fredrikson and Matthews 1990; Lawler et al. 1981). Individuals in westernized societies experience both multiple life stressors and diets with high salt content. Stress and sodium intake are often considered to independently raise blood pressure; however, studies are discussed below showing stressor-induced sodium intake. Stressors do not necessarily deplete animals of sodium, but activate many of the same hormone systems that are stimulated by manipulations that deplete the body of sodium.

Stress-induced sodium intake has been associated with a variety of stressors in many species. Rats exposed to social stress increase sodium intake, and this increase is attenuated by administration of clonidine or reserpine (Bourjeili et al. 1995). Denton and colleagues noted that wild rabbits exposed to a psychological stressor drastically increased sodium intake (Denton et al. 1984), and mice exposed to restraint stress or food deprivation increased salt consumption (Denton et al. 1999; Kuta et al. 1984).

Additional studies have examined the neuromodulatory and hormonal basis for stress-induced sodium appetite. ICV administration of corticotropin releasing factor, a central peptide involved in the stress response, increases sodium intake in mice and rabbits (Denton et al. 1999). In rabbits the increase in sodium intake is also accompanied by enhanced sodium excretion (Tarjan et al. 1991). Adrenocorticotropin stimulating hormone enhances sodium intake and sodium retention through a peripheral action (Denton et al. 1984, 1999). In addition, the effect of adrenocorticotropin stimulating hormone on sodium appetite is abolished by adrenalectomy in rats and sheep (Weisinger et al. 1978, 1980). The α2-adrenoreceptor antagonist and anxiogenic drug yohimbine also elicits a robust sodium appetite (Zardetto-Smith et al. 1994). In a recent study, we found that yohimbine-induced sodium appetite, but not furo/cap-induced sodium appetite, is blocked by pretreatment with the beta-adrenergic receptor antagonist propranolol, but not atenolol (Figure 14.4). These data implicate central β-receptors in the genesis of stress-induced sodium appetite.

14.3.4 Pregnancy, Perinatal Events, and Sensitization of Sodium Appetite

The mechanisms involved in the process of sensitization of sodium appetite may not remain consistent throughout the life span of a species, and there are likely to be periods during ontogeny when the same degree of disruption in sodium homeostasis has more or less severe consequences. A growing body of research indicates that during the perinatal period, both mothers and their offspring are more sensitive to conditions that alter sodium homeostasis. In the case of the offspring, disrupting sodium balance will manifest lifelong alterations in physiological and behavioral responses to sodium deficits or dietary excess. Four lines of research have examined this topic: (1) the effect of pregnancy on sodium intake of dams, (2) the effects of sodium depletion in dams on offspring's sodium intake, (3) the effects of dam dietary sodium excesses or insufficiencies on offspring, and (4) early life stress.

During gestation, significant amounts of sodium are necessary for the development of healthy offspring. The sodium requirement of developing fetuses causes a sodium deficit in dams. Dams increase sodium intake and retain more sodium than nonpregnant rats; this is an effect that appears around day 8 of pregnancy (Barelare and Richter 1937; Churchi-l et al. 1980; Pike and Yao 1971). During pregnancy, females increase genital licking, a behavior partially explained by the need to recycle excreted sodium, as this behavior is attenuated when hypertonic saline is accessible

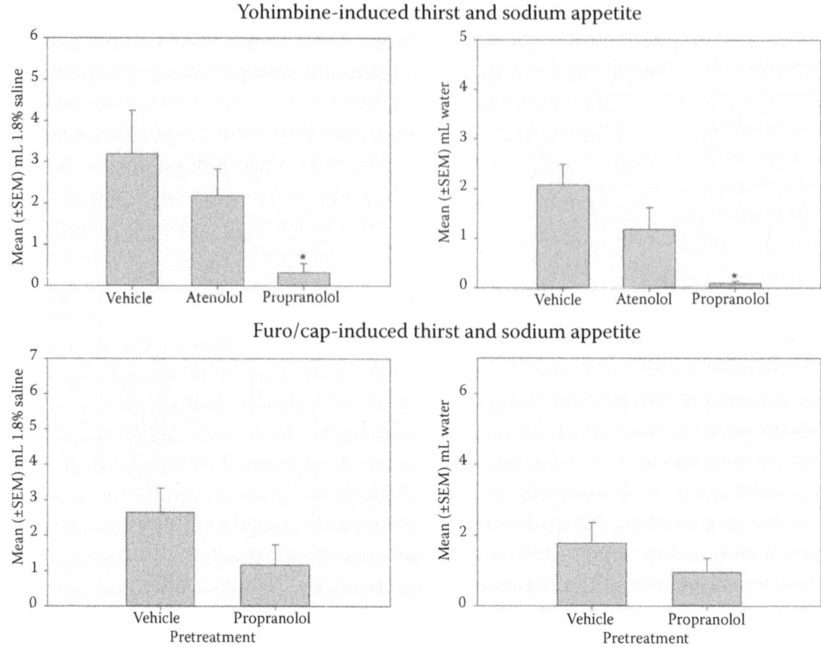

FIGURE 14.4 Propranolol pretreatment inhibits yohimbine induced thirst and sodium appetite, but has no effect on furo/cap induced thirst and sodium appetite. Top: rats were given injections of either the β-adrenergic receptor antagonist propranolol (10 mg/kg, IP), which crosses the blood–brain barrier, or the β-adrenergic receptor antagonist atenolol (10 mg/kg, IP), which does not cross the blood–brain barrier, before receiving yohimbine (3 mg/kg, SC). Only propranolol significantly reduced water and sodium intakes, suggesting a role for central β receptors in stress-induced sodium intake ($*p < 0.05$, Dunn's post-test after significant ANOVA on ranks). Bottom: rats were given furo/cap injections to exhibit rapid need-induced intake. Pretreatment with propranolol had no significant effect on thirst or sodium appetite.

(Steinberg and Bindra 1962). Despite elevated sodium intake and enhanced retention, dams have equivalent or less plasma sodium than nonpregnant females, and about 60% of retained sodium is found in the conceptus (Churchi-l et al. 1980). Gestation-induced sodium deficiency elevates activity of the classic (circulating) RAAS in rats and humans (Garland et al. 1987; Wilson et al. 1980). In humans, taste changes occur congruent with elevated RAAS activity and an active sodium appetite; pregnant women have greater difficulty discriminating between different concentrations of sodium chloride and prefer hypertonic saline solutions (Brown and Toma 1986). Sodium intake during pregnancy is critical as diastolic blood pressure falls in dams maintained on a low sodium diet (Beauséjour et al. 2003). Female rats also increase sodium intake during lactation (Schleifer 1990). The increased need-induced sodium intake in dams may utilize or evoke sodium appetite sensitization. Thus, pregnancy may be a natural process that takes advantage of sodium appetite sensitization.

In rats, the manipulations that consistently increase sodium appetite in the offspring are found in studies using sodium depletion in pregnant females. Initially it was shown that when females were given three polyethylene glycol injections during pregnancy, their offspring exhibited enhanced need-free sodium intake during adulthood (Nicolaidis et al. 1990). This phenomenon was extended by findings demonstrating that furosemide-induced sodium depletions elevated sodium appetite in the progeny. This effect was blocked by administering a high dose of angiotensin converting enzyme inhibitor to the dams to block the action of angiotensin II (Galaverna et al. 1995). When the offspring were sodium depleted as adults, both need and need-free sodium appetite were further potentiated in animals with a history of prenatal sodium depletions. Finally, dams that received partial aortic ligations, which increase RAAS activity and cause thirst and sodium appetite, give birth to progeny that exhibit elevated need-free intake of sodium solutions (Argüelles et al. 2000). These studies suggest that elevated RAAS activity during pregnancy has an organizational effect on adult sodium appetite in the progeny. It also appears that prenatal and adult sodium depletions have an additive effect on need-free and need-induced sodium intake (Galaverna et al. 1995). In humans, postnatal sodium depletions, such as infantile vomiting or diarrhea, increase adult sodium intake and preference (Leshem 1998). In addition, reports of frequent bouts of morning sickness in pregnant women serve as good predictors of high sodium intake in the offspring of such mothers (Crystal and Bernstein 1995).

Studies examining perinatal sodium diets have found equivocal effects on sodium intake in the offspring. In dams of Sprague–Dawley rats that were maintained on a 2.3% NaCl diet, progeny did not elevate sodium intake, and in the Wistar strain maintained on this diet, offspring decreased sodium intake (Myers et al. 1985). Another study found that sodium deprivation from days 2 or 3 through day 12 of pregnancy elevated water intake in progeny, but had no effect on sodium preference (Mouw et al. 1978). Sodium chloride deprivation from gestation day 3 to postnatal day 12 resulted in an enhanced preference for water (Hill et al. 1986). When these authors extended sodium deprivation from gestation day 3 to postnatal day 35, they observed decreased chorda tympani responses to sodium; however, sodium chloride preference was not measured in this protocol. Contreras and Kosten (1983) found that progeny exposed to a 3% sodium chloride diet from gestation day 1 to postnatal day 30 had an enhanced preference for sodium chloride and glucose solutions.

Finally, early life stress enhances need-free sodium intake in adulthood. Maternal separation, which is a significant stressor, increases adult need-free sodium intake (Leshem et al. 1996). These experimenters also found that providing the pups with a saline-soaked towel that allowed them to consume salt prevented enhanced adult need-free intake. Others found that rats exposed to early life stressors had elevated basal activity of both the hypothalamic–pituitary–adrenal axis and the RAAS system as adults (Edwards et al. 1999). Considering that mineralocorticoids and glucocorticoids enhance sodium appetite, elevated activities of these systems may be critical for enhanced sodium intake in rats exposed to early life stress.

14.4 CONCLUSIONS AND FURTHER CONSIDERATIONS

Sodium appetite sensitization is most likely mediated by CNS plasticity. Currently, the mechanisms mediating sodium appetite sensitization are unclear. However, we can propose several possibilities. Changes in numerous receptors, neurotransmitters, and second messengers mediate plasticity in many phenomena (Bliss and Collingridge 1993; Freeman and Steinmetz 2011; Kauer and Malenka 2007; Kelley 2004; Ruscheweyh et al. 2011; Wolf et al. 2004). In sodium appetite sensitization, up-regulation of dopamine and changes in opioid function in limbic brain areas are likely. AT_1 receptor up-regulation may also play an important role in sensitization. Anatomical substrates of interest include areas involved in body fluid homeostasis and limbic areas important for goal directed behavior. Future work that targets these systems will help elucidate the mediators of sodium appetite sensitization.

As originally formulated by Alan Epstein and his colleagues (Epstein 1991, 1982; Sakai et al. 1989), behavioral sensitization and related neuroplasticity associated with maintaining body fluid homeostasis may serve an adaptive purpose as sensitized animals in sodium-deficient environments will seek and ingest more sodium when sodium depleted. Fessler (2003) has proposed that sodium appetite sensitization also serves a long-term prophylactic effect by providing a buffer to prevent sudden dehydration imposed by pathogen exposure. Fessler also raises the point that sodium appetite sensitization comes with a caloric cost of seeking and ingesting more sodium, creating a trade-off between energy expenditure and protecting body fluid homeostasis. Sodium depletion during gestation leads to an organizational effect in the form of enhanced sodium consumption by the offspring. This allows progeny to be better equipped to face the challenges of a sodium deficient environment. It is possible that increased intake of sodium in need-free states may also help animals remember the location of environmental sodium through latent learning.

Sodium appetite sensitization provides a good example of processes related to the concepts of *allostasis* (Sterling and Eyer 1988) and *allostatic load* (McEwen and Stellar 1993). Allostasis is associated with changes in the operating range of physiological endpoints (e.g., body fluids, blood pressure, body temperature) and the ability of the body to increase or decrease vital functions to achieve a new steady state in response to challenges (McEwen and Stellar 1993; Sterling and Eyer 1988). Although a change in a physiological endpoint (e.g., the elevation of blood pressure) and the activation of physiological and/or behavioral mechanisms (e.g., increased sympathetic tone; increased salt intake to expand blood volume) to defend the new level of function may in the short-term enhance survival, the sustained or repeated activation of such countermeasures can have negative health effects. The strain or wear and tear on organs and tissues resulting from the body using allostasic defenses is the consequence of an individual's allostatic load (McEwen and Stellar 1993). In the case of sodium appetite, the increased salt intake as a consequence of sensitization produced by earlier challenges to body fluid homeostasis or cardiovascular function may add to the allostatic load of an individual. Excessive salt intake has been linked to disorders such as hypertension, heart failure, and renal disease. Future research determining the mechanisms underlying the sensitization of need-induced and need-free salt intake is likely to generate findings with important translational relevance as they may aid in explaining the nature of salt craving and the causes of high, excess sodium intake in humans (Morris et al. 2008).

ACKNOWLEDGMENTS

Research support to the authors' laboratory and/or support for the preparation of this chapter was provided by NIH Grants from the National Heart, Lung, and Blood Institute HL14388 and HL 098207, the National Institute of Diabetes and Digestive and Kidney Diseases DK 66086, and the National Institutes of Mental Health MH 80241, and National Institutes of Aging AG 025465.

REFERENCES

Acerbo, M. J., and A. K. Johnson. 2011. Behavioral cross-sensitization between DOCA-induced sodium appetite and cocaine-induced locomotor behavior. *Pharmacol Biochem Behav* 98:440–8.

Aguilera, G., A. Kiss, and X. Luo. 1995. Increased expression of type 1 angiotensin II receptors in the hypothalamic paraventricular nucleus following stress and glucocorticoid administration. *J Neuroendocrinol* 7:775–83.

Antelman, S. M., and L. A. Chiodo. 1983. Amphetamine as a stressor. *Stimulants: Neurochemical, Behavioral and Clinical Perspectives,* 269–299. New York: Raven Press.

Antelman, S. M., A. J. Eichler, C. A. Black, and D. Kocan. 1980. Interchangeability of stress and amphetamine in sensitization. *Science* 207:329–31.

Argüelles, J., J. I. Brime, P. Lopez-Sela, C. Perillan, and M. Vijande. 2000. Adult offspring long-term effects of high salt and water intake during pregnancy. *Horm Behav* 37:156–62.

Bare, J. K. 1949. The specific hunger for sodium chloride in normal and adrenalectomized white rats. *J Comp Physiol Psychol* 42:242–53.

Barelare, B., and C. P. Richter. 1937. Increased sodium chloride appetite in pregnant rats. *American Journal of Physiology—Legacy Content* 121:185.

Beauséjour, A., K. Auger, J. St-Louis, and M. Brochu. 2003. High-sodium intake prevents pregnancy-induced decrease of blood pressure in the rat. *Am J Physiol Heart Circ Physiol* 285:H375–83.

Berridge, K. C. 1996. Food reward: Brain substrates of wanting and liking. *Neurosci Biobehav Rev* 20:1–25.

Berridge, K. C., F. W. Flynn, J. Schulkin, and H. J. Grill. 1984. Sodium depletion enhances salt palatability in rats. *Behav Neurosci* 98:652–60.

Berridge, K. C., T. E. Robinson, and J. W. Aldridge. 2009. Dissecting components of reward: Liking, wanting, and learning. *Curr Opin Pharmacol* 9:65–73.

Bliss, T. V. P., and G. L. Collingridge. 1993. A synaptic model of memory: Long-term potentiation in the hippocampus. *Nature* 361:31–39.

Bourjeili, N., M. Turner, J. Stinner, and D. Ely. 1995. Sympathetic nervous system influences salt appetite in four strains of rats. *Physiol Behav* 58:437–43.

Brown, J. E., and R. B. Toma. 1986. Taste changes during pregnancy. *Am J Clin Nutr* 43:414–8.

Bryant, R. W., A. N. Epstein, J. T. Fitzsimons, and S. J. Fluharty. 1980. Arousal of a specific and persistent sodium appetite in the rat with continuous intracerebroventricular infusion of angiotensin II. *J Physiol* 301:365–82.

Bunnemann, B., A. Lippoldt, J. A. Aguirre, A. Cintra, and R. Metzger. 1993. Glucocorticoid regulation of angiotensinogen gene expression in discrete areas of the male rat brain. *Neuroendocrinology* 57:856–62.

Castrén, E., and J. M. Saavedra. 1989. Angiotensin II receptors in paraventricular nucleus, subfornical organ, and pituitary gland of hypophysectomized, adrenalectomized, and vasopressin-deficient rats. *Proc Nat Acad Sci USA* 86:725–9.

Churchi-l, S. E., H. H. Bengele, and E. A. Alexander. 1980. Sodium balance during pregnancy in the rat. *Am J Physiol Regul Integr Comp Physiol* 239:R143–8.

Clark, J. J., and I. L. Bernstein. 2004. Reciprocal cross-sensitization between amphetamine and salt appetite. *Pharmacol Biochem Behav* 78:691–8.

Clark, J. J., and I. L. Bernstein. 2006a. A role for D2 but not D1 dopamine receptors in the cross-sensitization between amphetamine and salt appetite. *Pharmacol Biochem Behav* 83:277–84.

Clark, J. J., and I. L. Bernstein. 2006b. Sensitization of salt appetite is associated with increased "wanting" but not "liking" of a salt reward in the sodium-deplete rat. *Behav Neurosci* 120:206–10.

Contreras, R. J. 1977. Changes in gustatory nerve discharges with sodium deficiency: A single unit analysis. *Brain Res* 121:373–8.

Contreras, R. J., and T. Kosten. 1983. Prenatal and early postnatal sodium chloride intake modifies the solution preferences of adult rats. *J Nutrition* 113:1051–62.

Cooper, S. J., and D. B. Gilbert. 1984. Naloxone suppresses fluid consumption in tests of choice between sodium chloride solutions and water in male and female water-deprived rats. *Psychopharmacology* 84:362–7.

Cotman, C. W., and N. C. Berchtold. 2002. Exercise: A behavioral intervention to enhance brain health and plasticity. *Trends Neurosci* 25:295–301.

Crystal, S. R., and I. L. Bernstein. 1995. Morning sickness: Impact on offspring salt preference. *Appetite* 25:231–40.

Dahl, L. K. 1961. Possible role of chronic excess salt consumption in the pathogenesis of essential hypertension. *Am J Cardiol* 8:571–5.

Dahl, L. K. 1972. Salt and hypertension. *Am J Clin Nutr* 25:231–44.

Dayas, C. V., K. M. Buller, J. W. Crane, Y. Xu, and T. A. Day. 2001. Stressor categorization: Acute physical and psychological stressors elicit distinctive recruitment patterns in the amygdala and in medullary noradrenergic cell groups. *Eur J Neurosci* 14:1143–52.

De Luca, L. A., Jr., D. T. B. Pereira-Derderian, R. C. Vendramini, R. B. David, and J. V. Menani. 2010. Water deprivation-induced sodium appetite. *Physiol Behav* 100:535–44.

De Luca, L. A., Jr., Z. Xu, G. H. Schoorlemmer et al. 2002. Water deprivation-induced sodium appetite: Humoral and cardiovascular mediators and immediate early genes. *Am J Physiol Regul Integr Comp Physiol* 282:R552–9.

Denton, D. A., J. R. Blair-West, M. I. McBurnie, J. A. P. Miller, R. S. Weisinger, and R. M. Williams. 1999. Effect of adrenocorticotrophic hormone on sodium appetite in mice. *Am J Physiol Regul Integr Comp Physiol* 277:R1033–40.

Denton, D. A., J. P. Coghlan, D. T. Fei et al. 1984. Stress, ACTH, salt intake and high blood pressure. *Clin Exp Hypertens* 6:403–15.

Edwards, E., J. A. King, and J. C. Fray. 1999. Increased basal activity of the HPA axis and renin-angiotensin system in congenital learned helpless rats exposed to stress early in development. *Int J Dev Neurosci* 17:805–12.

Epstein, A. N. 1982. Mineralocorticoids and cerebral angiotensin may act together to produce sodium appetite. *Peptides* 3:493–4.

Epstein, A. N. 1991. Neurohormonal control of salt intake in the rat. *Brain Res Bull* 27:315–20.

Falk, J. L. 1965. Water Intake and NaCl appetite in sodium depletion. *Psychol Rep* 16:315–25.

Falk, J. L. 1966. Serial sodium depletion and NaCl solution intake. *Physiol Behav* 1:75–7.

Fessler, D. M. T. 2003. An evolutionary explanation of the plasticity of salt preferences: Prophylaxis against sudden dehydration. *Med Hypotheses* 61:412–5.

Fitts, D. A., and D. B. Masson. 1989. Forebrain sites of action for drinking and salt appetite to angiotensin or captopril. *Behav Neurosci* 103:865–72.

Fitts, D. A., and R. L. Thunhorst. 1996. Rapid elicitation of salt appetite by an intravenous infusion of angiotensin II in rats. *Am J Physiol* 270:R1092–8.

Fitzsimons, J. T. 1961. Drinking by rats depleted of body fluid without increase in osmotic pressure. *J Physiol* 159:297–309.

Fluharty, S. J., and A. N. Epstein. 1983. Sodium appetite elicited by intracerebroventricular infusion of angiotensin II in the rat: II. Synergistic interaction with systemic mineralocorticoids. *Behav Neurosci* 97:746–58.

Frankmann, S. P., D. M. Dorsa, R. R. Sakai, and J. B. Simpson. 1986. A single experience with hyperoncotic colloid dialysis persistently alters water and sodium intake. In *NATO ASI series Series A, Life Sciences*, ed. G. De Caro, A. N. Epstein, and M. Massi. New York: Plenum Press.

Fredrikson, M., and K. A. Matthews. 1990. Cardiovascular responses to behavioral stress and hypertension: A meta-analytic review. *Ann Behav Med* 12:30–9.

Freeman, J. H., and A. B. Steinmetz. 2011. Neural circuitry and plasticity mechanisms underlying delay eyeblink conditioning. *Learn Mem* 18:666–77.

Fregly, M. J., and N. E. Rowland. 1985. Role of renin–angiotensin–aldosterone system in NaCl appetite of rats. *Am J Physiol Regul Integr Comp Physiol* 248:R1–11.

Galaverna, O., L. A. De Luca Jr., J. Schulkin, S. Yao, and A. N. Epstein. 1992. Deficits in NaCl ingestion after damage to the central nucleus of the amygdala in the rat. *Brain Res Bull* 28:89–98.

Galaverna, O., S. Nicolaidis, S. Z. Yao, R. R. Sakai, and A. N. Epstein. 1995. Endocrine consequences of prenatal sodium depletion prepare rats for high need-free NaCl intake in adulthood. *Am J Physiol Regul Integr Comp Physiol* 269:R578–83.

Ganesan, R., and C. Sumners. 1989. Glucocorticoids potentiate the dipsogenic action of angiotensin II. *Brain Res* 499:121–30.

Garland, H. O., J. C. Atherton, C. Baylis, M. R. A. Morgan, and C. M. Milne. 1987. Hormone profiles for progesterone, oestradiol, prolactin, plasma renin activity, aldosterone and corticosterone during pregnancy and pseudopregnancy in two strains of rat: Correlation with renal studies. *J Endocrinol* 113:435–44.

Gosnell, B. A., and M. J. Majchrzak. 1990. Effects of a selective mu opioid receptor agonist and naloxone on the intake of sodium chloride solutions. *Psychopharmacology* 101:66–71.

Grill, H. J., and R. Norgren. 1978. The taste reactivity test: I. Mimetic responses to gustatory stimuli in neurologically normal rats. *Brain Res* 143:263–79.

Herman, J. P., H. Figueiredo, N. K. Mueller et al. 2003. Central mechanisms of stress integration: Hierarchical circuitry controlling hypothalamo–pituitary–adrenocortical responsiveness. *Front Neuroendocrinol* 24:151–80.

Hill, D. L., C. M. Mistretta, and R. M. Bradley. 1986. Effects of dietary NaCl deprivation during early development on behavioral and neurophysiological taste responses. *Behav Neurosci* 100:390–8.

Hubbell, C. L., and N. B. McCutcheon. 1993. Opioidergic manipulations affect intake of 3% NaCl in sodium-deficient rats. *Pharmacol Biochem Behav* 46:473–6.

Jacobs, K. M., G. P. Mark, and T. R. Scott. 1988. Taste responses in the nucleus tractus solitarius of sodium-deprived rats. *J Physiol* 40:393–410.

Jalowiec, J. E. 1974. Sodium appetite elicited by furosemide: Effects of differential dietary maintenance. *Behav Biol* 10:313–27.

Jalowiec, J. E., and E. M. Stricker. 1973. Sodium appetite in adrenalectomized rats following dietary sodium deprivation. *J Comp Physiol Psychol* 82:66–77.

Karst, H., S. Berger, M. Turiault, F. Tronche, G. Schütz, and M. Joëls. 2005. Mineralocorticoid receptors are indispensable for nongenomic modulation of hippocampal glutamate transmission by corticosterone. *Proc Nat Acad Sci USA* 102:19204–7.

Kauer, J. A., and R. C. Malenka. 2007. Synaptic plasticity and addiction. *Nat Rev Neurosci* 8:844–58.

Kelley, A. E. 2004. Memory and addiction: Shared neural circuitry and molecular mechanisms. *Neuron* 44:161–79.

King, S. J., J. W. Harding, and K. E. Moe. 1988. Elevated salt appetite and brain binding of angiotensin II in mineralocorticoid-treated rats. *Brain Res* 448:140–9.

Koob, G. F., and M. Le Moal. 2005. Plasticity of reward neurocircuitry and the 'dark side' of drug addiction. *Nat Neurosci* 8:1442–4.

Kuta, C. C., H. U. Bryant, J. E. Zabik, and G. K. Yim. 1984. Stress, endogenous opioids and salt intake. *Appetite* 5:53–60.

Lawler, J. E., G. F. Barker, J. W. Hubbard, and R. G. Schaub. 1981. Effects of stress on blood pressure and cardiac pathology in rats with borderline hypertension. *Hypertension* 3:496–505.

Leshem, M. 1998. Salt preference in adolescence is predicted by common prenatal and infantile mineralofluid loss. *Physiol Behav* 63:699–704.

Leshem, M., M. Maroun, and S. Del Canho. 1996. Sodium depletion and maternal separation in the suckling rat increase its salt intake when adult. *Physiol Behavior* 59:199–204.

Li, H. Y., A. Ericsson, and P. E. Sawchenko. 1996. Distinct mechanisms underlie activation of hypothalamic neurosecretory neurons and their medullary catecholaminergic afferents in categorically different stress paradigms. *Proc Nat Acad Sci USA* 93:2359–64.

Liedtke, W. B., M. J. McKinley, L. L. Walker et al. 2011. Relation of addiction genes to hypothalamic gene changes subserving genesis and gratification of a classic instinct, sodium appetite. *Proc Nat Acad Sci USA* 108:12509–14.

Lucas, L. R., C. A. Grillo, and B. S. McEwen. 2003. Involvement of mesolimbic structures in short-term sodium depletion: In situ hybridization and ligand-binding analyses. *Neuroendocrinology* 77:406–15.

Lucas, L. R., C. A. Grillo, and B. S. McEwen. 2007. Salt appetite in sodium-depleted or sodium-replete conditions: Possible role of opioid receptors. *Neuroendocrinology* 85:139–47.

Lucas, L. R., P. Pompei, and B. S. McEwen. 2000. Salt appetite in salt-replete rats: Involvement of mesolimbic structures in deoxycorticosterone-induced salt craving behavior. *Neuroendocrinology* 71:386–95.

Ma, L. Y., B. S. McEwen, R. R. Sakai, and J. Schulkin. 1993. Glucocorticoids facilitate mineralocorticoid-induced sodium intake in the rat. *Horm Behav* 27:240–50.

Masson, D. B., and D. A. Fitts. 1989. Subfornical organ connectivity and drinking to captopril or carbachol in rats. *Behav Neurosci* 103:873–80.

McEwen, B. S., and E. Stellar. 1993. Stress and the individual. Mechanisms leading to disease. *Arch Intern Med* 153:2093–101.

Moe, K. E., M. L. Weiss, and A. N. Epstein. 1984. Sodium appetite during captopril blockade of endogenous angiotensin II formation. *Am J Physiol Regul Integr Comp Physiol* 247:R356–65.

Morris, M. J., E. S. Na, and A. K. Johnson. 2008. Salt craving: The psychobiology of pathogenic sodium intake. *Physiol Behavior* 94:709–21.

Morris, M. J., E. S. Na, A. J. Grippo, and A. K. Johnson. 2006. The effects of deoxycorticosterone-induced sodium appetite on hedonic behaviors in the rat. *Behav Neurosci* 120:571–8.

Morris, M. J., E. S. Na, and A. K. Johnson. 2010. Mineralocorticoid receptor antagonism prevents hedonic deficits induced by a chronic sodium appetite. *Behav Neurosci* 124:211–24.

Mousseau, M. C., S. N. Thornton, F. P. Martial, F. Lienard, and S. Nicolaidis. 1996. Neuronal responses to iontophoretically applied angiotensin II, losartan and aldosterone, as well as gustatory stimuli, in non-anesthetized control and desoxycorticosterone acetate-pretreated rats. *Regul Pept* 66:51–4.

Mouw, D. R., A. J. Vander, and J. Wagner. 1978. Effects of prenatal and early postnatal sodium deprivation on subsequent adult thirst and salt preference in rats. *Am J Physiol Renal Physiol* 234:F59–63.

Myers, J. B., V. J. Smidt, S. Doig, R. D. Nicolantonio, and T. O. Morgan. 1985. Blood pressure, salt taste and sodium excretion in rats exposed prenatally to high salt diet. *Clin Exp Pharmacol Physiol* 12:217–20.

Na, E. S., M. J. Morris, and A. K. Johnson. 2012. Opioid mechanisms that mediate the palatability of and appetite for salt in sodium replete and deficient states. *Physiol Behavior* 106:164–70.

Na, E. S., M. J. Morris, and A. K. Johnson. 2009. Behavioral cross-sensitization between morphine-induced locomotion and sodium depletion-induced salt appetite. *Pharmacol Biochem Behav* 93:368–74.

Na, E. S., M. J. Morris, R. F. Johnson, T. G. Beltz, and A. K. Johnson. 2007. The neural substrates of enhanced salt appetite after repeated sodium depletions. *Brain Research* 1171:104–10.

Nicolaidis, S., O. Galaverna, and C. H. Metzler. 1990. Extracellular dehydration during pregnancy increases salt appetite of offspring. *Am J Physiol Regul Integr Comp Physiol* 258:R281–3.

Ortiz, J., L. W. Fitzgerald, S. Lane, R. Terwilliger, and E. J. Nestler. 1996. Biochemical adaptations in the mesolimbic dopamine system in response to repeated stress. *Neuropsychopharmacology* 14:443–52.

Pereira, D. T. B., J. V. Menani, and L. A. De Luca Jr. 2010. FURO/CAP: A protocol for sodium intake sensitization. *Physiol Behav* 99:472–81.

Pike, R. L., and C. Yao. 1971. Increased sodium chloride appetite during pregnancy in the rat. *J Nutr* 101:169–75.

Rice, K. K., and C. P. Richter. 1943. Increased sodium chloride and water intake of normal rats treated with desoxycorticosterone acetate. *Endocrinology* 33:106.

Richter, C. P. 1936. Increased salt appetite in adrenalectomized rats. *Am J Physiol Legacy Content* 115:155–161.

Roitman, M. F., E. Na, G. Anderson, T. A. Jones, and I. L. Bernstein. 2002. Induction of a salt appetite alters dendritic morphology in nucleus accumbens and sensitizes rats to amphetamine. *J Neurosci* 22:RC225.

Roitman, M. F., G. E. Schafe, T. E. Thiele, and I. L. Bernstein. 1997. Dopamine and sodium appetite: Antagonists suppress sham drinking of NaCl solutions in the rat. *Behav Neurosci* 111:606–11.

Ruscheweyh, R., O. Wilder-Smith, R. Drdla, X. G. Liu, and J. Sandkühler. 2011. Long-term potentiation in spinal nociceptive pathways as a novel target for pain therapy. *Mol Pain* 7:20.

Sakai, R. R., W. B. Fine, A. N. Epstein, and S. P. Frankmann. 1987. Salt appetite is enhanced by one prior episode of sodium depletion in the rat. *Behav Neurosci* 101:724–31.

Sakai, R. R., S. P. Frankmann, W. B. Fine, and A. N. Epstein. 1989. Prior episodes of sodium depletion increase the need-free sodium intake of the rat. *Behav Neurosci* 103:186–92.

Sakai, R. R., B. S. McEwen, S. J. Fluharty, and L. Y. Ma. 2000. The amygdala: Site of genomic and nongenomic arousal of aldosterone-induced sodium intake. *Kidney Int* 57:1337–45.

Sakai, R. R., S. Nicolaidis, and A. N. Epstein. 1986. Salt appetite is suppressed by interference with angiotensin II and aldosterone. *Am J Physiol Regul Integr Comp Physiol* 251:R762–8.

Schleifer, L. A. 1990. The effects of reproductive status on sodium chloride intake and preference curves in rats, Department of Psychology, Concordia University, Montreal.

Schoorlemmer, G. H. M., A. K. Johnson, and R. L. Thunhorst. 2001. Circulating angiotensin II mediates sodium appetite in adrenalectomized rats. *Am J Physiol Regul Integr Comp Physiol* 281:R723–9.

Shelat, S. G., S. J. Fluharty, and L. M. Flanagan-Cato. 1998. Adrenal steroid regulation of central angiotensin II receptor subtypes and oxytocin receptors in rat brain. *Brain Res* 807:135–46.

Shelat, S. G., J. L. King, L. M. Flanagan-Cato, and S. J. Fluharty. 2000. Mineralocorticoids and glucocorticoids cooperatively increase salt intake and angiotensin II receptor binding in rat brain. *Neuroendocrinology* 69:339–51.

Smith, D. F., and E. M. Stricker. 1969. The influence of need on the rat's preference for dilute NaCl solutions. *Physiol Behav* 4:407–10.

Steinberg, J., and D. Bindra. 1962. Effects of pregnancy and salt-intake on genital licking. *J Comp Physiol Psychol* 55:103–6.

Sterling, P., and J. Eyer. 1988. Allostasis: A new paradigm to explain arousal pathology. In *Handbook of Life Stress, Cognition and Health*, ed. S. Fisher, and J. Reason, 629–49. New York: Wiley.

Stricker, E. M. 1981. Thirst and sodium appetite after colloid treatment in rats. *J Comp Physiological Psychol* 95:1–25.

Stricker, E. M., E. Thiels, and J. G. Verbalis. 1991. Sodium appetite in rats after prolonged dietary sodium deprivation: A sexually dimorphic phenomenon. *Am J Physiol Regul Integr Comp Physiol* 260:R1082–8.

Takamata, A., G. W. Mack, C. M. Gillen, and E. R. Nadel. 1994. Sodium appetite, thirst, and body fluid regulation in humans during rehydration without sodium replacement. *Am J Physiol Regul Integr Comp Physiol* 266:R1493–502.

Tarjan, E., D. A. Denton, and R. S. Weisinger. 1991. Corticotropin-releasing factor enhances sodium and water intake/excretion in rabbits. *Brain Res* 542:219–24.

Thornton, S. N., and S. Nicolaidis. 1994. Long-term mineralocorticoid-induced changes in rat neuron properties plus interaction of aldosterone and ANG II. *Am J Physiol Regul Integr Comp Physiol* 266:R564–71.

Thunhorst, R. L., T. G. Beltz, and A. K. Johnson. 2007. Glucocorticoids increase salt appetite by promoting water and sodium excretion. *Am J Physiol Regul Integr Comp Physiol* 293:R1444–51.

Thunhorst, R. L., and D. A. Fitts. 1994. Peripheral angiotensin causes salt appetite in rats. *Am J Physiol Regul Integr Comp Physiol* 267:R171–7.

Thunhorst, R. L., and A. K. Johnson. 1994. Renin–angiotensin, arterial blood pressure, and salt appetite in rats. *Am J Physiol Regul Integr Comp Physiol* 266:R458–65.

Thunhorst, R. L., M. Morris, and A. K. Johnson. 1994. Endocrine changes associated with a rapidly developing sodium appetite in rats. *Am J Physiol Regul Integr Comp Physiol* 267:R1168–73.

Trujillo, K. A., and H. Akil. 1991. Inhibition of morphine tolerance and dependence by the NMDA receptor antagonist MK-801. *Science* 251(4989):85–87.

Ulrich-Lai, Y. M., and J. P. Herman. 2009. Neural regulation of endocrine and autonomic stress responses. *Nat Rev Neurosci* 10:397–409.

Weisinger, R. S., D. A. Denton, M. J. McKinley, and J. F. Nelson. 1978. ACTH induced sodium appetite in the rat. *Pharmacol Biochem Behav* 8:339–42.

Weisinger, R. S., J. P. Coghlan, D. A. Denton et al. 1980. ACTH-elicited sodium appetite in sheep. *Am J Physiol Endocrinol Metab* 239:E45–50.

Wilson, K. M., C. Sumners, S. Hathaway, and M. J. Fregly. 1986. Mineralocorticoids modulate central angiotensin II receptors in rats. *Brain Res* 382:87–96.

Wilson, M., A. A. Morganti, I. Zervoudakis et al. 1980. Blood pressure, the renin–aldosterone system and sex steroids throughout normal pregnancy. *Am J Med* 68:97–104.

Wolf, M. E., S. L. Dahlin, X. T. Hu, C. J. Xue, and K. White. 1995. Effects of lesions of prefrontal cortex, amygdala, or fornix on behavioral sensitization to amphetamine: Comparison with *N*-methyl-D-aspartate antagonists. *Neuroscience* 69:417–39.

Wolf, M. E., X. Sun, S. Mangiavacchi, and S. Z. Chao. 2004. Psychomotor stimulants and neuronal plasticity. *Neuropharmacology* 47:61–79.

Woolf, C. J., and M. W. Salter. 2000. Neuronal plasticity: Increasing the gain in pain. *Science* 288:1765–9.

Yang, Z., and A. N. Epstein. 1991. Blood-borne and cerebral angiotensin and the genesis of salt intake. *Horm Behav* 25:461–76.

Zardetto-Smith, A. M., T. G. Beltz, and A. K. Johnson. 1994. Role of the central nucleus of the amygdala and bed nucleus of the stria terminalis in experimentally-induced salt appetite. *Brain Res* 645:123–34.

15 Homeostasis and Body Fluid Regulation

An End Note

Laurival A. De Luca Jr., Richard B. David, and José V. Menani

CONTENTS

15.1 HOMEOSTASIS: A CRITICIZED CONCEPT

Homeostasis is accepted universally as a synonym of equilibrium or stability in biological systems and commonly used to describe activities of cells, organs, individuals, and society (Abbott 2003; Cannon 1929; Palagi and Mancini 2011). Tradition and universal acceptance easily preclude more explanations about the concept of homeostasis, being unlikely that the reading of this book has caused much doubt about its meaning.

However, the universal use and attribution of a control mechanism based exclusively on negative feedback have raised objections to the term homeostasis (e.g., Berridge 2004; Sterling 2012). It is not our intent to review and discuss here the foundations of the many objections raised about homeostasis. We would like to recall the original concept instead, showing how appropriate and valid it is in the context of neural and endocrine mechanisms that regulate the composition of body fluid and produce homeostasis in mammals.

15.2 HOMEOSTASIS AS DEFINED BY CANNON

We may identify the article of Walter Cannon published in *Physiological Reviews* (1929) as a landmark to the original definition of homeostasis. Cannon based on Claude Bernard considered homeostasis as steady states maintained by mechanisms peculiar to living beings. He also emphasized that the existence of such states derives from a constant internal environment. Bernard defined internal environment as the fluid matrix made of blood and lymph or, by extension, the extracellular fluid (ECF) (Cannon 1929; Guyton and Hall 2000; Takei 2000). It should be clear from Cannon's (1929) article, and from the reports of other authors, who later adopted the original concept (e.g., Guyton and Hall 2000; Takei 2000), that the constancy of the ECF constitutes the essence of the concept of homeostasis. Cannon referred specifically to the constancy of physical–chemical constituents of the ECF as varying within narrow limits. He called those constituents homeostatic categories. Among such categories, we may find temperature, water, osmolarity, and sodium, all related to mechanisms discussed in previous chapters.

Cannon considered a living being as an open system with automatic reactive mechanisms. Reactive mechanisms were assumed by Cannon to correct large fluctuations, thereby maintaining each homeostatic category within narrow and constant limits. Let us take sodium as an example. Its normal concentration in the human blood varies about 7% within a range of 138–146 mmol/L (Guyton and Hall 2000). As illustrated throughout this book, several autonomic, neuroendocrine, and behavioral mechanisms are associated with the homeostasis of this ion.

15.3 HOMEOSTASIS VERSUS HOMEOSTATIC MECHANISMS: EXAMPLE OF SODIUM

Powerful multiple homeostatic mechanisms maintain sodium concentration and osmolarity of the ECF of most bony fishes and amphibians, and birds and mammals, to about one-third that of the sea. Such mechanisms operate to avoid either desiccation or overhydration, and their appearance in evolution was important for vertebrate conquest of the terrestrial environment (Guyton and Hall 2000; Kültz 2012; Takei et al. 2000).*

Responses to sodium loss or gain provide an example of how different types of mechanisms contribute to body fluid homeostasis. Homeostatic mechanisms in general may operate in a reaction mode, usually involving negative feedback, or through a combination of reaction with prediction or anticipation (Carpenter 2004; Moore-Ede 1986).

The concept of reactive or negative feedback mechanisms regulating sodium availability in body fluids is ingrained in the field. Studies that began with Curt Richter about the same time Cannon defined homeostasis (Moran and Schulkin 2000) inspired pioneer work searching for brain reactive mechanisms that control both ingestion and renal excretion of sodium (Covian et al. 1975). Such paradigm of reactive mechanisms is the foundation of productive investigation about mechanisms of body fluid homeostasis as shown by many examples throughout this book.

Another paradigm that has been increasingly investigated involves anticipation. Anticipatory mechanisms that control sodium intake were initially indicated by the finding that history of repeated sodium depletions enhances sodium intake (Falk 1966). Later it was demonstrated that angiotensin II and aldosterone, hormones

* This contrasts with the flexible osmotic concentration of the ECF found in aquatic invertebrates (e.g., Freire et al. 2008)—which hardly fits Cannon's definition of homeostasis.

secreted or produced in response to sodium depletion, also act by modifying the behavior, increasing the intensity of both need-induced and need-free sodium intake (Sakai et al. 1989).

The implications of the effects of those hormones for neural and behavioral plasticity, particularly for the behavior of sodium intake, were already discussed in previous chapters (e.g., Chapters 4, 13, and 14). We would like to call attention to the fact that changes in sodium intake as a function of episodic sodium depletion must result from plastic changes in the mechanisms that produce the behavior. Such changes are not predicted from mechanisms that only react to alterations in body sodium, but have been interpreted as an anticipated response to future dehydration (Epstein 1991; Fessler 2003).

Now, the fact that the neural and behavioral mechanisms that produce sodium homeostasis are plastic does not mean that the normal sodium range is necessarily modified. For example, whereas repeated episodes of sodium depletion enhances sodium intake, they alter neither plasma sodium concentration nor indicators of blood volume in rats (Sakai et al. 1989). Another example to illustrate the same point derives from congenital adrenal hyperplasia associated with salt waste. Adult humans bearing this pathology have an enhanced sodium appetite positively correlated with history of perinatal hyponatremia, but they also have normal serum sodium concentration (Kochli et al. 2005).

So, we may conclude that sodium or body fluid homeostasis occurs in the face of plastic neural and behavioral alterations. This teaches us that homeostatic mechanisms (reactive, anticipatory, or both) should not be confounded with the product of their operation (the maintenance of the homeostatic categories within narrow values).

15.4 MINERAL INTAKE AND HOMEOSTASIS

In this section, we further discuss how clear-cut observations that support homeostasis and reactive mechanisms may evolve to more complex situations that only apparently contradict homeostasis. The contradiction disappears if we take into account Cannon's definition and control mechanisms that involve more than simple negative feedback. In order to show complex situations compatible with homeostasis, we will use two major examples of mechanisms and behaviors associated with the regulation of osmotic and ionic concentration of the ECF.

Consider first what is called osmometric theory of thirst as a major example. Cellular and behavioral studies suggest that hypertonicity of the ECF activate osmoreceptors located outside the blood–brain barrier by shrinkage of their intracellular volume [intracellular fluid (ICF)] (Bourque 2008; Fitzsimons 1961; Kutscher 1966; Liedtke 2007; Schoorlemmer et al. 2000). A linear function between water intake and the osmotic load, and the termination of drinking by the exact amount of water necessary to dilute the load (Fitzsimons 1961) suggests a direct negative feedback in operation. The theory elegantly suggests that a cellular-dehydrated animal behaves like an osmometer, by drinking water as a linear function of the amount of the osmotic load detected by the osmoreceptors—interestingly, the termination of drinking begins in advance to the correction of the ECF tonicity (Carpenter 2004; Stricker and Hoffmann 2006). As a corollary to the theory, if the cellular-dehydrated rat behaves strictly like an osmometer, it should ingest only water when allowed choices of fluids. However, as we will see next, it does not ingest only water.

When confronted in the laboratory with a free choice of multiple bottles, one bottle containing water and the others containing different mineral solutions, the cellular-dehydrated rat, similar to the extracellular-dehydrated rat, ingests less water and selects to ingest solutions containing minerals (Constancio et al. 2011; David et al. 2008a; Pereira et al. 2005). The chosen minerals are preferentially sodium—mainly in the form of sodium bicarbonate—or potassium.

Such nuances in mineral intake may derive from mechanisms evolved in the wild where water with dissolved minerals, and not pure water, is apparently the norm (Blake et al. 2011). Yet, whatever the actual reason there is for a cellular-dehydrated animal to ingest sodium or potassium, recall that ion excess is dealt with through increased excretion by normal kidneys (Fitzsimons 1961; Kutscher 1966; Schoorlemmer et al. 2000). So, osmotic concentration of the ECF is readjusted to predisturbance levels by a combination of behavior and renal function.

This is why we should not be misled by looking separately at the intake or at the renal output of an ion and forgetting what happens to the ECF. For example, circadian urine potassium excretion of humans may differ about 60–70% from low to high peak levels (Moore-Ede 1986). In general human population, however, the concentration of potassium in the ECF varies within a range of 25% variation (Guyton and Hall 2000), far smaller than the variations in renal potassium excretion. Moreover, if we examine potassium concentration in the ECF under laboratory-controlled conditions we may find signs of tighter regulation. For example, in either pregnant or nonpregnant goats, circadian plasma potassium concentration may differ about 7–8% from low to high peak levels, even considering larger concurrent changes in hormones that control renal potassium excretion (Skotnicka 2003).

A second major example suggests that the control of osmotic and sodium homeostasis is more complex than a simple negative feedback when it occurs along with mechanisms that regulate the volume of ECF. An introduction to such complexity involving the simultaneous regulation of both ICF and ECF in terms of control theory may be found elsewhere (Carpenter 2004). Here, we only would like to recall some neural and endocrine mechanisms to illustrate the same point.

What would an animal that has double dehydration, of both ICF and ECF, do first—drink water or ingest salt? If your answer is to drink water, you may be right. The powerful dipsogen angiotensin II, mentioned throughout this book, is produced in response to dehydration of the ECF. The same angiotensin II also induces sodium appetite (Sakai et al. 1989). However, there is a delay in the production of sodium appetite, operationally defined as increased ingestion of hypertonic NaCl, related with the time length of dehydration. The dilution hypothesis suggests that angiotensin II produces water intake (thirst) first; sodium appetite comes next as the ingested water dilutes the ECF and removes an inhibition produced by angiotensin II itself (Stricker and Verbalis 1990). According to the hypothesis, the mechanism seems to be protecting an already dehydrated animal from further cellular dehydration by preventing excess ingestion of osmolytes. A protective mechanism against cellular dehydration is also suggested by several studies with the lateral parabrachial nucleus (LPBN; Chapters 3, 9, 10, and 11).

Initial work established in the mid-1990s that a dipsogenic dose of angiotensin II injected in the lateral cerebral ventricles also immediately produces hypertonic NaCl intake, in a two-bottle choice test, in rats that received bilateral injection of a serotonergic antagonist in the LPBN region (Menani et al. 1996). A similar result was obtained in the early phase of sodium depletion when sodium appetite is still absent and usually only thirst is present (Menani et al. 2000). Later, it was shown that serotonergic antagonism in the LPBN of rats with quickly induced sodium appetite, as produced by a combination of furosemide plus low dose of captopril (see Chapter 14 for different procedures to induce sodium appetite), enhanced only hypertonic NaCl intake, with no effect on isotonic NaCl intake (David et al. 2008b). Although these results suggest a protection mechanism selective to hypertonic NaCl, the mechanism is perhaps not that simple because recent preliminary data suggest that the blockade of serotonin receptors in the LPBN may also enhance isotonic NaCl intake of cellular-dehydrated rats (David et al., unpublished data).

The main point here is to emphasize that a behavioral mechanism may apparently involve "errors"—for example, animals' choice for dissolved minerals when cellular-dehydrated—not predicted by the osmometric theory, and perhaps not even predicted if we assume that homeostasis results from a simple, direct, negative feedback. The mechanism to achieve osmotic, sodium or potassium homeostasis and the aforementioned examples naturally involve not only brain pattern generators that command the behavior to initiate and terminate the ingestion of minerals and water, but also effector systems such as the kidney. The brain and the kidney may operate through feedback and feedforward mechanisms that result in homeostasis. Therefore, it is not too much to say again that, although the homeostatic mechanism as proposed by Cannon (1929) fits mostly a reactive, direct negative feedback, there are other mechanisms that make homeostasis possible to occur (Carpenter 2004; Moore-Ede 1986). The definition of homeostasis has its own validity independently from the type of animal machinery that produces it.

15.5 CONCLUSION

We have drawn some examples from body fluid homeostasis to address the concept of homeostasis under the definition proposed originally by Walter Cannon (1929). The definition refers particularly to the constancy of physical–chemical constituents (homeostatic categories) of the ECF in a changing external environment. Whatever control mechanism ingrained in the animal machinery, reactive or predictive, to produce it, homeostasis is valid as long as we apply Cannon's definition. In this context, homeostasis remains a valid concept to describe the state of body fluids of mammals as regulated by neural and endocrine mechanisms.

ACKNOWLEDGMENTS

Research of the authors and the preparation of this book were supported by CAPES, CNPq, FAPESP, FAPESP-PRONEX, FUNDUNESP, PROPe-UNESP, and PROPG-UNESP.

REFERENCES

Abbott, L. F. 2003. Balancing homeostasis and learning in neural circuits. *Zoology* 106:365–71.

Berridge, K. C. 2004. Motivation concepts in behavioral neuroscience. *Physiol Behav* 81:179–209.

Blake, J. G., D. Mosquera, J. Guerra, B. A. Loiselle, D. Romo, and K. Swing. 2011. Mineral licks as diversity hotspots in lowland forest of eastern Ecuador. *Diversity* 3:217–34.

Bourque, C. W. 2008. Central mechanisms of osmosensation and systemic osmoregulation. *Nat Rev Neurosci* 9:519–31.

Cannon, W. B. 1929. Organization for physiological homeostasis. *Physiol Rev* 9:399–431.

Carpenter, R. H. S. 2004. Homeostasis: A plea for a unified approach. *Adv Physiol Educ* 28:S180–7.

Constancio, J., D. T. B. Pereira-Derderian, J. V. Menani, and L. A. De Luca Jr. 2011. Mineral intake independent from gastric irritation or pica by cell-dehydrated rats. *Physiol Behav* 104:659–65.

Covian, M. R., J. Antunes-Rodrigues, C. G. Gentil, W. A. Saad, L. A. A. Camargo, and C. R. Silva Neto. 1975. Central control of salt balance. In *Neural Integration of Physiological Mechanisms and Behaviour*, ed. G. J. Mogenson, and F. R. Calaresu, 267–82. Toronto: Univ. Toronto Press.

David, R. B., J. V. Menani, and L. A. De Luca Jr. 2008a. Central angiotensin II induces sodium bicarbonate intake in the rat. *Appetite* 51:82–9.

David, R. B., J. V. Menani, and L. A. De Luca Jr. 2008b. Serotonergic receptor blockade in the lateral parabrachial nucleus: Different effects on hypertonic and isotonic NaCl intake. *Brain Res* 1187:137–45.

Epstein, A. N. 1991. Neurohormonal control of salt intake in the rat. *Brain Res Bull* 27:315–20.

Falk, J. L. 1966. Serial sodium depletion and NaCl solution intake. *Physiol Behav* 1:75–7.

Freire, C. A., E. M. Amado, L. R. Souza et al. 2008. Muscle water control in crustaceans and fishes as a function of habitat, osmoregulatory capacity, and degree of euryhalinity. *Comp Biochem Physiol Part A* 149:435–46.

Fitzsimons, J. T. 1961. Drinking by nephrectomized rats injected with various substances. *J Physiol* 155:563–79.

Fessler, D. M. T. 2003. An evolutionary explanation of the plasticity of salt preferences: Prophylaxis against sudden dehydration. *Med Hypotheses* 61:412–5.

Guyton, A. C., and J. E. Hall. 2000. *Textbook of Medical Physiology*. Philadelphia: W.B. Saunders Company.

Kültz, D. 2012. The combinatorial nature of osmosensing in fishes. *Physiology* 27:259–75.

Kochli, A., Y. Tenenbaum-Rakover, and M. Leshem. 2005. Increased salt appetite in patients with congenital adrenal hyperplasia 21-hydroxylase deficiency. *Am J Physiol Regul Integr Comp Physiol* 288:R1673–81.

Kutscher, C. L. 1966. An osmometric analysis of drinking in salt injected rats. *Physiol Behav* 1:79–83.

Liedtke, W. 2007. Role of TRPV ion channels in sensory transduction of osmotic stimuli in mammals. *Exp Physiol* 92:507–12.

Menani, J. V., R. L. Thunhorst, and A. K. Johnson. 1996. Lateral parabrachial nucleus and serotonergic mechanisms in the control of salt appetite in rats. *Am J Physiol Regul Integr Comp Physiol* 270:R162–8.

Menani, J. V., L. A. De Luca Jr., R. L. Thunhorst, and A. K. Johnson. 2000. Hindbrain serotonin and the rapid induction of sodium appetite. *Am J Physiol Regul Integr Comp Physiol* 279:R126–31.

Moore-Ede, M. C. 1986. Physiology of the circadian timing system: Predictive versus reactive homeostasis. *Am J Physiol Regul Integr Comp Physiol* 250:R737–52.

Moran, T. H., and J. Schulkin. 2000. Curt Richter and regulatory physiology. *Am J Physiol Regul Integr Comp Physiol* 279:R357–63.

Palagi, E., and G. Mancini. 2011. Playing with the face: Playful facial "chattering" and signal modulation in a monkey species (*Theropithecus gelada*). *J Comp Psychol* 125:11–21.

Pereira D. T. B., R. B. David, R. C. Vendramini, J. V. Menani, and L. A. De Luca Jr. 2005. Potassium intake during cell dehydration. *Physiol Behav* 85:99–106.

Sakai, R. R., S. P. Frankmann, W. B. Fine, and A. N. Epstein. 1989. Prior episodes of sodium depletion increase the need-free sodium intake of the rat. *Behav Neurosci* 103:186–92.

Schoorlemmer, G. H. M., A. K. Johnson, and R. L. Thunhorst. 2000. Effect of hyperosmotic solutions on salt excretion and thirst in rats. *Am J Physiol Regul Integr Comp Physiol* 278:R917–23.

Skotnicka, E. 2003. Circadian variations of plasma renin activity (PRA), aldosterone and electrolyte concentrations in plasma in pregnant and non-pregnant goats. *Comp Biochem Physiol Part C: Toxicol Pharmacol* 134:385–95.

Sterling, P. 2012. Allostasis: A model of predictive regulation. *Physiol Behav* 106:5–15.

Stricker, E. M., and J. G. Verbalis. 1990. Sodium appetite. In *Neurobiology of food and fluid intake*, ed. E. M. Stricker, 387–419. New York: Plenum Press.

Stricker, E. M., and M. L. Hoffmann. 2006. Control of thirst and salt appetite in rats: Early inhibition of water and NaCl ingestion. *Appetite* 46:234–7.

Takei, Y. 2000. Comparative physiology of body-fluid regulation in vertebrates with special reference to thirst regulation. *Jpn J Physiol* 50:171–86.

Index

Page numbers followed by f and t indicate figures and tables, respectively.

F

G

H

I

O

P

T - #0371 - 071024 - C2 - 234/156/15 - PB - 9780367379414 - Gloss Lamination